# APPROXIMATION THEORY

PROCEEDINGS OF THE CONFERENCE JOINTLY ORGANIZED
BY THE MATHEMATICAL INSTITUTE OF THE
POLISH ACADEMY OF SCIENCES AND
THE INSTITUTE OF MATHEMATICS
OF THE ADAM MICKIEWICZ UNIVERSITY
HELD IN POZNAŃ, 22–26 AUGUST, 1972

*Edited by*

ZBIGNIEW CIESIELSKI
*Mathematical Institute of the Polish Academy of Sciences, Gdańsk, Poland*

*and*

JULIAN MUSIELAK
*Institute of Mathematics of the Adam Mickiewicz University, Poznań, Poland*

D. REIDEL PUBLISHING COMPANY
DORDRECHT-HOLLAND / BOSTON-U.S.A.

PWN-POLISH SCIENTIFIC PUBLISHERS
WARSZAWA

Library of Congress Catalog Card Number 74-80524

ISBN 90 277 0483 X

Distributors for Albania, Bulgaria, Chinese People's Republic, Czechoslovakia, Cuba, German Democratic Republic, Hungary, Mongolia, Poland, Rumania, Democratic Republic of Vietnam, the U.S.S.R. and Yugoslavia
ARS POLONA—RUCH
Krakowskie Przedmieście 7, 00-068 Warszawa, Poland

Distributors for the U.S.A., Canada and Mexico
D. REIDEL PUBLISHING COMPANY, INC.
306 Dartmouth Street, Boston, Mass. 02116, U.S.A.

Distributors for all other countries
D. REIDEL PUBLISHING COMPANY
P.O. Box 17, Dordrecht, Holland



Printed in Poland by D.R.P.

# PREFACE

The Conference on Approximation Theory held in Poznań, August 22–26, 1972, was organized jointly by the Mathematical Institute of the Polish Academy of Sciences and the Institute of Mathematics of the Adam Mickiewicz University.

These proceedings contain part of the papers presented during the conference.

This volume can be regarded in a sense as a continuation of the Proceedings of the Conference on Constructive Function Theory, Varna 1970, and the Proceedings of the Conference on Constructive Theory of Functions, Budapest 1969.

Z. CIESIELSKI and J. MUSIELAK

# LIST OF PARTICIPANTS

| | | |
|---|---|---|
| 1. | J. Adamska | Poland |
| 2. | G. Albinus | GDR |
| 3. | J. Albrycht | Poland |
| 4. | A. Alexiewicz | Poland |
| 5. | W. Arend | Poland |
| 6. | V. V. Arestov | USSR |
| 7. | M.-B. A. Babaev | USSR |
| 8. | V. A. Baskakov | USSR |
| 9. | V. I. Belyj | USSR |
| 10. | V. I. Berdyshev | USSR |
| 11. | S. Bergman | USA |
| 12. | H.-P. Blatt | GFR |
| 13. | J. Bochenek | Poland |
| 14. | A. Borucka-Cieślewicz | Poland |
| 15. | D. Braess | GFR |
| 16. | W. W. Breckner | Rumania |
| 17. | B. Brosowski | GFR |
| 18. | Z. Ciesielski | Poland |
| 19. | R. Denchev | Bulgaria |
| 20. | S. Dimiev | Bulgaria |
| 21. | B. Dreseler | GFR |
| 22. | K. Fanta | Hungary |
| 23. | T. L. Fields | Canada |
| 24. | G. Freud | Hungary |
| 25. | P. M. Gauthier | USA |
| 26. | E. Görlich | GFR |
| 27. | M. v. Golitschek | GFR |
| 28. | L. Y. Hedberg | Sweden |
| 29. | T. Hedberg | Sweden |
| 30. | K. H. Hoffmann | GFR |
| 31. | J. Hojdar | ČSSR |
| 32. | T. Iwiński | Poland |
| 33. | A. Janicki | Poland |
| 34. | M. Jaroszewska | Poland |
| 35. | H. H. Kallioniemi | Sweden |
| 36. | M. Karpiński | Poland |
| 37. | O. Kis | Hungary |
| 38. | W. Meyer-König | GFR |

| | | |
|---|---|---|
| 39. | W. Köhnen | GFR |
| 40. | W. Kołwzan | Poland |
| 41. | G. Kozłowska | Poland |
| 42. | J. Krogulska | Poland |
| 43. | W. Kurc | Poland |
| 44. | I. Labuda | Poland |
| 45. | L. Leindler | Hungary |
| 46. | S. Lyttkens | Sweden |
| 47. | T. Markiewicz | Poland |
| 48. | S. Markov | Bulgaria |
| 49. | J. Maruşciac | Rumania |
| 50. | W. Matuszewska | Poland |
| 51. | M. Mikosz | Poland |
| 52. | H. Musielak | Poland |
| 53. | J. Musielak | Poland |
| 54. | R. J. Nessel | GFR |
| 55. | E. Neuman | Poland |
| 56. | G. P. Névai | Hungary |
| 57. | A. Opyrchał | Poland |
| 58. | W. Orlicz | Poland |
| 59. | T. Orłowska | Poland |
| 60. | K. I. Oskolkov | USSR |
| 61. | S. Pawelke | GFR |
| 62. | W. Pleśniak | Poland |
| 63. | R. V. Poliakov | USSR |
| 64. | V. Popov | Bulgaria |
| 65. | E. Popoviciu | Rumania |
| 66. | T. Popoviciu | Rumania |
| 67. | D. Przeworska-Rolewicz | Poland |
| 68. | A. Przeździecka | Poland |
| 69. | P. Pych | Poland |
| 70. | W. Raczyński | Poland |
| 71. | M. Radnay | Hungary |
| 72. | H. Ratajski | Poland |
| 73. | L. Rempulska | Poland |
| 74. | S. Rolewicz | Poland |
| 75. | P. Russev | Bulgaria |
| 76. | K. Scherer | GFR |
| 77. | W. Schempp | GFR |
| 78. | W. Schipp | Hungary |
| 79. | A. Sharma | Canada |
| 80. | B. Sendov | Bulgaria |
| 81. | J. Siciak | Poland |
| 82. | M. Sikorski | Poland |

| | |
|---|---|
| 83. W. Sippel | GFR |
| 84. M. Skowroński | Poland |
| 85. M. Skwarczyński | Poland |
| 86. Z. Stankiewicz | Poland |
| 87. S. Stańko | Poland |
| 88. E. L. Stark | GFR |
| 89. S. B. Stečkin | USSR |
| 90. S. Stoiński | Poland |
| 91. Yu. N. Subbotin | USSR |
| 92. J. Szabados | Hungary |
| 93. J. Szalay | Hungary |
| 94. J. Szelmeczka | Poland |
| 95. R. Taberski | Poland |
| 96. P. M. Tamrazov | USSR |
| 97. S. A. Teliakowskij | USSR |
| 98. S. Troianski | Bulgaria |
| 99. F. Unger | GDR |
| 100. A. K. Varma | USA |
| 101. P. Vértesi | Hungary |
| 102. A. Waszak | Poland |
| 103. H. Weber | GFR |
| 104. P. Weyman | Poland |
| 105. M. Wiegner | GFR |
| 106. T. Winiarska | Poland |
| 107. T. Winiarski | Poland |
| 108. Z. Wronicz | Poland |
| 109. J. Ziemánek | ČSSR |

CONTENTS

# ON THE BEST APPROXIMATION OF THE OPERATORS OF DIFFERENTIATION AND RELATED QUESTIONS

V. V. ARESTOV

*Swerdlovsk*

**1.** We exploit the following notation: $I$ is the real line or the half-line $[0, \infty)$; $p, q, r$ are parameters from the interval $1 \leqslant p, q, r \leqslant \infty$; $k, n$ are integers with $0 \leqslant k < n$; $L_p = L_p(I)$ for $1 \leqslant p < \infty$ is the space of functions $x$ integrable in $p$-th power, $L_\infty$ is the space of essentially bounded functions; the norm is defined by

$$\|x\|_p = \|x\|_{L_p(I)} = \begin{cases} \left\{\int_I |x(t)|^p dt\right\}^{1/p} & \text{for } 1 \leqslant p < \infty, \\ \operatorname{ess\,sup}_{t\in I} |x(t)| & \text{for } p = \infty. \end{cases}$$

Let, further, $W_p^n$, $1 < p \leqslant \infty$, denote the set of functions $x$ possessing a locally absolutely continuous derivative of order $n-1$ and such that $x^{(n)} \in L_p$; $W_1^n$ denotes the set of functions $x$ of the form

$$x(t) = \sum_{i=0}^{n-1} c_i t^i + \frac{1}{(n-1)!} \int_0^t (t-\tau)^{n-1} d\xi(\tau),$$

where $c_i$ are real numbers and $\xi = x^{(n-1)}$ has bounded variation: $V\xi < \infty$. For $x \in W_1^n$ we put $\|x^{(n)}\|_1 = Vx^{(n-1)}$. Finally, we denote by $W_{p,r}^n$ the intersection of the spaces $L_r$ and $W_p^n$, and by $Q = Q_{p,r}^n$ the set of functions $x \in W_{p,r}^n$ with the property $\|x^{(n)}\|_p \leqslant 1$.

For any pair of real numbers $\mu_0 > 0$, $\mu_n > 0$ consider the quantity

$$\mu_k = \sup \|x^{(k)}\|_q,$$

where the supremum is taken over all functions $x \in W_{p,r}^n$ with $\|x\|_r \leqslant \mu_0$, $\|x^{(n)}\|_p \leqslant \mu_n$. The following formula expressing the dependence of $\mu_k$ on $\mu_0, \mu_n$ can be easily derived from the norm homogeneity property (see [15], [7])

$$\mu_k = D\mu_0^\alpha \mu_n^\beta,$$

where

$$\alpha = \frac{n-k-p^{-1}+q^{-1}}{n-p^{-1}+r^{-1}}, \qquad \beta = 1-\alpha,$$

and $D$ is a constant independent of $\mu_0$, $\mu_n$.

It hence follows that the functions of class $W^n_{p,r}$ satisfy the inequality

$$\|x^{(k)}\|_q \leqslant D\|x\|_r^\alpha \|x^{(n)}\|_p^\beta. \tag{1}$$

Inequalities of such type have been first considered by Hardy and Littlewood [12], Landau [14], Hadamard [11]. Further interesting results concerning this topic have been obtained by Kolmogorov [13] and Sz.-Nagy [15]. In recent years the constant $D$ was found for many specific values of $k, n, p, q, r, I$ as well as the set of extreme functions for which inequality (1) turns into an equality (see references in [3], [7], [9]). Full conditions for the finiteness of $D$ were obtained by Gabushin [7] who proved that $D$ is finite if and only if

$$\frac{n-k}{r}+\frac{k}{p} \geqslant \frac{n}{q}. \tag{2}$$

We now formulate the problem first studied by Stečkin [17], [16] concerning the best approximation (on the set of $n$ times differentiable functions) to the operator of differentiation of order $k$ by bounded linear operators. Put

$$U(T) = \sup_{x\in Q^n_{p,r}} \|x^{(k)}-Tx\|_q, \tag{3}$$

$$E(N) = \inf_{\|T\|^{L_q}_{L_r}\leqslant N} U(T), \tag{4}$$

where $T$ is a linear operator from $L_r$ to $L_q$. Here we assume that if $r$ or $q$ equals $\infty$, then $L_r$ resp. $L_q$ is replaced by the space $C = C(I)$ of continuous functions. The study of the quantity $E(N)$ includes the problem of the existence, uniqueness and other properties of an extremal operator $T$, i.e. an operator with the properties

$$\|T\| \leqslant N, \qquad U(T) = E(N). \tag{5}$$

The dependence of $E(N)$ on $N$ is given by the formula (see [17] and also [3])

$$E(N) = N^{-\nu}E(1), \quad \nu = \frac{\alpha}{\beta} = \frac{n-k-p^{-1}+q^{-1}}{k+r^{-1}-q^{-1}} \geqslant 0. \tag{6}$$

Stečkin [17] pointed out the connection between the problem (4) and inequality (1). This connection is expressed by the formula

(7) $$E(N) \geqslant G(N),$$

where

$$G(N) = \sup_{x \in Q}(\|x^{(k)}\|_q - N\|x\|_r) = \beta\alpha^{\alpha/\beta}D^{1/\beta}N^{-\alpha/\beta}.$$

Inequality (7) gives a lower estimate for $E(N)$ by the constant $D$, thus automatically also by any lower estimate for $D$, which can be obtained by means of any function $x \in W^n_{p,r}$. On the other hand, for a given operator $T$ with $\|T\| \leqslant N$ the quantity $U(T)$ is an upper estimate for $E(N)$, thus also for the constant $D$. If a function $x$ and an operator $T$ are chosen in such a way, that in the chain of inequalities

$$(\|x^{(k)}\|_q - N\|x\|_r)\|x^{(n)}\|_p^{-1} \leqslant G(N) \leqslant E(N) \leqslant U(T)$$

the first and the last term coincide, then a solution of both problems is found; then the function $x$ is extremal in inequality (1) and the operator $T$ is extremal in the problem (4).

According to this scheme solutions of the latter problem for specific values of $k, n, p, q, r, I$ are given in the papers [16], [17], [1]–[3], [5], [8]–[10], [19], [21]. In some of those cases the corresponding precise inequality (1) has been known formerly. In the other ones (see [5], [8], [21]) this was the way to find the least constant in (1). We remark that the paper of Domar [6] gives, with use of a result of Kolmogorov [13], the solution of the problem (4) for $p = q = r = \infty$, $I = (-\infty, \infty)$ and arbitrary $k, n$ (for $n = 2, 3$ cf. [17], for $n = 4, 5$ cf. [1]).

In the case $q = \infty$ $E(N)$ coincides with the quantity (see [2], [3], [8]–[10], [21])

(8) $$e(N) = \inf_{\|T\|_{L_r^*} \leqslant N} \sup_{x \in Q^n_{p,r}} (x^{(k)}(0) - Tx)$$

of the best approximation to the operator of differentiation at the point $t = 0$ by bounded functionals. Gabushin [10] has shown that

(9) $$e(N) = G(N) = \beta\alpha^{\alpha/\beta}D^{1/\beta}N^{-\alpha/\beta} \quad \text{for } q = \infty.$$

The equality $E(N) = G(N)$, except of the case $q = \infty$, takes place e.g. for $p = q = r = 2$, $I = (-\infty, \infty)$ (see [19]). In the general case, however, (7) is a strict inequality, though a corresponding example cannot be given here.

**2.** A connection exists not only between the problem (4) and inequality (1); these problems are also related to an approximation to a certain class of differentiable functions by another similar class.

Let $1 \leqslant p', q', r' \leqslant \infty$; $0 < m \leqslant n$; $B^n_{r'}(N)$ is the set of functions $\varphi \in W^n_{r'}$ with $\|\varphi^{(n)}\|_{r'} \leqslant N$; $B^m_{q'} = B^m_{q'}(1)$; $\mathfrak{M}(N)$ is the set of linear (i.e. homogeneous and additive) operators $S$ from $W^m_{q'}$ to $W^n_{r'}$ such that

(10) $$\|(S\psi)^{(n)}\|_{r'} \leqslant N\|\psi^{(m)}\|_{q'}$$

for $\psi \in W^m_{q'}$. Then

(11) $$F(N) = F(N, I) = \sup_{\psi \in B^m_{q'}} \inf_{\varphi \in B^n_{r'}(N)} \|\psi - \varphi\|_{p'}$$

is the approximation in the metric $L_{p'}$ to the class $B^m_{q'}$ by the class $B^n_{r'}(N)$, and

(12) $$F_L(N) = F_L(N, I) = \inf_{S \in \mathfrak{M}(N)} \sup_{\psi \in B^m_{q'}} \|\psi - S\psi\|_{p'}$$

is the corresponding linear approximation. Clearly $F(N) \leqslant F_L(N)$.

The quantities $F(N)$ and $F_L(N)$ have been dealt with in papers [4], [18], [19], [20]. In particular, it is known [4], that

(13) $$F(N) = N^{-\nu}F(1), \qquad F_L(N) = N^{-\nu}F_L(1),$$

where

$$\nu = \left(m + \frac{1}{p'} - \frac{1}{q'}\right)\left(n - m + \frac{1}{q'} - \frac{1}{r'}\right)^{-1}$$

and $F(N)$ is finite if [20] and only if [4]

$$\frac{m}{r'} + \frac{n-m}{p'} \leqslant \frac{n}{q'}.$$

Under some assumptions $F_L(N)$, thus also $F(N)$, can be estimated from above by $E(N)$ (see [4], [18], [19]). Namely, suppose that

(14) $$m = n - k, \quad p = r = q', \quad p' = r' = q,$$

and that the extremal operator $T$ with respect to (4), is defined on the set $W^n_p$, is linear, commutes with the operation of differentiation of order $n$, and

(15) $$\sup_{x \in B^n_p} \|x^{(k)} - Tx\|_q = E(N).$$

We shall show that then

(16) $$F_L(N) \leqslant E(N),$$

and consequently

(17) $$F(N) \leqslant F_L(N) \leqslant E(N).$$

To every function $\psi \in W_p^{n-k}$ we assign a function $x \in W_p^n$ such that $x^{(k)} = \psi$. Define the operator $S$ by

$$S\psi = Tx \quad \text{for} \quad \psi \in W_p^{n-k}.$$

The functions $x$ can be chosen so as to make $S$ linear. We have

$$(S\psi)^{(n)} = (Tx)^{(n)} = Tx^{(n)} = T\psi^{(n-k)},$$

whence follows that $S$ fulfils (10). Further, condition (15) yields

$$\|\psi - S\psi\|_q = \|x^{(k)} - Tx\|_q \leqslant E(N)\|x^{(n)}\|_p = E(N)\|\varphi^{(n-k)}\|_p$$

and (16) follows.

We now formulate one more problem of an approximation to a class by a class. Let $w_{r'}^n$ be the set of functions $x \in W_{r'}^n$ with $x^{(i)}(0) = 0$ for $i = 0, 1, \ldots, n-1$; let $b_{r'}^n(N)$ be the set of functions $x \in w_{r'}^n$ with $\|x^{(n)}\|_{r'} \leqslant N$; $b_{q'}^m = b_{q'}^m(1)$; and let $\mathfrak{M}'(N)$ be the set of linear operators $S$ from $w_{q'}^m$ to $w_{r'}^m$ satisfying condition (10). Finally, put

(18) $$f(N) = f(N, I) = \sup_{\psi \in b_{q'}^m} \inf_{\varphi \in b_{r'}^n(N)} \|\psi - \varphi\|_{p'},$$

(19) $$f_L(N) = f_L(N, I) = \inf_{S \in \mathfrak{M}'(N)} \sup_{\psi \in b_{q'}^m} \|\psi - S\psi\|_{p'}.$$

Repeating the argument of [4] one can prove formulae (13) with $f$ in place of $F$.

Since

$$F(N) = \sup_{\psi \in b_{q'}^m} \inf_{\varphi \in B_{r'}^n(N)} \|\psi - \varphi\|_{p'}$$

and $b_{q'}^m \subset B_{q'}^m$, thus

(20) $$F(N) \leqslant f(N).$$

The inequality

(21) $$F_L(N) \leqslant f_L(N)$$

also holds, since the formula

$$S\psi = \psi - \psi_0 + S'\psi_0,$$

where

$$\psi_0(t) = \psi(t) - \sum_{i=0}^{m-1} \psi^{(i)}(0)\frac{t^i}{i!}, \qquad \psi \in W_{q'}^m,$$

assigns to every operator $S' \in \mathfrak{M}'(N)$ an operator $S \in \mathfrak{M}(N)$ in such a way that the operators $S$ and $S'$ give equal approximations of corresponding classes.

The quantities $F(N)$ and $f(N)$ are connected with the constant $D$ in (1) as follows.

THEOREM 1. *Let*

$$(22) \quad m = n-k, \qquad \frac{1}{p}+\frac{1}{p'} = 1, \qquad \frac{1}{q}+\frac{1}{q'} = 1, \qquad \frac{1}{r}+\frac{1}{r'} = 1.$$

*Then*

$$(23) \qquad G(N) = \beta\alpha^{\alpha/\beta}D^{1/\beta}N^{-\alpha/\beta} = \begin{cases} F(N) & \text{for } I = (-\infty, \infty), \\ f(N) & \text{for } I = [0, \infty). \end{cases}$$

This theorem follows rather simply from Lemma 1 of the paper [10] by Gabushin.

A function $\psi$ will be called *maximal* with respect to the problem (11) if $\psi \in B_{q'}^m$ and the supremum in (11) is attained for that function. A maximal function with respect to the problem (18) is defined analogously.

Together with equalities (23) we may assert that in the case $q = \infty$ the function $\theta$ defined by

$$\theta(t) = \frac{t_+^{n-k-1}}{(n-k-1)!}, \qquad \text{where} \qquad t_+ = \begin{cases} t & \text{for } t > 0, \\ 0 & \text{for } t \leqslant 0, \end{cases}$$

is maximal; and in the case $1 \leqslant q < \infty$, under the condition that there exists an extremal function $z$ with respect to (1), normed by

$$\|z^{(n)}\|_p = 1, \qquad \|z\|_r = \left(\frac{\alpha D}{N}\right)^{1/\beta},$$

any function $\eta$ such that

$$\eta^{(n-k)} = \|z^{(k)}\|_q^{1-q}|z^{(k)}|^{1/(q-1)}\operatorname{sign} z^{(k)}$$

is maximal.

THEOREM 2. *Let conditions* (22) *be fulfilled. Then, in the case* $I = (-\infty, \infty)$, *for any values of the remaining parameters we have*

$$E(N) = F_L(N), \tag{24}$$

*and in the case* $I = [0, \infty)$, *for* $1 < q \leqslant \infty$, *we have*

$$E(N) = f_L(N). \tag{25}$$

Under assumptions of Theorem 2 we have, by (25), (24), (23), (21) and (20),

$$E(N) \geqslant F_L(N) \geqslant F(N),$$

whereby conditions (22) are different from (14) in inequalities (17).

Equalities (9), (23)–(25) and the maximality of $\theta$ for $q = \infty$ imply the following statement.

COROLLARY. *Let*

$$m = n-k, \quad q = \infty, \quad q' = 1, \quad p' = \frac{p}{p-1}, \quad r' = \frac{r}{r-1}.$$

*Then*

$$F_L(N) = F(N) = e(N) = \inf_{\varphi \in B^n_{r'}(N)} \|\theta - \varphi\|_{p'} \quad \textit{for } I = (-\infty, \infty),$$

$$f_L(N) = f(N) = e(N) = \inf_{\varphi \in b^n_{r'}(N)} \|\theta - \varphi\|_{p'} \quad \textit{for } I = [0, \infty).$$

**3.** The quantities (11), (12), (18), (19) make sense for the half-axis as well as for the axis.

Subbotin [18] obtained lower estimate coinciding for $F(N, [0, \infty))$ and $F(N, (-\infty, \infty))$ for $p' = q' = r' = \infty$ and $p' = q' = r' = 1$; he also proved that this estimate is precise for $n \leqslant 5$. In the case $p' = q' = r' = \infty$ for an arbitrary $n$ an upper estimate for those quantities is obtained in [4]; this estimate coincides with the estimate from [18]. Consequently, in those cases $F(N)$ does not depend on $I$. This fact is valid for any values of the parameters.

THEOREM 3. *The following equalities hold*:

$$F(N, [0, \infty)) = F(N, (-\infty, \infty)),$$
$$f(N, [0, \infty)) = f(N, (-\infty, \infty)).$$

Concerning the problems of linear approximation to a class by a class, it can be shown (for particular cases cf. [18]), that

$$F_L(N, [0, \infty)) \leqslant F_L(N, (-\infty, \infty)),$$
$$f_L(N, [0, \infty)) \leqslant f_L(N, (-\infty, \infty)),$$

and if $q' \leqslant p'$, $q' \leqslant r'$, then

$$f_L(N, [0, \infty)) = f_L(N, (-\infty, \infty)).$$

For $q = \infty$, according to Theorem 3 and corollary, we have

$$F(N, [0, \infty)) \leqslant F_L(N, [0, \infty)) \leqslant F_L(N, (-\infty, \infty)) = F(N, (-\infty, \infty)) = F(N, [0, \infty)),$$

i.e. the quantities $F(N)$ and $F_L(N)$ do not depend on $I$ and are equal. Similarly we can verify that for $q = \infty$ also the quantities $f(N)$ and $f_L(N)$ do not depend on $I$ and are equal.

## REFERENCES

[1] V. V. Arestov, 'On the Best Approximation to the Operators of Differentiation', *Mat. Zametki* **1** (1967) 149–154.

[2] V. V. Arestov, 'On the Best Uniform Approximation to the Operators of Differentiation', *ibidem* **5** (1969) 273–284.

[3] V. V. Arestov, 'On the Best Approximation to the Operators of Differentiation in the Uniform Metric', Author's Thesis, Moskva 1969.

[4] V. V. Arestov and V. N. Gabushin, 'On the Approximation to Classes of Differentiable Functions', *Mat. Zametki* **9** (1971) 105–112.

[5] V. I. Berdyshev, 'On the Best $L[0, \infty)$ Approximation to the Operator of Differentiation', *ibidem* **9** (1971) 477–481.

[6] Y. Domar, 'An Extremal Problem Related to Kolmogoroff's Inequality for Bounded Functions', *Arkiv för Mat.* **7** (1968) 433–441.

[7] V. N. Gabushin, 'Inequalities for the Norms of a Function and Its Derivative in the $L_p$ Metrics', *Mat. Zametki* **1** (1967) 291–298.

[8] V. N. Gabushin, 'On the Best Approximation to the Operator of Differentiation on the Half-Axis', *ibidem* **6** (1969) 573–582.

[9] V. N. Gabushin, 'Inequalities for the Norms of a Function and Its Derivative, and Their Applications', Author's Thesis, Moskva 1970.

[10] V. N. Gabushin, 'The Best Approximation to Functionals on Certain Sets', *Mat. Zametki* **8** (1970) 551–562.

[11] J. Hadamard, 'Sur le module maximum d'une fonction et de ses dérivées', *Soc. Math. de France, Comptes Rendus des Scinces* (1914) 68–72.

[12] G. H. Hardy and J. E. Littlewood, 'Contribution to the Arithmetic Theory of Series', *Proc. London Math. Soc.* **(2) 11** (1912) 411–478.

[13] A. N. Kolmogorov, 'On Inequalities Between the Least Upper Bounds of Successive Derivatives of an Arbitrary Function on an Infinite Interval', *Učebn. Zap. Mosk. Univ.* **30**, *Matematika* **3** (1939) 3–16.

[14] E. Landau, 'Einige Ungleichungen für zweimal differentierbare Funktionen', *Proc. London Math. Soc.* **(2) 13** (1913) 43–49.

[15] B. Sz.-Nagy, 'Über Integralungleichungen zwischen Funktion und ihrer Ableitung', *Acta Univ. Szeged, Sect. Sci. Math.* **10** (1941) 64–74.

[16] S. B. Steĉkin, 'Inequalities Between the Norms of the Derivatives of an Arbitrary Function', *ibidem* **26** (1965) 225–230.

[17] S. B. Steĉkin 'The Best Approximation to Linear Operators', *Mat. Zametki* **1** (1967) 149–154.

[18] Yu. N. Subbotin, 'The Best Approximation to a Function Class by Another Class, *ibidem* **2** (1967) 495–504.

[19] Yu. N. Subbotin and L. V. Taikov. 'On the Best Approximation to the Operator of Differentiation in the Space $L_2$', *ibidem* **3** (1968) 257–264.

[20] S. B. Steĉkin, 'The Connection Between Spline Approximations and the Problem of an Approximation to a Class by a Class', *ibidem* **9** (1971) 501–510.

[21] L. V. Taĭkov, 'On Kolmogorov Type Inequalities and the Best Formulae of Numerical Differentiation, *ibidem* **4** (1963) 233–238.

# A GENERALIZATION OF THEOREMS OF KOROVKIN ON CONDITIONS FOR AND THE ORDER OF CONVERGENCE OF A SEQUENCE OF POSITIVE OPERATORS

V. A. BASKAKOV

*Moskva*

We consider a sequence of linear operators

$$\mathscr{L}_n(f; x) = \int_a^b f(t)\, d\varphi_n(x, t), \qquad n = 1, 2, 3, \ldots, \tag{1}$$

where for every $n$ and for every fixed $x \in [a, b]$ the function $\varphi_n(x, t)$ has bounded variation with respect to $t \in [a, b]$, and $f(x) \in C[a, b]$.

If the functions $\varphi_n(x, t)$ are non-decreasing in $t$, then the operators (1) are positive. The two following fundamental theorems on conditions for and the order of convergence of sequences of positive operators are due to Korovkin ([2], Theorems 3 and 17):

1. *If the operators* (1) *are positive and if*

$$\mathscr{L}_n(t^i; x) \rightrightarrows x^i, \qquad i = 0, 1, 2, \tag{2}$$

*uniformly on the segment* $[a, b]$, *then for any function* $f(x) \in C[a, b]$ *we have*

$$\mathscr{L}_n(f, x) \rightrightarrows f(x)$$

*uniformly on* $[a, b]$.

2. *If the values of the positive operators* (1) *are polynomials of degree* $\leqslant n$, *then for at least one of the three functions* $1, x, x^2$ *the order of the approximation* (2) *does not exceed* $O(n^{-2})$.

The aim of the present paper is to obtain analogous results for a more general class of operators.

We denote by $E_{2k}, k \geqslant 1$, the class of operators (1) such that for every fixed $x \in [a, b]$ the integrals

$$\mathscr{I}^{(1)}_{2k,n}(t) = \int_a^t dt_1 \int_a^{t_1} dt_2 \ldots \int_a^{t_{2k-1}} d\varphi_n(x, t_{2k}), \qquad a \leqslant t < x,$$

$$\mathscr{I}^{(2)}_{2k,n}(t) = \int_t^b dt_1 \int_{t_1}^b dt_2 \ldots \int_{t_{2k-1}}^b d\varphi_n(x, t_{2k}), \qquad x \leqslant t \leqslant b, n = 1, 2, \ldots,$$

have a constant sign for all $t \in [a, b]$; this sign can depend on $n$. Obviously, the class $E_{2k}$ for any $k \geqslant 1$ contains the class of positive operators (1).

THEOREM 1. *If a sequence of operators* (1) *of class* $E_{2k}, k \geqslant 1$, *satisfies the conditions*

$$\mathscr{L}_n(t^i; x) \rightrightarrows x^i, \qquad i = 0, 1, \ldots, 2k,$$

*uniformly in* $x \in [a, b]$, *then for any function* $f(x)$, $2k$ *times continuously differentiable on* $[a, b]$ *we have*

$$\mathscr{L}_n(f; x) \rightrightarrows f(x)$$

*uniformly on* $[a, b]$.

COROLLARY 1. *If a sequence of operators* (1) *satisfies the conditions of Theorem* 1 *and if the norms of these operators are uniformly bounded, i.e. if*

$$\int_a^b |d\varphi_n(x, t)| \leqslant C_1, \qquad n = 1, 2, \ldots,$$

*where* $C_1$ *is an absolute constant, then for any function* $f(x) \in C[a, b]$ *we have*

$$\mathscr{L}_n(f; x) \rightrightarrows f(x)$$

*uniformly on* $[a, b]$.

The proof of this corollary follows from Theorem 1 and the theorem of Banach.

THEOREM 2. *If the values of the operators* (1) *are polynomials of degree* $\leqslant n$ *and satisfy the conditions of Theorem* 1, *then for at least one* $i$ *from the set* $i = 0, 1, \ldots, 2k$ *we have*

$$n^{2k} \max_{x \in [a, b]} |\mathscr{L}_n(t^i; x) - x^i| \nrightarrow 0 \qquad \text{as } n \to \infty.$$

Note that for $k = 1$ Corollary 1 and Theorem 2 are direct extensions of the above quoted Korovkin's theorems for positive operators onto a wider class of operators $E_2$.

Analogous theorems hold also for operators of class $E_{2k}$ concerning the convergence not to the functions in question, but to their derivatives of an even order:

THEOREM 3. *If a sequence of operators* (1) *of class* $E_{2k}$, $k > 1$, *satisfies the conditions*

$$\mathscr{L}_n(t^i; x) \rightrightarrows (x^i)^{(2l)}, \qquad i = 0, 1, \ldots, 2k,$$

*uniformly on* $[a, b]$ ($l$ *fixed*, $< k$), *then for any function* $f(x)$, $2k$ *times continuously differentiable on* $[a, b]$, *we have*

$$\mathscr{L}_n(f; x) \rightrightarrows f^{(2l)}(x)$$

*uniformly on* $[a, b]$.

COROLLARY 2. *If a sequence of operators* (1) *of class* $E_{2k}$, $k > 1$, *satisfies the conditions of Theorem* 3 *and the condition*

$$\int_a^x |\mathscr{I}^{(1)}_{2l,n}(t)|\, dt + \int_x^b |\mathscr{I}^{(2)}_{2l,n}(t)|\, dt \leqslant C_2$$

*for all* $x \in [a, b]$, *where* $C_2$ *is an absolute constant, then for any function* $f(x)$, $2l$ *times continuously differentiable on* $[a, b]$, *we have*

$$\mathscr{L}_n(f; x) \rightrightarrows f^{(2l)}(x)$$

*uniformly on* $[a, b]$.

THEOREM 4. *If the operators* (1) *are polynomials of degree* $\leqslant n$ *and satisfy the conditions of Theorem* 3, *then for at least one* $i$ *from the set* $i = 0, 1, \ldots, 2k$ *we have*

$$n^{2k-2l} \max_{x \in [a,b]} |\mathscr{L}_n(t^i; x) - (x^i)^{(2l)}| \nrightarrow 0 \qquad \text{as } n \to \infty.$$

Observe that Theorems 2 and 4 express the fact, that polynomial operators of class $E_{2k}$, $k \geqslant 1$, as well as positive operators, are not a good tool for approximation of analytic functions or functions of a high order of differentiability.

It is not difficult to see that if the first order derivatives of the Bernstein polynomials, the singular Weierstrass integrals and certain other positive operators are written in the form (1), then the functions $\varphi_n(x, t)$

for every $n$ and every fixed $x \in [a, b]$ are decreasing in the interval $[a, t(n, x)]$ and increasing in $[t(n, x), b]$. For the Bernstein polynomials and the singular Weierstrass integrals we have $t(n, x) = x$. Thus the kernel $d\varphi_n(x, t)$ of the integrals (1) changes sign only at one point $t = t(n, x)$. The class of operators of this type is called by Korovkin [3] the class $S_1$. It can be also easily seen that the $m$-th order derivatives of the Bernstein polynomials and of certain other positive operators are operators of class $S_m$, in Korovkin's terminology [3]. Korovkin has studied conditions for and the order of convergence of operators of class $S_m$ to functions. According to what is done above, it seems natural to study conditions for and the order of convergence of operators of class $S_m$ also to the derivatives of $m$-th order. Concerning this topic, the following theorems were proved in the author's paper [1]:

THEOREM A. *If a sequence of operators* (1) *of class* $S_1$ *satisfies the conditions*

$$\mathscr{L}_n(t^i, x) \rightrightarrows (x^i)', \quad i = 0, 1, 2, 3,$$

*uniformly on* $[a, b]$, *then for any function* $f(x)$ *continuously differentiable on* $[a, b]$ *we have*

$$\mathscr{L}_n(f; x) \rightrightarrows f'(x)$$

*uniformly on* $[a, b]$.

The following theorem has a local character:

THEOREM B. *If a sequence of operators* (1) *of class* $S_1$ *with* $t(n, x) = x$, $n = 1, 2, \ldots$, *satisfies the condition*

$$\mathscr{L}_n(t^i, x) \to (x^i)', \quad i = 0, 1, 2, 3,$$

*at the point* $t = x$, *then for any bounded function differentiable at the point* $x$ *we have*

$$\mathscr{L}_n(f; x) \to f'(x).$$

THEOREM C. *If the operators* (1) *of class* $S_m$, $m \geqslant 1$, *satisfy the condition*

$$\mathscr{L}_n(t^i, x) \rightrightarrows (x^i)^{(m)}, \quad i = 0, 1, 2, \ldots, m+2,$$

*uniformly on* $[a, b]$, *then for any function* $f(x)$, $m$ *times continuously differentiable on* $[a, b]$ *we have*

$$\mathscr{L}_n(f; x) \rightrightarrows f^{(m)}(x)$$

*uniformly on* $[a, b]$.

We remark here that Theorems A and B have been obtained also by Tihomirov [6].

The following results provide an extension of Theorem A.

We denote by $E_{2k+1}$ the class of operators (1) whose kernels are such that for every $x \in [a, b]$ the integrals,

$$\mathscr{I}^{(1)}_{2k+1,n}(t) = \int_a^t dt_1 \int_a^{t_1} dt_2 \ldots \int_a^{t_{2k}} d\varphi_n(x, t_{2k+1}), \qquad a \leqslant t < x,$$

$$\mathscr{I}^{(2)}_{2k+1,n}(t) = \int_t^b dt_1 \int_{t_1}^b dt_2 \ldots \int_{t_{2k}}^b d\varphi_n(x, t_{2k+1}), \qquad x \leqslant t \leqslant b, \quad n = 1, 2, \ldots,$$

have constant, but mutually opposite signs.

THEOREM 5. *If operators* (1) *of class* $E_{2k+1}$, $k \geqslant 1$, *satisfy the condition*

$$\mathscr{L}_n(t^i, x) \rightrightarrows (x^i)^{(2l+1)}, \qquad i = 0, 1, \ldots, 2k+1,$$

*uniformly on* $[a, b]$ ($l$ *fixed*, $< k$), *then for any function* $f(x)$, $2k+1$ *times continuously differentiable on* $[a, b]$, *we have*

$$\mathscr{L}_n(f; x) \rightrightarrows f^{(2l+1)}(x)$$

*uniformly on* $[a, b]$.

COROLLARY 3. *If operators* (1) *satisfy the conditions of Theorem* 5 *and the condition*

$$\int_a^x |\mathscr{I}^{(1)}_{2l+1,n}(t)|\, dt + \int_x^b |\mathscr{I}^{(2)}_{2l+1,n}(t)|\, dt \leqslant C_3$$

*for all* $x \in [a, b]$, *where* $C_3$ *is an absolute constant, then for any function* $f(x)$, $2l+1$ *times continuously differentiable on* $[a, b]$, *we have*

$$\mathscr{L}_n(f; x) \rightrightarrows f^{(2l+1)}(x)$$

*uniformly on* $[a, b]$.

THEOREM 6. *If the values of the operators* (1) *are polynomials of degree* $\leqslant n$ *and satisfy the conditions of Theorem* 5, *then for at least one* $i$ *from the set* $i = 0, 1, \ldots, 2k+1$ *we have*

$$n^{2k-2l} \max_{x \in [a,b]} |\mathscr{L}_n(t^i, x) - (x^i)^{(2l+1)}| \nrightarrow 0 \qquad \text{as } n \to \infty.$$

We remark that in Theorems 1, 3, 5 the system of monomials $\{x^i\}$ can be replaced by Chebyshev's systems of sufficiently many times differentiable functions.

We also remark that Theorem 5 for $k = 1$ and $l = 0$ coincides with Theorem 5 of Korovkin's paper [2].

A similar generalization of Korovkin's theorems on positive operators can also be obtained for the approximation of periodic functions by trigonometric polynomials. We give here some of these results.

Let now

$$\mathscr{L}_n(f; x) = \frac{1}{\pi}\int_{-\pi}^{\pi} f(x+t)\mathscr{U}_n(t)\,dt, \qquad n = 1, 2, \dots, \tag{3}$$

where

$$\mathscr{U}_n(t) = \tfrac{1}{2} + \sum_{k=1}^{n} \lambda_k^{(n)} \cos kt.$$

If $\mathscr{U}_n(t) \geqslant 0$ for $t \in [0, \pi]$, then the operators (3) are positive. Concerning operators of type (3), the following theorems were proved by Korovkin ([2], Theorems 4 and 17):

1. *If* $\lim_{n\to\infty} \lambda_1^{(n)} = 1$, *then for any function* $f(x) \in C_{2\pi}$

$$\lim_{n\to\infty} \mathscr{L}_n(f, x) = f(x) \tag{4}$$

*uniformly in* $x$.

2. *The difference* $1 - \lambda_1^{(n)}$ *has order not greater than* $O(n^{-2})$.

It follows from the last statement, that the order of the approximation (4) for $f(x) = \sin x$ and $f(x) = \cos x$ cannot exceed $O(n^{-2})$.

The subsequent theorems show that the above Korovkin's results hold for a more general class of operators, containing the class of positive operators.

We denote by $F$ the class operators (3) such that the integrals

$$\int_{t}^{\pi/2} dt_1 \int_{t_1}^{\pi} U_n(t_2)\,dt_2 \qquad \text{for } 0 \leqslant t < \frac{\pi}{2},$$

$$\int_{\pi/2}^{t} dt_1 \int_{t_1}^{\pi} U_n(t_2)\,dt_2 \qquad \text{for } \frac{\pi}{2} \leqslant t \leqslant \pi$$

are non-negative. Obviously, the class $F$ contains the class of positive operators.

THEOREM 7. *If a sequence of operators* (3) *of class F satisfies the conditions*

(a) $$\lim_{n\to\infty} \lambda_1^{(n)} = 1,$$

(b) $$\|\mathcal{L}_n\| = \frac{1}{\pi}\int_{-\pi}^{\pi} |\mathcal{U}_n(t)|\,dt \leqslant C_4, \qquad n = 1, 2, \ldots,$$

*where $C_4$ is an absolute constant, then for any function $f(x) \in C^2$ we have*

$$\lim_{n\to\infty} L_n(f; x) = f(x)$$

*uniformly in $x$.*

THEOREM 8. *If the operators* (3) *are of class F and their norms are uniformly bounded, then the difference $1-\lambda_1^{(n)}$ converges to zero not more rapidly than $O(n^{-2})$.*

This theorem shows that operators of class $F$, as well as positive operators, cannot approximate functions with good regularity properties in an order higher than $O(n^{-2})$.

We now give one more result concerning the order of approximation of twice differentiable functions by operators of class $F$.

Let $W^{(2)}M$ denote the class of periodic functions with period $2\pi$ which have an absolutely continuous derivative and whose second order derivative fulfils the inequality $|f''(x)| \leqslant M$ at all points of its existence.

Write

$$\mathcal{E}_n W^{(2)}M = \sup_{f\in W^{(2)}M} \max_x |L_n(f; x) - f(x)|.$$

THEOREM 9. *If the operators* (3) *are of class F, then*

$$E_n W^{(2)}M = \frac{4M}{\pi}\left[\left(\frac{\pi^3}{32} - \sum_{k=1}^{n} \frac{\sin k\pi/2}{k^3}\right) + \sum_{k=1}^{n} (1-\lambda_k^{(n)})\frac{\sin k\pi/2}{k^3}\right].$$

Since positive operators belong to the class $F$, thus Theorem 9 can be applied e.g. to the Fejér operators. We obtain:

COROLLARY 4 (S. M. Nikol'skii [5]). *For the Fejér operators we have*

$$E_n W^{(2)}M = \frac{4M}{\pi n}\sum_{k=1}^{\infty} \frac{(-1)^k}{(2k+1)^2} + O(n^{-2}).$$

We now pass to another sort of results concerning the convergence of operators (1) and (3). Put

$$\varphi(x,t) = \begin{cases} 0 & \text{for } a \leqslant t < x, \\ 1 & \text{for } x \leqslant t \leqslant b, \end{cases} \qquad x \in (a, b],$$

$$\varphi(a,t) = \begin{cases} 0 & \text{for } t = a, \\ 1 & \text{for } a < t \leqslant b. \end{cases}$$

THEOREM 10. *A sequence of operators* (1) *with* $\varphi_n(x, a) = 0$ *converges to* $f(x)$ *uniformly on* $[a, b]$ *for every* $f(x) \in C[a, b]$ *if and only if*:

(i) $\mathscr{L}_n(1; x) \rightrightarrows 1$ *uniformly on* $[a, b]$;

(ii) *the sequence* $\{\varphi_n(x, t)\}_1^\infty$ *converges in mean to* $\varphi(x, t)$ *uniformly with respect to* $x \in [a, b]$;

(iii) *the total variations of the differences* $\varphi_n(x, t) - \varphi(x, t)$ *are bounded uniformly in n and x.*

An analogous result for linear summation methods of Fourier series (3) can be given a specific form. Let $\varphi(t)$ be the odd periodic function with period $2\pi$ defined by $\varphi(t) = \frac{1}{2}(\pi - t)$ for $t \in (0, \pi]$ and let

$$T_n(t) = \sum_{k=1}^{n} \frac{\lambda_k^{(n)}}{k} \sin kt.$$

THEOREM 11. *A sequence of operators* (3) *converges to* $f(x)$ *uniformly in x for every* $f(x) \in C_{2\pi}$ *if and only if*:

(i) *the sequence* $\{T_n(t)\}_1^\infty$ *converges in mean to* $\varphi(t)$;

(ii) *the total variations of the differences* $T_n(t) - \varphi(t)$ *are bounded uniformly in n.*

THEOREM 12. *If the sequence of polynomials* $T_n(t)$ *converges in mean to* $\varphi(t)$ *and the convergence is of order* $O(n^{-1})$, *then the norms of operators* (3) *are bounded uniformly in n.*

It is not difficult to obtain estimates for the order of approximation of various classes of functions by operators (1) and (3). Those estimates can be expressed in terms of certain constructive characteristic of the function class in question and by the order of convergence $\varphi_n(x, t) \to \varphi(x, t)$ in mean ($T_n(t) \to \varphi(t)$ in mean in the periodic case).

We give just one of such estimates. Let $E_n = E_n(f)$ be the best approximation of a function $f(x)$ by trigonometric polynomials of degree

$\leqslant n$; let $V_n$ be the total variation of the difference $T_n(t)-\varphi(t)$ on $[-\pi, \pi]$, and let

$$\beta_n = \int_{-\pi}^{\pi} |T_n(t)-\varphi(t)|\,dt.$$

THEOREM 13. *If* $f(x) \in C_{2\pi}$, *then*

$$|\mathcal{L}_n(f; x)-f(x)| \leqslant \frac{1}{\pi}(1+V_n)E_n+\frac{12}{\pi}\beta_n \sum_{k=0}^{n-1} E_k.$$

## REFERENCES

[1] V. A. Baskakov, 'On Certain Sequences of Linear Operators, Convergent in the Space of Continuous Functions', Author's Thesis, Moskva 1955.

[2] P. P. Korovkin, 'Linear Operators and the Approximation Theory', Fizmatgiz, Moskva 1959.

[3] P. P. Korovkin, 'Convergent Sequences of Linear Operators', *Usp. Mat. Nauk* **17**, 4 (1962) 147–152.

[4] P. P. Korovkin, On the Uniqueness Conditions for the Moment Problem and for the Convergence of Sequences of Linear Operators', *Učebn. Zap. Kalinin. Ped. Inst.* **26** (1958) 95–102.

[5] S. M. Nikol'skii, 'On the Approximation of Periodic Functions by Trigonometric Polynomials', *Trudy Mat. Inst. Steklov* **15** (1945).

[6] N. B. Tihomirov, 'On the Approximation by Linear Operators to Derivatives of Functions', *Trudy Centr. Zon. Ob'ed. Mat. Katedr. Funkc. Analiz. i Teoria Funkcii* **2**, *Kalinin* 1871, 162–175.

# ON THE UNIFORM CONTINUITY OF THE OPERATOR OF BEST APPROXIMATION

V. I. BERDYSHEV

*Sverdlovsk*

We introduce the following notation: for a Banach space $X$, $X^*$ is its conjugate; for $x, y \in X$ let $xy = \|x-y\|$; for a given set $M \subset X$ and an element $x \in X$ let $xM = \inf\{xz: z \in M\}$, $x_M = \{z \in M: xz = xM\}$; for two sets $M_1, M_2 \subset X$ write

$$M_1 M_2 = \min\{\sup_{x \in M_2} xM_1,\ \sup_{x \in M_1} xM_2\}.$$

A set $M \subset X$ is called a *set of the existence of a best approximation*, short: a *set of existence* (cf. [4]), if $x_M$ is non-empty for any $x \in X$. If, for any $x \in X$, the set $x_M$ consists of a single element, then $M$ is called a *Chebyshev set* (cf. [4]). The following fact is well known: every closed convex subset $M \subset X$ is a set of existence (a Chebyshev set) only in the case when $X$ is reflexive (resp. reflexive and strictly convex). If $M$ is a set of existence, we can define the operator of best approximation $x \to x_M$ (the metric projection). The problem of the continuity of the operator of best approximation has been extensively studied; a survey of known results can be found in [13]. Phelps [14] proved that the Hilbert space $X$ is characterized by the following property: for every closed convex subset $M \subset X$ the inequality $x_M y_M \leqslant xy$ holds true. Holmes and Kripke [9] investigated conditions for $x_M y_M \leqslant K \cdot xy$, where $K$ may depend on $x$ and $M$.

Let $\mathscr{M}$ denote the class of convex sets of existence $M \subset X$, let $\mathscr{L}$ denote the class of all subspaces of existence $L \subset X$, and $\mathscr{L}_k$ the class of subspaces $L \in \mathscr{L}$ of dimension $k$, $k = 1, 2, \ldots$, $\mathscr{L}^k$ the class of subspaces $L \in \mathscr{L}$ of codimension $k$. For a set $M \in \mathscr{M}$ put

(1) $\omega(t, M) = \omega(t, M)_X = \sup\{x_M y_M: xy \leqslant t,\ yM \leqslant 1\}, \quad 0 \leqslant t \leqslant 1,$

and let

(2) $$\omega(t) = \omega(t)_X = \sup\{\omega(t, L): L \in \mathscr{L}\}$$

be the modulus of continuity of the operator of best approximation in the space $X$. Write

(3) $$\omega_k(t) = \sup\{\omega(t, L): L \in \mathscr{L}_k\}, \quad \omega^k(t) = \sup\{\omega(t, L): L \in \mathscr{L}^k\}.$$

The following equality has been established in [1]:

(4) $$\omega(t) = \sup\{\omega(t, M): M \in \mathscr{M}\}.$$

As can be seen from examples given by Bernstein [2] and Stečkin [7], $\omega_2(t)_X \underset{t\to 0}{\nrightarrow} 0$ if $X = C[a, b]$ is the space of functions continuous on $[a, b]$ with the sup-norm. The following example shows that also $\omega_1(t)$ need not converge to 0 as $t \to 0$.

EXAMPLE 1. Let $X = L[-1, 1]$ be the space of functions $x(t)$ integrable on $[-1, 1]$ with the norm $\|x(t)\| = \int_{-1}^{1} |x(t)|\,dt$, and let $L$ denote the one-dimensional subspace of constant functions; let $\varepsilon \in (0, 1)$ and put:

$$x = x(t) = \begin{cases} -1, & -1 \leqslant t < \varepsilon, \\ 1, & \varepsilon \leqslant t \leqslant 1, \end{cases}$$

$$y = y(t) = \begin{cases} -1, & -1 \leqslant t < -\varepsilon, \\ 1, & -\varepsilon \leqslant t \leqslant 1, \end{cases}$$

We then have $xy = 4\varepsilon$, $x_L = -1$, $y_L = +1$, $x_L y_L = 4$ and thus $\omega(t)_X \geqslant 4$ for $0 < t \leqslant 1$.

Stečkin has posed the problem of characterization of spaces $X$ in which the operator of best approximation is uniformly continuous, i.e. $\lim_{t\to 0} \omega(t)_X = 0$. This problem can be solved in the terms of sections of the unit ball $V$ of $X$.

**1.** Let $d(M)$ denote the diameter of a set $M \subset X$. For a set $M$ contained in a hyperplane $P \subset X$ denote (cf. [6]):

(5) $$s(M)_P = \inf\{\sup_{x\in M-p} f(x) - \inf_{x\in M-p} f(x); f \in (P-P)^*, \|f\| = 1\}, \quad p \in P.$$

Further, write

(6) $$\Omega(t) = \Omega(t)_X = \sup d(V \cap P),$$

where the supremum is taken with respect to all hyperplanes $P \subset X$ such that $s(V \cap P)_P \leqslant t$.

Most of the subsequent statements are given without proofs. The proofs can be found in [1].

THEOREM 1. *The following conditions are equivalent*:

(a) $$\lim_{t \to 0} \omega(t) = 0;$$

(b) $$\lim_{t \to 0} \omega_k(t) = 0, \quad k = 2, 3, \ldots;$$

(c) $$\lim_{t \to 0} \sup_{k=1,2,\ldots} \omega_k(t) = 0;$$

(d) $$\lim_{t \to 0} \omega^k(t) = 0, \quad k = 2, 3, \ldots;$$

(e) $$\lim_{t \to 0} \Omega(t) = 0.$$

The relation $\lim_{t \to 0} \omega_1(t) = 0$, and thus also $\lim_{t \to 0} \omega(t) = 0$ need not hold even in a three-dimensional non-strictly convex space:

EXAMPLE 2. Let $X = \{x = (a_1, a_2, a_3)\colon -\infty < a_i < \infty, i = 1, 2, 3\}$ with the norm $\|x\| = \max_{i=1,2,3} |a_i|$, $x = (1-\varepsilon, 0, 1+\varepsilon)$, $y = (1+\varepsilon, 0, 1-\varepsilon)$, $\varepsilon > 0$, $L = \{\lambda(-\varepsilon, 1, \varepsilon)\colon |\lambda| < \infty\}$. We have $xy = 2\varepsilon$, $x_L = (-\varepsilon, 1, \varepsilon)$, $y_L = (\varepsilon, -1, -\varepsilon)$, $x_L y_L = 2$, hence $\omega(t)_X \geqslant 2$ for $0 < t \leqslant 1$.

We denote by *UC* the class of reflexive strictly convex spaces such that $\lim_{t \to 0} \omega(t) = 0$. Let *UR* denote the class of uniformly convex spaces, i.e. spaces with strictly positive modulus of convexity

(7) $$\delta(t) = \delta(t)_X = \inf_{\substack{\|x_1\| = \|x_2\| = 1 \\ \|x_1 - x_2\| = t}} \left(1 - \frac{\|x_1 + x_2\|}{2}\right),$$

for $t \in (0, 2)$. Finally, let $E$ denote the class of spaces such that for every hyperplane $P \subset X$, every sequence $\{x_n\} \subset P$ satisfying the condition $\lim_{n \to \infty} \|x_n\| = \inf\{\|x\| : x \in P\}$ necessarily converges. Clearly, every space $X \in E$ is reflexive and strictly convex. The class $E$ has been introduced by Ky Fan and Glicksberg [12]. They also have shown that in a space $X \in E$ the operator of best approximation is continuous. It is known that $UR \subset E$.

THEOREM 2. *The following inclusions*

$$UR \subset UC \subset E \tag{8}$$

*hold and are proper*: $UR \neq UC \neq E$.

We shall give examples of reflexive strictly convex spaces $X^{(1)}$, $X^{(2)}$ such that $X^{(1)} \in UC \setminus UR$, $X^{(2)} \in E \setminus UC$.

Denote $V_r(x) = \{z : xz \leqslant r\}$, $\theta$ is the zero element of the space in question.

EXAMPLE 3. Let $X = l_2$ be the space of sequences $x = (\xi_1, \xi_2, \ldots)$ with $\|x\| = \sqrt{\sum_{i=1}^{\infty} \xi_i^2} < \infty$; $e_1 = (1, 0, 0, \ldots)$, $e_2 = (0, 1, 0, \ldots)$, ...; let $\beta$ be a fixed number, $\frac{1}{\sqrt{2}} < \beta < 1$. We define $X^{(1)}$ as the space $X$ with the norm induced by the unit ball $V^{(1)} = V_1(\theta) \cap [\bigcap_{k=1}^{\infty} (V_{r_k}(ke_k) \cap \cap V_{r_k}(-ke_k))]$, where $r_k = \sqrt{(k+\beta)^2 + 1 - \beta^2}$.

EXAMPLE 4. Let $X = l_2$, $x_i^j = \frac{1+\beta}{2} e_i + (-1)^j \gamma \cdot e_{i+1}$, $j = 1, 2$, where $e_i$ are defined as in the preceding example, $\gamma = \sqrt{1 - \left(\frac{1+\beta}{2}\right)^2}$ and $\beta \in \left(\frac{1}{\sqrt{2}}, 1\right)$ is chosen so that $\|x_i^j\| = 1$. Let $M^i$ denote the intersection of all balls with radii $\leqslant i$ containing the elements $\pm x_i^j$, $j = 1, 2$ and the set $\{x \in V_1(\theta) : (e_i, x) \leqslant \beta\}$. The unit ball of the space $X^{(2)}$ is defined as $V^{(2)} = \bigcap_{i=1}^{\infty} M^i$.

**2.** Let $\Omega_k(t)$ and $\Omega^k(t)$ denote supremum (6) taken with respect to planes $P \subset X$ with $\dim P = k$, resp. $\operatorname{codim} P = k$.

We introduce the functions

$$\begin{aligned} \tilde{\omega}(t, L) &= \sup\{x_L y_L, xy \leqslant t, yy_L = 1\}, \\ \tilde{\omega}(t) &= \sup\{\tilde{\omega}(t, L) : L \in \mathscr{L}\}, \end{aligned} \tag{9}$$

$$\begin{aligned} \tilde{\Omega}(t, L) &= \sup\{xy : xL = \|x\| = yL = \|y\| = 1,\ (x-y)L \leqslant t\}, \\ \tilde{\Omega}(t) &= \sup\{\tilde{\Omega}(t, L) : L \in \mathscr{L}\}. \end{aligned} \tag{10}$$

The modulus of convexity $\delta(t)$ (see (7)) of $X$ is a function continuous and strictly increasing on the interval $[\max_{\delta(t)=0} t, 2]$. Thus the function $\varepsilon(t)$ inverse to $\delta(t)$ is well defined on the interval $[0, \delta(2)]$.

We list some properties of the introduced functions:

($\omega^1$) $\omega(t, L) = \sup\{x_L y_L : xy \leqslant t, y\theta \leqslant 1\}$
$= \sup\{x_L y_L : xy \leqslant t, yy_L = y\theta \leqslant 1\}$;

($\omega^2$) $\omega(t) = \sup \omega(t, P)$, the supremum taken with respect to all hyperplanes $P \subset X$;

($\omega^3$) functions $\omega(t, L), \omega(t)$ are continuous non-decreasing on $(0, 1)$, $\omega(0, L) = \omega(0) = 0$;

($\omega^4$) $t \leqslant \omega(t) \leqslant 2(1+t)$; for any hypersubspace $L \subset X$, $x_L y_L \leqslant 2 \cdot xy$ and consequently $\omega^1(t) \leqslant 2t$;

($\omega^5$) (a) $\omega(s) \geqslant s \dfrac{\omega(t)}{t}$ for $0 \leqslant s \leqslant t$, i.e. $\dfrac{\omega(t)}{t}$ is non-increasing;

(b) let $c_t$ be the infimum of numbers $c$, $0 < c \leqslant 1$ with the property: for any $\varepsilon > 0$ there exist a subspace $L$ and elements $x, y$ such that $xy = t$, $\min\{xx_L, yy_L\} \leqslant c$, $x_L y_L \geqslant \omega(t) - \varepsilon$; then $\omega(s) = s \dfrac{\omega(t)}{t}$ for $t \leqslant s \leqslant \dfrac{t}{c}$;

(c) if $c_t < 1$ for all $t$ from some interval $(t_0, t_1)$, $0 \leqslant t_0 < t_1 \leqslant 1$, then $\omega(t) = \dfrac{\omega(t_0)}{t_0} t$ for $t_0 \leqslant t \leqslant t_1$; the analogues of these properties hold for $\omega(t, L)$;

($\omega^6$) $\omega^i(t) = \omega_k(t) = \omega(t)$, $k = 1, 2, \ldots, \infty$, $i = 2, \ldots$, $0 \leqslant t \leqslant 1$, $X$ being strictly convex;

($\Omega^1$) the function $\Omega(t)$ is non-decreasing in $[0, 2]$ and continuous in $(0, 2]$; $\Omega(t) \geqslant t$ for $0 \leqslant t \leqslant 2$, $\Omega(2) = 2$; the analogues of these properties hold for $\tilde{\Omega}(t)$;

($\Omega^2$) $\tilde{\Omega}(t, L) = \sup\{\sup_{x^i \in \theta P_i} x^1 x^2 : P_1 P_2 \leqslant t, P_i - P_i = L, \theta P_i = 1, i = 1, 2\}$;

($\Omega^3$) if $X$ is reflexive strictly convex, then $\tilde{\Omega}(t, L) \leqslant \Omega(t)$;

($\Omega^4$) $\Omega^i(t) = \Omega_k(t) = \Omega(t), i = 1, 2, \ldots, k = 1, 2, \ldots, \infty, 0 \leqslant t \leqslant 2$;

($\varepsilon^1$) $\lim_{t\to 0} t/\varepsilon(t) = 0$;

($\varepsilon^2$) $\Omega(t) \leqslant \varepsilon(t/2)$, $0 \leqslant t \leqslant 2$.

We shall prove some of the above properties. If $x, y \in X$, $yy_L \leqslant 1$, $y_1 \in y_L$, then putting $x' = x - y_1$, $y' = y - y_1$ we have: $x'_L = x_L - y_1$, $\theta \in y'_L$, $x'y' = xy$, $x'_L y'_L = x_L y_L$, $\theta y' \leqslant 1$. Thus ($\omega^1$) holds.

The inequality $\omega^1(t) \leqslant 2t$ (see ($\omega^4$)) is a consequence of the following lemma. In the sequel aff$M$ denotes the affine envelope of the set $M$.

LEMMA 1. *If $x, x', y \in X$ and $P$ is a plane in $X$ such that $x_P \neq \emptyset$, $x' \in x_P$, $y \in \operatorname{aff}(P \cup \{x\})$, then $y_P \neq \emptyset$, the element $y' = y + \dfrac{yP}{xx'}(x' - x)$ belongs to $y_P$ and $x'y' \leqslant 2xy$.*

Only the last inequality requires a proof. Let e.g. $xx' \geqslant yy'$. Write $z = x' + (y - y')$. Then

$$x'y' = zy \leqslant xy + xz = xy + xx' - yy' \leqslant xy + xy' - yy' \leqslant 2xy.$$

We now prove ($\omega^5$) assuming, for simplicity, that $X$ is strictly convex. For any $\varepsilon > 0$ there exist a subspace $L$ and elements $x, y$ such that $xy = t$, $yy_L \leqslant c_t + \varepsilon$, $x_L y_L \geqslant \omega(t) - \varepsilon$. For $0 < \lambda < \dfrac{1}{c_t + \varepsilon}$ we have

$$(\lambda x)(\lambda y) = \lambda t, \qquad (\lambda y)(\lambda y)_L = \lambda \cdot yy_L \leqslant \lambda(c_t + \varepsilon) \leqslant 1,$$

$$\omega(\lambda t) \geqslant (\lambda x)_L (\lambda y)_L = \lambda \cdot x_L y_L \geqslant \lambda\big(\omega(t) - \varepsilon\big).$$

Hence (a) $\omega(\lambda t) \geqslant \lambda \omega(t)$ for $0 < \lambda < 1/c_t$. In particular, $\omega(\mu T) \geqslant \mu \omega(T)$ for $c_t \leqslant \mu \leqslant 1$ provided $c_t < 1$. Putting $\lambda = 1/\mu$, $t = T/\mu$ we get (b) $\lambda \omega(t) \geqslant \omega(\lambda t)$ for $1 \leqslant \lambda \leqslant 1/c_t$.

In view of (a) we have $\omega(t) \leqslant \dfrac{\omega(t_0)}{t_0} t$ for $t_0 \leqslant t \leqslant t_1$. Suppose there exists a number $t_2$, $t_0 < t_2 \leqslant t_1$, such that $\omega(t_2) < \dfrac{\omega(t_0)}{t_0} t_2$. Let $\dfrac{\omega(t_2)}{t_2} < K < \dfrac{\omega(t_0)}{t_0}$, $t_3 = \max\{t \leqslant t_2 : \omega(t) \leqslant Kt\}$. Since $c_{t_3} < 1$, thus, by (b), for all $t$ with $t_3 \leqslant t \leqslant t_3/c_{t_3}$ we have $\omega(t) = Kt$, a contradiction.

From the monotonicity of $\omega(t)$ and by ($\omega^5$) (a) we get for $0 < s < t < 1$ the inequality

$$0 \leqslant \omega(t) - \omega(s) \leqslant (1 - s/t)\omega(t),$$

hence easily follows the continuity of $\omega(t)$.

We now prove $(\Omega^3)$. Let $x, y \in X$, $xL = \|x\| = yL = \|y\| = 1$, $(x-y)L \leqslant t$. Put $P' = L+x$, $P'' = L+y$, $P = \operatorname{aff}(P' \cup P'')$. Then $d(V \cap P) \geqslant xy$, $P'P'' \leqslant t$. Since $X$ is strictly convex, the set $V \cap P$ lies between the planes $P', P''$, hence $s(V \cap P) \leqslant P'P''$.

Property $(\varepsilon^1)$ follows from the convergence $\lim_{t\to 0} \frac{\delta(t)}{t} = 0$, which holds already in any two-dimensional space.

**3.** We shall give a condition for the uniform continuity of the operator of best approximation from a fixed subspace. We shall employ the following

LEMMA 2. *Let $L$ be a subspace of existence,* $\operatorname{codim} L \neq 1$; *let $x, y \notin L$, $y \notin \operatorname{aff}(L \cup \{x\})$. Then for any $r$ with* $\min(xL, yL) \leqslant r \leqslant \max(xL, yL)$ (*short*: $r \in [xL, yL]$), *there exist planes $P_i$, $i = 1, 2$, such that* $P_i - P_i = L$, $\theta P_i = 1$, $i = 1, 2$,

$$P_1 P_2 \leqslant 2\frac{xy}{r}, \qquad x_L y_L \leqslant r(\theta_{P_1}\theta_{P_2}) + 2\cdot xy.$$

The proof of this lemma in the case $\dim L = 1$ may be found in [1] and can be fully applied also in the general case.

COROLLARY. *Let $L \subset X$ be a subspace of existence, and let $x, y \notin L$, $r \in [xL, yL]$. Then*

$$x_L y_L \leqslant r \cdot \tilde{\Omega}\left(2\frac{xy}{r}, L\right) + 2\cdot xy. \tag{11}$$

The proof results in applying Lemma 1 if $y \in \operatorname{aff}(L \cup \{x\})$ and Lemma 2 in the opposite case.

THEOREM 3. *For a Chebyshev subspace $L \subset X$ the conditions*

(a) $\lim_{t\to 0} \omega(t, L) = 0$,

(b) $\lim_{t\to 0} \tilde{\Omega}(t, L) = 0$

*are equivalent.*

PROOF. If $x, y \in X$, $\|x\| = \|y\| = xL = yL = 1$, $(x-y)L = t$, then, putting $y' = x_{L+y}$, we get $xy' = t$. Hence, since $x_L = \theta$, $y' - y = y'_L$, thus $xy \leqslant xy' + y'y = xy' + \theta(y'-y) = xy' + x_L y'_L$. So (a) implies (b). The converse implication can be easily obtained by applying (11) and the obvious inequality $x_L y_L \leqslant 2(xy + r)$.

**4.** We shall give a condition for $\omega(t) \to 0$, $t \to 0$. Let $0 < t < 1$. Let $c^t$ denote the supremum of number $c$ with the following property: for any $\varepsilon > 0$ there exist a subspace $L$ and elements $x, y$ such that $xy = t$, $x_L y_L \geqslant \omega(t) - \varepsilon$, $\min(xL, yL) \leqslant 1$, $\max(xL, yL) \geqslant c$. Clearly (see ($\omega^5$)) $c_t \leqslant c^t$.

LEMMA 3. *The following inequalities hold true*:

$$\omega(t) \leqslant r\tilde{\Omega}\left(\frac{2t}{r}\right) + 2t, \tag{12}$$

$$\omega(t) \leqslant r\Omega\left(\frac{2t}{r}\right) + 2t, \tag{12'}$$

*for* $r = c^t$, $r = \min(1, c^t)$.

PROOF. For any $\varepsilon > 0$ we can find $L \subset X$, $x, y \in X$, such that $xy = t$, $\omega(t) \leqslant x_L y_L + \varepsilon$, $\min(xL, yL) \leqslant 1$, $\max(xL, yL) \geqslant c^t - \varepsilon$. Then, by (11), we have for $r \in [xL, yL]$

$$\omega(t) \leqslant r\tilde{\Omega}\left(\frac{2t}{r}, L\right) + 2t + \varepsilon.$$

Inequality (12′) is proved similarly.

REMARK. Analogous inequalities hold for $\omega(t, L)$, and $\tilde{\Omega}(t, L)$.

THEOREM 4. *If* $\lim_{t\to 0} \tilde{\Omega}(t) = 0$, *then* $\lim_{t\to 0} \omega(t) = 0$.

PROOF. If $\lim_{t\to 0} t/c^t = 0$, the assertion follows from (12). So suppose that there exists a sequence of positive numbers $t_i$ with $\lim t_i = 0$ and such that $t_i/c^{t_i} \geqslant \mu > 0$. Then $c_{t_i} \to 0$, and, by ($\omega^5$), $\omega(s) = s\omega(t_i)/t_i$ for $t_i \leqslant s < t_i/c_{t_i}$, $i = 1, 2, \ldots$ Hence follows $\omega(s) = s\omega(t_1)/t_1$ for $0 \leqslant s \leqslant \mu$.

We shall give some estimates for $\omega(t)$. Let $\varphi(t)$ be a function defined on $[0, 1]$ and let $0 < T \leqslant 1$. Denote by $\varphi(t)^*$ the smallest concave majorant of $\varphi(t)$ on $[0, T]$ and put $\varphi(t)_* = \max_{0 \leqslant x \leqslant T} \varphi(t, x)$, where

$$\varphi(t, x) = \begin{cases} \dfrac{\varphi(x)}{x} t & \text{for } 0 \leqslant t \leqslant x, \\ \varphi(x) & \text{for } x < t \leqslant T. \end{cases}$$

THEOREM 5. *Let* $T = \max\{t: \omega(t) \leqslant \Omega(2t)+2t, 0 \leqslant t \leqslant 1\}$. *Then*

$$\omega(t) \leqslant (\Omega(2t)+2t)_* \leqslant (\Omega(2t)+2t)^*, \quad 0 \leqslant t \leqslant T;$$

(13)

$$\omega(t) = \frac{\omega(T)}{T}t, \quad T \leqslant t \leqslant 1.$$

PROOF. Put $\varphi(t) = \Omega(2t)+2t$. If $c_t = 1$, then, by (12′) and the inequality $c_t \leqslant c^t$, we have $\omega(t) \leqslant \varphi(t)$. Consequently, if $t_0, t_1, t_0 < t_1$, are such that $\omega(t_0) = \varphi(t_0)$, $\omega(t_1) = \varphi(t_1)$ or $t_1 = 1$, and $\omega(t) > \varphi(t)$ for $t_0 < t < t_1$, then $c_t < 1$ for $t_0 < t < t_1$. By ($\omega^5$) (c) we have $\omega(t) \leqslant \varphi(t)_*$ for $0 \leqslant t \leqslant T$ and $\omega(t) = \dfrac{\omega(T)}{T}t$ for $T \leqslant t \leqslant 1$. The inequality $\varphi(t)_* \leqslant \varphi(t)^*$ is obvious.

COROLLARY. $\omega(t) \leqslant 2\Omega(2t)^*$ *for* $0 \leqslant t \leqslant T$.

THEOREM 6. *There exists* $T$, $0 < T \leqslant 1$, *such that*

(14) $$\omega(t) \leqslant 2\varepsilon(t)^*, \quad 0 \leqslant t \leqslant T.$$

PROOF. Let $\omega(t) > \Omega(2t)+2t$ for $t > 0$. Then by (13) $\omega(t) = Kt$ for $0 \leqslant t \leqslant 1$ and the theorem follows from ($\varepsilon^1$). In the opposite case inequality ($\varepsilon^2$) should be applied.

Note that for certain spaces, e.g. for $L_p[a, b]$, $1 < p < \infty$ the equality $\varepsilon(t) = \varepsilon(t)^*$ holds (see [8]).

From (12), ($\Omega^1$) and ($\Omega^2$) follows

THEOREM 7. *If the space X is reflexive and strictly convex, then*

(15) $$\hat{\omega}(t) \leqslant 2\tilde{\Omega}(2t), \quad 0 \leqslant t \leqslant 1.$$

**5.** By an analogy to (1), for a set $M \subset X$, $M \neq X$, we define the quantity

(16) $$\mu(t, M) = \sup\{d(M \cap V_{xM+t}(x)): xM \leqslant 1\}.$$

It is natural to call a set $M$ a uniformly approximately compact set (see [5]) if $\lim_{t\to 0} \mu(t, M) = 0$. In the case when $M$ is convex set of existence the quantity $\mu(t, M)$ coincides with the supremum (16) taken with respect to all $x$ such that $xM = 1$. In fact, if $xM < 1$ and if $P = \{z: f(z) = c,\ f \in X^*\}$ is a hyperplane separating the sets $M$ and

$V_{xM}(x)$, $\inf f(M) \geqslant c$, then, writing $y = x' + \dfrac{x-x'}{xx'}$, $x' \in x_M$, we have $yM = yP = yx' = 1$,

$$\{z\colon zy \leqslant 1+t, f(z) \geqslant c\} \supset \{z\colon zx \leqslant xM+t, f(z) \geqslant c\}$$

and consequently $M \cap V_{xM+t}(x) \subset M \cap V_{yM+t}(y)$.

Denote

$$\hat{\varepsilon}(t) = \sup_{f \in X^*, \|f\|=1} d\{x \in V\colon f(x) \geqslant 1-t\}. \tag{17}$$

Let $M$ be convex set of existence. Having used a hyperplane separating the set $M$ and $V_{xM}(x)$, $x \notin M$, we have

$$\mu(t, M) \leqslant (1+t)\hat{\varepsilon}\left(\frac{t}{1+t}\right). \tag{18}$$

Let $x, y \notin M$. For any $y' \in y_M$ $xy' \leqslant xM + 2xy$, i.e. $y_M \subset M \cap \cap V_{xM+2xy}(x)$, hence

$$x'y' \leqslant \mu(2xy, M), \quad \forall x' \in x_M, \quad y' \in y_M. \tag{19}$$

By the way we have proved

$$x'y' \leqslant (xM+2xy)\tilde{\varepsilon}\left(\frac{2xy}{xM+2xy}\right), \quad \forall x' \in x_M, \quad y' \in y_M, \tag{20}$$

and

$$\mu(t) \overset{\text{def}}{=} \sup_{M \in \mathscr{V}} \mu(t, M) = (1+t)\tilde{\varepsilon}\left(\frac{t}{1+t}\right), \tag{21}$$

where $\mathscr{V}$ is a class of all convex subsets $M \subset X$.

THEOREM 8. *The following conditions are equivalent*:

(a) $\lim_{t \to 0} \mu(t, M) = 0$ *for every* $M \in V$,

(b) $X$ *is uniformly convex.*

PROOF. It is known that a space is uniformly convex only if $\lim_{t \to 0} \tilde{\varepsilon}(t) = 0$. It remains to apply (18).

Inequality (20) for convex Chebyshev set and Theorem 8 were kindly delivered by Vlasov.

We shall estimate $x_M y_M$ by the modulus of convexity of the space.

THEOREM 9. *If $M \subset X$ is a convex set of existence, then for all $x, y \in X$ and $x' \in x_M \setminus y_M$, $y' \in y_M \setminus x_M$*

$$(22)\quad x'y' \leqslant xM \cdot \varepsilon\left(\frac{2xy}{xM+2xy}\right)+xy, \qquad x'y' \leqslant xM \cdot \varepsilon\left(\frac{xy}{xM}\right)+xy.$$

PROOF. Write

$$R = xM+2\cdot xy, \qquad x_1 = x+\frac{R}{xM}\cdot(x'-x),$$

$$y_1 = x+\frac{R}{xy'}\cdot(y_M-x), \qquad y_2 = x+\frac{xM}{xy'}(y'-x),$$

$$v = \frac{x_1+y_1}{2}, \qquad w = \frac{x'+y_2}{2}.$$

The segments $[x_1, y_1]$ and $[x', y_2]$ are parallel, the point $y'$ belongs to $[y_1, y_2]$, the point $z = [v, x] \cap [x', y']$ belongs to $[v, w]$ and, as is easy to see, $wz \leqslant vw/2$; moreover, $(x', y') \cap V_{xM}(x) = \varnothing$. Applying the inequalities $xv \geqslant xz \geqslant xM$ we obtain

$$(23)\quad xM-xw = (R-xv)\frac{xM}{R} \leqslant (R-xM)\frac{xM}{R} = 2\cdot xy\cdot\frac{xM}{R},$$

$$(24)\quad xM-xw \leqslant xz-xw = zw \leqslant \tfrac{1}{2}\cdot vw \leqslant \tfrac{1}{2}(R-xM) = xy.$$

Then (see (7))

$$\delta\left(\frac{x'y_2}{xM}\right) \leqslant \frac{xM-xw}{xM}, \qquad \text{i.e.} \qquad x'y_2 \leqslant xM\cdot\varepsilon\left(\frac{xM-xw}{xM}\right).$$

Hence

$$(25)\quad x'y' \leqslant x'y_2+y_2y' \leqslant xM\cdot\varepsilon\left(\frac{xM-xw}{xM}\right)xy.$$

Substituting (23), resp. (24), into (25) we obtain the first, resp. the second, of inequalities (22).

Applying (20) or (22) we can also obtain the inclusion $UR \subset UC$.

The author is grateful to S. B. Stečkin for the formulation of the problem.

## REFERENCES

[1] V. I. Berdyshev, 'On the Modulus of Continuity of the Operator of Best Approximation', *Mat. Zametki* **15** (1974) 797–808.

[2] S. N. Bernstein, 'Extremal Properties of Polynomials', *ONTI*, 1937.

[3] J. A. Clarkson, 'Uniformly Convex Spaces', *Trans. Amer. Math. Soc.* **40**, 3 (1936) 396–414.

[4] N. V. Efimov and S. B. Stečkin, 'Certain Properties of Chebyshev Sets', *Dokl. Akad. Nauk SSSR* **118**, (1958) 17–19.

[5] N. V. Efimov, 'Approximative Compactness and Chebyshev Sets', *ibidem* **140**, (1961) 522–524.

[6] H. G. Eggleston, *Convexity*, Cambridge 1958.

[7] P. V. Galkin, 'On the Modulus of Continuity of the Operator of Best Approximation for the Space of Continuous Functions', *Mat. Zametki* **10** (1971) 601–613.

[8] O. Hanner, 'On the Uniform Convexity of $L^p$ and $l^p$', *Arkiv för Math.* **3**, 3 (1956) 236–244.

[9] R. Holmes and B. Kripke, 'Smoothness of Approximation', *Michigan Math. J.* **15** (1968) 225–248.

[10] V. K. Ivanov, 'On a Certain Type of Incorrect Linear Equations in Topological Vector Spaces', *Sibirsk. Mat. V.* **6** (1965), 832–839.

[11] V. L. Klee, 'Some Characterization of Reflexivity', *Rev. Ci.* (Lima) **52** (1950) 15–23.

[12] Ky Fan and J. Glicksberg, 'Some Geometric Properties of the Spheres in a Normed Linear Space', *Duke Math. J.* **25** (1958) 553–568.

[13] E. V. Oshman, 'On the Continuity of the Metric Projection and Geometric Properties of the Unit Sphere in a Banach Space', Author's Thesis, Sverdlovsk 1970.

[14] R. P. Phelps, 'Convex Sets and Nearest Points', *Proc. Amer. Math. Soc.* **8** (1957) 790–797.

# STETIGKEITSEIGENSCHAFTEN VON OPTIMIERUNGSAUFGABEN UND LINEARE TSCHEBYSCHEFF-APPROXIMATION

HANS-PETER BLATT

*Erlangen*

*Abstract.* We consider the problem how the feasible region, the set of solutions and the minimum value are changing under perturbations of the non-linear programming problem. The constraints are always convex, but the objective function needs only to be continuous. As an application we prove the convergence of a gradient method for linear Chebyshev approximation.

Wir untersuchen die Frage, wie sich bei Optimierungsaufgaben die zulässigen Punkte, die Lösungen und der Extremalwert der Zielfunktion bei einer Änderung des Problems verhalten. Dantzig, Folkman und Shapiro [1] untersuchten Stetigkeitseigenschaften der Minimallösungen einer konvexen Funktion über einem Bereich des $\boldsymbol{R}^n$, wenn sie diese Funktion und ihren Definitionsbereich änderten. Als Anwendung betrachteten sie lineare Optimierungsaufgaben. Kürzlich zeigte Krabs [4] für abstrakte Optimierungsaufgaben, daß sich der Minimalwert der Zielfunktion bei einer stetigen Änderung des Problems auch stetig ändert. Als wesentliches Hilfsmittel benutzt er dabei die duale Optimierungsaufgabe. Bei unserer Untersuchung brauchen nur die Restriktionen konvex zu sein, dagegen setzen wir von der Zielfunktion nur die Stetigkeit voraus. Als Anwendung der Theorie zeigen wir die Konvergenz eines Gradientenverfahrens für lineare Tschebyscheff-Approximation.

**1. Stetigkeitseigenschaften von Optimierungsaufgaben.** $Q$ sei eine kompakte, konvexe Teilmenge des $\boldsymbol{R}^n$ und $|\cdot|$ eine Norm in $\boldsymbol{R}^n$. $\Lambda$ sei ein metrischer Raum mit Metrik $d$. Für jedes $\lambda \in \Lambda$ seien $m$ $(m < n)$ affinlineare Funktionen

$$g_{\lambda,\mu}: \boldsymbol{R}^n \to \boldsymbol{R}$$
$$x \to \langle a_{\lambda,\mu}, x\rangle + h_{\lambda,\mu}$$

gegeben. Dabei sind $a_{\lambda,\mu} \in \boldsymbol{R}^n$, $b_{\lambda,\mu} \in \boldsymbol{R}$ für $\mu = 1, \ldots, m$, und $\langle\ ,\ \rangle$ bezeichnet das Skalarprodukt in $\boldsymbol{R}^n$. Außerdem seien $a_{\lambda,1}, \ldots, a_{\lambda,m}$ linear unabhängig.

Für jedes $\lambda \in \Lambda$ bezeichne $\Omega(\lambda)$ eine Indexmenge und

$$f_{\lambda,\mu}: Q \to \boldsymbol{R}$$

sei für alle $\mu \in \Omega(\lambda)$ eine stetige, konvexe Funktion.

Weiterhin geben wir eine Funktion

$$F: \Lambda \times Q \to \boldsymbol{R}$$

vor, die für festes $\lambda$ bezüglich $x \in Q$ stetig ist, und betrachten für jedes $\lambda \in \Lambda$ folgendes Optimierungsproblem $(O_\lambda)$: Minimiere die Funktion $F(\lambda, x)$ unter den Nebenbedingungen

(1.1) $$x \in Q,$$

(1.2) $$g_{\lambda,\mu}(x) = 0 \quad \text{für } \mu = 1, \ldots, m,$$

(1.3) $$f_{\lambda,\mu}(x) \leqslant 0 \quad \text{für } \mu \in \Omega(\lambda).$$

Die Menge der Restriktionen, die als Gleichungen auftreten, bezeichnen wir mit

(1.4) $$R_\lambda^{(1)} := \{g_{\lambda,\mu} \mid \mu = 1, \ldots, m\},$$

und die als Ungleichungen auftreten mit

(1.5) $$R_\lambda^{(2)} := \{f_{\lambda,\mu} \mid \mu \in \Omega(\lambda)\}.$$

Setzen wir

(1.6) $$Z_\lambda^{(1)} := \{x \in Q \mid g_{\lambda,\mu}(x) = 0 \text{ für } \mu = 1, \ldots, m\}$$

und

(1.7) $$Z_\lambda^{(2)} := \{x \in Q \mid f_{\lambda,\mu}(x) \leqslant 0 \text{ für } \mu \in \Omega(\lambda)\},$$

dann gilt für die Menge der zulässigen Punkte $Z_\lambda$ unseres Optimierungsproblems $(O_\lambda)$:

(1.8) $$Z_\lambda = Z_\lambda^{(1)} \cap Z_\lambda^{(2)}.$$

Die Mengen $Z_\lambda^{(1)}$, $Z_\lambda^{(2)}$ und $Z_\lambda$ sind konvex und kompakt. Außerdem setzen wir sie für jedes $\lambda \in \Lambda$ als nicht leer voraus.

Die Lösungsmenge von $(O_\lambda)$ ist

(1.9) $$L_\lambda = \{x \in Z_\lambda \mid F(\lambda, x) = \min_{z \in Z_\lambda} F(\lambda, z)\}$$

und ist kompakt.

Um die Stetigkeitseigenschaften von $Z_\lambda$ und $L_\lambda$ zu untersuchen, brauchen wir geeignete Stetigkeitsbegriffe für mengenwertige Abbildungen:

Es seien $X$ und $Y$ metrische Räume und $\mathscr{K}(Y)$ die Menge der kompakten Teilmengen von $Y$, die nicht leer sind.

DEFINITION (Hahn [2]). Eine Abbildung $\psi: X \to \mathscr{K}(Y)$ heißt *oberhalbstetig* (bzw. *unterhalbstetig*) in $x_0 \in X$, falls es zu jeder in $Y$ offenen Menge $G$ mit $\psi(x_0) \subset G$ (bzw. $\psi(x_0) \cap G \neq \emptyset$) eine Umgebung $U$ von $x_0$ gibt mit $\psi(x) \subset G$ (bzw. $\psi(x) \cap G \neq \emptyset$) für alle $x \in U$.

Sind $A$ und $B$ zwei nichtleere Teilmengen des metrischen Raumes $X$ mit Metrik $\bar{d}$, so bezeichnen wir mit

(1.10) $$\operatorname{dist}(A, B) := \max\bigl(\sup_{a\in A} \inf_{b\in B} \bar{d}(a, b), \sup_{b\in B} \inf_{a\in A} \bar{d}(a, b)\bigr)$$

die Hausdorff-Distanz von $A$ zu $B$.

Wir versehen $C(Q)$ mit der Tschebyscheff-Norm $\|\cdot\|$ und betrachten kompakte Teilmengen von $C(Q)$ stets bezüglich dieser Topologie. Demgemäß ist im folgenden bei $\operatorname{dist}(R_\lambda^{(i)}, R_{\bar{\lambda}}^{(i)})$ $(\lambda, \bar{\lambda} \in \Lambda$ und $i = 1, 2)$ stets in (1.10) die durch die Tschebyscheff-Norm induzierte Metrik zu nehmen.

Im Beweis der Sätze brauchen wir folgendes Hilfsmittel:

HILFSSATZ 1. *Sei $C$ eine konvexe Menge des linearen Raumes $L$. Ist $x$ aus* Int $C$ *(Inneres von $C$) und $y$ aus der abgeschlossenen Hülle von $C$, so liegt die Menge*

(1.11) $$\{\alpha x + \beta y \mid \alpha > 0, \beta > 0, \alpha + \beta = 1\}$$

*in* Int $C$.

BEWEIS: vgl. Valentine [7], S. 20.

HILFSSATZ 2. *Sei $\bar{\lambda} \in \Lambda$ und* $\lim_{\lambda \to \bar{\lambda}} \operatorname{dist}(R_\lambda^{(1)}, R_{\bar{\lambda}}^{(1)}) = 0$. *Dann ist die Abbildung*

$$\psi_1 : \Lambda \to \mathscr{K}(Q)$$
$$\lambda \to Z_\lambda^{(1)}$$

*oberhalbstetig in $\bar{\lambda}$. Falls zusätzlich $\psi_1(\bar{\lambda}) \cap \operatorname{Int} Q \neq \emptyset$ oder $\psi_1(\lambda) = \psi_1(\bar{\lambda})$ für alle $\lambda$ aus einer Umgebung von $\bar{\lambda}$ ist, dann ist $\psi_1$ unterhalbstetig in $\bar{\lambda}$.*

BEWEIS. (1) Oberhalbstetigkeit: Nehmen wir an, $\psi_1$ wäre in $\bar{\lambda}$ nicht oberhalbstetig: Es existiert also eine offene Umgebung $G$ von $\psi_1(\bar{\lambda})$ und eine Folge $\lambda_i \overset{i\to\infty}{\to} \bar{\lambda}$ mit $\psi_1(\lambda_i) \not\subset G$. Sei nun $x_i \in \psi_1(\lambda_i)$ und $x_i \notin G$ $(i = 1, 2, \ldots)$. Da $\{x_i\}$ beschränkt ist, existiert eine konvergente Teilfolge. Wir können also o.B.d.A. annehmen, daß $x_i \overset{i\to\infty}{\to} x \notin G$ konvergiert. Weiter seien die $g_{\lambda,\mu}$ $(\mu = 1, \ldots, m)$ so nummeriert, daß

$$g_{\lambda_i,\mu} \overset{\lambda_i\to\bar{\lambda}}{\to} g_{\bar{\lambda},\mu}$$

für $\mu = 1, \ldots, m$.

Dann folgt aus

$$|g_{\bar{\lambda},\mu}(x)| \leqslant |g_{\lambda_i,\mu}(x_i)| + |g_{\lambda_i,\mu}(x_i) - g_{\bar{\lambda},\mu}(x_i)| + |g_{\bar{\lambda},\mu}(x_i) - g_{\bar{\lambda},\mu}(x)|,$$

daß $x \in \psi_1(\bar{\lambda})$, im Widerspruch zu $x \notin G$.

(2) Unterhalbstetigkeit: Wir können uns auf den Fall $\psi_1(\lambda) \cap \mathrm{Int}\,Q \neq \emptyset$ beschränken. Zu jedem $\lambda \in \Lambda$ gibt es nun Punkte $x_i^\lambda \in \mathbf{R}^n$ $(i = 0, \ldots, n-m)$ mit $|x_i^\lambda - x_i^{\bar{\lambda}}| \overset{\lambda\to\bar{\lambda}}{\to} 0$, so daß sich jedes $x^\lambda \in \psi_1(\lambda)$ darstellen läßt als

$$x^\lambda = x_0^\lambda + \sum_{i=1}^{n-m} \varrho_i x_i^\lambda \qquad (\varrho_i \in \mathbf{R}).$$

Sei nun $G$ eine offene Menge mit $x \in G \cap \psi_1(\bar{\lambda})$. Da $\psi_1(\bar{\lambda}) \cap \mathrm{Int}\,Q \neq \emptyset$ und $G$ offen ist, gibt es nach Hilfssatz 1 einen Punkt

$$z = x_0^{\bar{\lambda}} + \sum_{i=1}^{n-m} \alpha_i x_i^{\bar{\lambda}} \in G \cap \psi_1(\bar{\lambda}) \cap \mathrm{Int}\,Q.$$

Nun gibt es ein $\varepsilon > 0$, so daß für $d(\lambda, \bar{\lambda}) < \varepsilon$

$$z^\lambda = x_0^\lambda + \sum_{i=1}^{n-m} \alpha_i x_i^\lambda \in G \cap \psi_1(\lambda).$$

Damit ist $\psi_1$ in $\bar{\lambda}$ unterhalbstetig.

BEMERKUNG. Falls $\psi_1(\bar{\lambda}) \cap \mathrm{Int}\,Q = \emptyset$ und es keine Umgebung von $\bar{\lambda}$ gibt, in der $\psi_1$ konstant ist, dann braucht $\psi_1$ in $\bar{\lambda}$ nicht unterhalbstetig zu sein, wie man an folgendem Beispiel sieht.

BEISPIEL 1. Im $\mathbf{R}^2$ betrachten wir $Q := \{(x, y) | -1 \leqslant x \leqslant +1, 0 \leqslant y \leqslant 1\}$ und

$$g_\lambda(x, y) = y + \lambda x \qquad (\lambda \in \mathbf{R}).$$

Dann ist $\psi_1$ in $\lambda = 0$ nicht unterhalbstetig.

Im folgenden brauchen wir noch eine Eigenschaft der Optimierungsaufgabe $(O_\lambda)$, die man in der Optimierungstheorie auch benötigt, um notwendige Bedingungen für eine Lösung zu erhalten (Mangasarian [5]).

DEFINITION. $Z_\lambda$ (bzw. $Z_\lambda^{(2)}$) erfüllt die *Slater-Bedingung*, falls $R_\lambda^{(2)}$ kompakt ist und ein Punkt $x \in Z_\lambda$ (bzw. $x \in Z_\lambda^{(2)}$) existiert mit $f_{\lambda,\mu}(x) < 0$ für alle $\mu \in \Omega(\lambda)$.

HILFSSATZ 3. *Sei* $\bar{\lambda} \in \Lambda$ *und* $\lim_{\lambda \to \bar{\lambda}} \operatorname{dist}(R_\lambda^{(2)}, R_{\bar{\lambda}}^{(2)}) = 0$. *Dann ist die Abbildung*

$$\psi_2 : \Lambda \to \mathscr{K}(Q)$$
$$\lambda \to Z_\lambda^{(2)}$$

*oberhalbstetig im Punkt* $\bar{\lambda}$.

*Falls zusätzlich* $Z_{\bar{\lambda}}^{(2)}$ *die Slater-Bedingung erfüllt oder falls* $\psi_2(\lambda) = \psi_2(\bar{\lambda})$ *in einer Umgebung von* $\bar{\lambda}$ *ist, dann ist* $\psi_2$ *im Punkt* $\bar{\lambda}$ *unterhalbstetig.*

BEWEIS. (1) Oberhalbstetigkeit: Nehmen wir an, es gäbe eine offene Menge $G \supset Z_{\bar{\lambda}}^{(2)}$ und eine Folge $\lambda_i \overset{i\to\infty}{\to} \bar{\lambda}$ mit $\psi_2(\lambda_i) \not\subset G$.

Sei nun $x_i \in \psi_2(\lambda_i)$ und $x_i \notin G$. Da die Folge $\{x_i\}$ beschränkt ist, können wir o.B.d.A. annehmen, daß $x_i \overset{i\to\infty}{\to} x \notin G$ konvergiert. Zu beliebigem $f_{\bar{\lambda},\bar{\mu}} \in R_{\bar{\lambda}}^{(2)}$ gibt es nach Voraussetzung eine Folge

$$f_{\lambda_i,\mu_i} \in R_{\lambda_i}^{(2)} \quad \text{mit} \quad \lim_{i\to\infty} \|f_{\bar{\lambda},\bar{\mu}} - f_{\lambda_i,\mu_i}\| = 0.$$

Also folgt aus

$$|f_{\bar{\lambda},\bar{\mu}}(x) - f_{\lambda_i,\mu_i}(x_i)| \leqslant |f_{\bar{\lambda},\bar{\mu}}(x) - f_{\bar{\lambda},\bar{\mu}}(x_i)| + |f_{\bar{\lambda},\bar{\mu}}(x_i) - f_{\lambda_i,\mu_i}(x_i)|$$

auch

$$f_{\bar{\lambda},\bar{\mu}}(x) \leqslant 0.$$

Damit ist $x \in Z_{\bar{\lambda}}^{(2)}$ im Widerspruch zu $x \notin G$.

(2) Unterhalbstetigkeit: Wir können uns auf den Fall, daß die Slater-Bedingung erfüllt ist, beschränken.

Sei $G$ offen in $Q$ mit $y \in \psi_2(\bar{\lambda}) \cap G$. Außerdem sei $x \in Z_{\bar{\lambda}}^{(2)}$ mit $f_{\bar{\lambda},\bar{\mu}}(x) < 0$ für alle $\bar{\mu} \in \Omega(\bar{\lambda})$. Da $R_{\bar{\lambda}}^{(2)}$ kompakt ist, existiert ein $\varepsilon_1 > 0$, so daß $f_{\bar{\lambda},\bar{\mu}}(x) \leqslant -\varepsilon_1$ für alle $\bar{\mu} \in \Omega(\bar{\lambda})$. Deshalb erfüllen auch alle Punkte $z$ aus (1.11) wegen der Konvexität von $f_{\bar{\lambda},\bar{\mu}}$ wieder die Relation

$$f_{\bar{\lambda},\bar{\mu}}(z) < 0 \quad \text{für } \bar{\mu} \in \Omega(\bar{\lambda}).$$

Also gibt es einen Punkt $z_0 \in G$ und ein $\varepsilon > 0$, so daß

$$f_{\bar{\lambda},\bar{\mu}}(z_0) \leqslant -2\varepsilon \quad \text{für alle } \bar{\mu} \in \Omega(\bar{\lambda}).$$

Sei nun

$$G_1 := \{x \in Q \mid f_{\bar{\lambda},\bar{\mu}}(x) \leqslant -\varepsilon \text{ für alle } \bar{\mu} \in \Omega(\bar{\lambda})\},$$

dann ist $G_2 := G \cap \operatorname{Int} G_1 \neq \emptyset$ und ebenfalls $G_2 \subset Z_{\bar{\lambda}}^{(2)}$.

Wir wählen eine Umgebung $U_1$ von $\bar{\lambda}$ so, daß $\operatorname{dist}(R_\lambda^{(2)}, R_{\bar{\lambda}}^{(2)}) < \varepsilon$ für alle $\lambda \in U_1$. Dann gilt für $x \in G_2$ und $f_{\lambda,\mu} \in R_\lambda^{(2)}$ ($\lambda \in U_1$) mit geeignetem $\bar{\mu} \in \Omega(\bar{\lambda})$:

$$f_{\lambda,\mu}(x) < \varepsilon + f_{\bar{\lambda},\bar{\mu}}(x) \leqslant 0.$$

Also gilt $G_2 \subset \psi_2(\lambda)$ für $\lambda \in U_1$, damit insbesondere

$$G \cap \psi_2(\lambda) \neq \emptyset \quad \text{für } \lambda \in U_1.$$

Bemerkung. Falls $Z_{\bar{\lambda}}^{(2)}$ nicht die Slater-Bedingung erfüllt und es keine Umgebung von $\bar{\lambda}$ gibt, in der $\psi_2$ konstant ist, dann braucht $\psi_2$ nicht unterhalbstetig in $\bar{\lambda}$ zu sein. Dazu im $\boldsymbol{R}^2$

Beispiel 2. Restriktionen:

$$\begin{aligned} f_{\lambda,1}(x, y) &= -x-1, \\ f_{\lambda,2}(x, y) &= x-1, \\ f_{\lambda,3}(x, y) &= -y, \\ f_{\lambda,4}(x, y) &= y+\lambda x \quad (\lambda \in \boldsymbol{R}). \end{aligned}$$

In $\lambda = 0$ ist $\psi_2$ oberhalb-, aber nicht unterhalbstetig.

Satz 1. *Sei $\bar{\lambda} \in \Lambda$ und $\lim_{\lambda \to \bar{\lambda}} \operatorname{dist}(R_\lambda^{(i)}, R_{\bar{\lambda}}^{(i)}) = 0$ für $i = 1, 2$. Dann ist die Abbildung*

$$\begin{aligned} \psi: \Lambda &\to \mathscr{K}(Q) \\ \lambda &\to Z_\lambda^{(1)} \cap Z_\lambda^{(2)} = Z_\lambda \end{aligned}$$

*oberhalbstetig im Punkt $\bar{\lambda}$.*

*Weiterhin ist $\psi$ unterhalbstetig in $\lambda$, falls zusätzlich*

(α) *$Z_{\bar{\lambda}}^{(1)} \cap \operatorname{Int} Q \neq \emptyset$ oder $Z_\lambda^{(1)} = Z_{\bar{\lambda}}^{(1)}$ in einer Umgebung von $\bar{\lambda}$,*

(β) *$Z_{\bar{\lambda}}$ die Slater-Bedingung erfüllt.*

Beweis. (1) Oberhalbstetigkeit: Wir nehmen an, $\psi$ wäre nicht oberhalbstetig in $\bar{\lambda}$. Dann gibt es eine offene Umgebung $G$ von $Z_{\bar{\lambda}}$ und eine Folge $\{\lambda_i\}$ mit $\lim_{i \to \infty} \lambda_i = \bar{\lambda}$ und $\psi(\lambda_i) \not\subset G$ ($i = 1, 2, \ldots$), d.h. es existiert ein $x_i \in \psi(\lambda_i)$ mit $x_i \notin G$. Da die Folge $\{x_i\}$ beschränkt ist, können wir diese Folge als konvergent gegen $x$ ansehen. $\psi_1$ und $\psi_2$ sind aber oberhalbstetig in $\bar{\lambda}$, also ist $x \in Z_{\bar{\lambda}}$. Dies steht im Widerspruch zu $x \notin G$.

(2) Unterhalbstetigkeit: Sei $G$ offen in $Q$ und $G \cap Z_{\bar{\lambda}} \neq \emptyset$. Wie im Beweis von Hilfssatz 3 gibt es dann eine offene Teilmenge $G_2$ von $G$ und eine Umgebung $U_1$ von $\bar{\lambda}$ mit der Eigenschaft:

$$G_2 \subset \psi_2(\lambda) \quad \text{für } \lambda \in U_1 \quad \text{und} \quad G_2 \cap \psi_1(\bar{\lambda}) \neq \emptyset.$$

Nach Hilfssatz 2 gibt es dann eine Umgebung $U_2$ von $\bar{\lambda}$ mit $\psi_1(\lambda) \cap \cap G_2 \neq \emptyset$ für alle $\lambda \in U_2$. Damit ist aber

$$\psi_1(\lambda) \cap \psi_2(\lambda) \cap G \neq \emptyset \quad \text{für alle } \lambda \in U_1 \cap U_2.$$

BEMERKUNG. Falls (β) nicht erfüllt ist, jedoch $\psi_2(\lambda) = \psi_2(\bar{\lambda})$ in einer Umgebung von $\bar{\lambda}$ gilt, dann ersetzen wir $Q$ durch $Q := Q \cap Z_{\bar{\lambda}}^{(2)}$ und betrachten die Optimierungsaufgaben mit $R_\lambda^{(2)} = \emptyset$.

Im nächsten Satz gebrauchen wir den Begriff „unterhalbstetige Funktion", wohl zu unterscheiden vom Begriff „unterhalbstetige Abbildung". Zu ihrer Definition vergleiche man Hahn [2].

SATZ 2. *Sei* $\bar{\lambda} \in \Lambda$ *und* $\lim_{\lambda \to \bar{\lambda}} \operatorname{dist}(R_\lambda^{(i)}, R_{\bar{\lambda}}^{(i)}) = 0$ *für* $i = 1, 2$. $F(\lambda, x)$ *sei in allen Punkten* $(\bar{\lambda}, x) \in \{\bar{\lambda}\} \times Q \subset \Lambda \times Q$ *stetig. Dann ist die Funktion*

$$\theta: \Lambda \to \boldsymbol{R}$$
$$\lambda \to \min_{x \in Z_\lambda} F(\lambda, x)$$

*unterhalbstetig im Punkt* $\bar{\lambda}$.

*Falls zusätzlich die Voraussetzungen* (α) *und* (β) *von Satz* 1 *gelten, dann ist die Abbildung*

$$\Phi: \Lambda \to \mathscr{K}(Q)$$
$$\lambda \to L_\lambda$$

*oberhalbstetig im Punkt* $\bar{\lambda}$ *und* $\theta$ *ist stetig in diesem Punkt.*

BEWEIS. (1) Unterhalbstetigkeit von $\theta$: Nehmen wir an, $\theta$ wäre nicht unterhalbstetig. Dann gibt es ein $\varepsilon > 0$, eine Folge $\lambda_i \overset{i\to\infty}{\to} \bar{\lambda}$ und Punkte $x_i \in Z_{\lambda_i}$ mit $F(\lambda_i, x_i) = \theta(\lambda_i) \leqslant \theta(\bar{\lambda}) - \varepsilon$. Wir können o.B.d.A. annehmen, daß die Folge $\{x_i\}$ konvergiert. Dann konvergiert sie aber nach Satz 1 gegen ein $z \in Z_{\bar{\lambda}}$. Dann folgt aber aus

$$\lim_{i\to\infty} F(\lambda_i, x_i) = F(\bar{\lambda}, z) \geqslant \theta(\bar{\lambda})$$

ein Widerspruch.

(2) Oberhalbstetigkeit von $\Phi$: Wir gehen wieder indirekt vor: Es existiere eine offene Umgebung $G$ von $L_{\bar{\lambda}}$ und eine Folge $\{\lambda_i\}$ mit $\lim \lambda_i = \bar{\lambda}$ und $\Phi(\lambda_i) \not\subset G$, d.h. es existiert ein $x_i \in \Phi(\lambda_i)$ mit $x_i \notin G$. Da die Folge $\{x_i\}$ beschränkt ist, können wir sie als konvergent gegen $x$ annehmen, und $x$ ist nach Satz 1 aus $Z_{\bar{\lambda}}$. Wir wollen nun zeigen, daß $x \in L_{\bar{\lambda}}$ ist, also $F(\bar{\lambda}, x) = \min_{z\in Z_{\bar{\lambda}}} F(\bar{\lambda}, z)$. Dazu nehmen wir indirekt an, daß für ein $x^* \in Z_{\bar{\lambda}}$ gelte:

$$F(\bar{\lambda}, x^*) = \min_{z\in Z_{\bar{\lambda}}} F(\bar{\lambda}, z) < F(\bar{\lambda}, x).$$

Da $\psi$ unterhalbstetig ist, existiert eine Folge $\{x_i^*\}$ mit $x_i^* \in Z_{\lambda_i}$ und $\lim_{i\to\infty} x_i^* = x^*$.

Aus dem Diagramm $(i \to \infty)$

$$\begin{array}{ccc} F(\lambda_i, x_i) & \to & F(\bar{\lambda}, x), \\ \rotatebox{90}{$\leqslant$} & & \\ F(\lambda_i, x_i^*) & \to & F(\bar{\lambda}, x^*) \end{array}$$

folgt:

$$F(\bar{\lambda}, x) \leqslant F(\bar{\lambda}, x^*).$$

Dies ist ein Widerspruch, also $x \in L_{\bar{\lambda}}$.

Nun ist aber $\lim_{i\to\infty} x_i = x \notin G$. Dies steht zu $L_{\bar{\lambda}} \subset G$ im Widerspruch.

(3) Stetigkeit von $\theta$: Sei $\varepsilon > 0$, dann gibt es zu jedem Punkt $z \in L_{\bar{\lambda}}$ eine offene Umgebung $U_z \times V_z$ von $(\bar{\lambda}, z) \in \Lambda \times Q$ mit

$$|F(\bar{\lambda}, z) - F(\lambda, x)| < \varepsilon$$

für alle $(\lambda, x) \in U_z \times V_z$. Da $L_{\bar{\lambda}}$ kompakt ist, existiert eine endliche Überdeckung $V := \bigcup_{i=1}^{k} V_{z_i}$ von $L_{\bar{\lambda}}$ ($z_i \in L_{\bar{\lambda}}$ für $i = 1, \ldots, k$).

Da $\Phi$ oberhalbstetig ist in $\bar{\lambda}$, gibt es eine Umgebung $U$ von $\bar{\lambda}$ mit $\Phi(\lambda) \subset V$ für $\lambda \in U$. Dann gilt aber für $\lambda \in U \cap \bigcap_{i=1}^{k} U_{z_i}$

$$|\theta(\lambda) - \theta(\bar{\lambda})| < \varepsilon,$$

also ist $\theta$ bei $\bar{\lambda}$ stetig.

BEMERKUNGEN. (1) Falls die Voraussetzungen (α) und (β) nicht erfüllt sind, dann braucht $\Phi$ nicht oberhalbstetig und $\theta$ nicht stetig zu sein. Führt man nämlich in den Beispielen 1 und 2 als Zielfunktionen $F(\lambda, x, y) = x$ ein, so sind diese Stetigkeitseigenschaften bei $\lambda = 0$ verletzt.

(2) $\Phi$ ist im allgemeinen nicht unterhalbstetig, dazu

BEISPIEL 3. $\Lambda = \boldsymbol{R}$. Restriktionen:

$$\begin{aligned} f_{\lambda,1}(x, y) &= -y, \\ f_{\lambda,2}(x, y) &= y+\lambda x-1, \\ f_{\lambda,3}(x, y) &= y-\lambda x-1. \end{aligned}$$

Zielfunktion:

$$F(\lambda, x, y) = -y-x.$$

Dann ist $L_1 = \{\alpha(1, 0)-(1-\alpha)(0, 1) | \, 0 \leqslant \alpha \leqslant 1\}$

$$L_\lambda = \begin{cases} \{(0, 1)\} & \text{für } \lambda > 1, \\ \left\{\left(\dfrac{1}{\lambda}, 0\right)\right\} & \text{für } \lambda < 1. \end{cases}$$

**2. Konvergenz eines Gradientenverfahrens für lineare Tschebyscheff-Approximation.** $B$ sei ein kompakter metrischer Raum, $C(B)$ der lineare Raum der auf $B$ stetigen, reellwertigen Funktionen mit der Norm

$$\|g\|_B := \max_{x \in B} |g(x)|$$

für $g \in C(B)$. $V = \operatorname{span}\{v^1, v^2, \ldots, v^n\}$ sei ein $n$-dimensionaler Teilraum von $C(B)$. Gesucht ist nun zu $f \in C(B)-V$ ein $v_0 \in V$ mit

$$\|f-v_0\|_B = \inf_{v \in V} \|f-v\|_B.$$

Für eine solche Minimallösung gilt folgender Charakterisierungssatz von Kolmogoroff [6]:

$v_0$ ist genau dann Minimallösung zu $f$, wenn

$$\min_{x \in M(v_0)} \big(f(x)-v_0(x)\big)v(x) \leqslant 0 \qquad \text{für alle } v \in V.$$

Dabei ist

(2.1) $$M(v_0) = \{x \in B | \, |f(x)-v_0(x)| = \Delta(v_0)\},$$

wobei wir zur Abkürzung $\Delta(v) := \|f-v\|_B$ für $v \in V$ setzen.

Um also nachzuprüfen, ob ein $v_0$ Minimallösung ist, braucht man nur das folgende lineare Optimierungsproblem $A(v_0, M(v_0))$ zu betrachten:

Maximiere die Zahl $\mu$ unter den Nebenbedingungen

$$\big(f(x)-v_0(x)\big)h(x)-\mu \geqslant 0 \qquad \text{für } x \in M(v_0),$$

$$|\alpha^i| \leqslant 1 \qquad \text{für } i = 1, 2, \ldots, n,$$

wobei die $\alpha^i$ die Koeffizienten von $h \in V$ in der Darstellung $h = \sum_{i=1}^{n} \alpha^i v^i$ bilden.

Ist $\mu(v_0, M(v_0))$ das gesuchte Maximum und ist $\mu(v_0, M(v_0)) = 0$, dann ist $v_0$ Minimallösung zu $f$. Anderenfalls kann man zu einer Lösung $\big(h_0, \mu(v_0, M(v_0))\big)$ des obigen Optimierungsproblems ein geeignetes $t > 0$ finden, so daß

$$\|f-v_0-th_0\|_B < \|f-v_0\|_B.$$

Auf diese Weise gelangt man schrittweise zu besseren Approximationen, jedoch ist die Konvergenz dieses Verfahrens keineswegs gesichert. Um diesen Mangel zu beseitigen, wollen wir eine Idee aufgreifen, die erstmals von Wittmeyer-Koch [8] angegeben wurde. Sie schlug grob gesagt vor, nicht nur die absoluten Extrema $M(v_0)$ für die Nebenbedingungen zu nehmen, sondern auch noch relative Extrema heranzuziehen, die sich nur wenig von absoluten unterscheiden. Ihre Vorgehensweise untermauerte sie durch numerische Ergebnisse. Später konnte Krabs [4] für ein ähnliches Verfahren im diskreten Fall, auch für gewisse nichtlineare Probleme, einen Konvergenzsatz beweisen.

Beschreibung des Verfahrens. Der $k$-te Schritt unseres Verfahrens verläuft so: Gegeben sind $v_k \in V$ und die kompakte Teilmenge $M_k \subset B$ mit $M(v_k) \subset M_k$.

Wir betrachten die Optimierungsaufgabe $A(v_k, M_k)$:

Maximiere $\mu$ unter den Nebenbedingungen

$$\text{(2.2)} \qquad \big(f(x)-v_k(x)\big)h(x)-\mu \geqslant 0 \qquad \text{für } x \in M_k,$$

$$|\alpha^i| \leqslant 1 \qquad \text{für } i = 1, \ldots, n,$$

$$\text{(2.3)} \qquad \text{mit} \qquad h = \sum_{i=1}^{n} \alpha^i v^i \in V.$$

Dieses Problem hat mindestens eine Lösung $(h_k, \mu_k) \in V \times \boldsymbol{R}$ mit $\mu_k = \mu(v_k, M_k) \geqslant 0$. Außerdem erfüllt für jedes solche Problem $A(v_k, M_k)$ die Menge der zulässigen Punkte die Slater-Bedingung, was eine wesentliche Voraussetzung von Satz 2 ist, um Stetigkeitseigenschaften heranziehen zu können.

Die Funktion

$$G(v_k, h_k; t) := \|f - v_k - th_k\|_B \tag{2.4}$$

ist eine bezüglich $t \in \boldsymbol{R}$ konvexe Funktion mit $\lim_{t\to\infty} G(v_k, h_k; t) = \infty$.

Wit machen zwei Fallunterscheidungen:

(a) $\mu_k = 0$: Ist $M_k = M(v_k)$, so ist $v_k$ Minimallösung zu $f$ und wir sind fertig. Anderenfalls setzen wir $v_{k+1} := v_k$.

(b) $\mu_k > 0$: Wir bestimmen ein $t_k > 0$, so daß $G(v_k, h_k; t)$ in $t_k$ für $t \geqslant 0$ das Minimum annimmt, und setzen

$$v_{k+1} := v_k + t_k h_k.$$

Nun haben wir noch $M_{k+1} \supset M(v_{k+1})$ geeignet zu wählen. Dies geschieht dadurch, daß wir von der Folge $\{M_k\}$ folgende Eigenschaft fordern:

Es existiere eine konvergente Teilfolge $\{M_{k_i}\}$ mit

$$\lim_{i\to\infty} v_{k_i} = v, \tag{2.5}$$

$$\lim_{i\to\infty} M_{k_i} = M \subset M(v), \tag{2.6}$$

$$\lim_{i\to\infty} \mu_{k_i} = 0. \tag{2.7}$$

Zur Definition einer konvergenten Mengenfolge vergleiche man Hahn [2].

Falls das Verfahren nicht abbricht, so erhalten wir von einer Startfunktion $v_1$ ausgehend, Approximationen $v_k$ mit

$$\Delta(v_1) \geqslant \Delta(v_2) \geqslant \ldots \geqslant \Delta(v_k) \geqslant \Delta(v_{k+1}) \geqslant \ldots$$

Da die Folge $\{\Delta(v_k)\}$ monoton fällt und beschränkt ist, konvergiert sie gegen einen Grenzwert $L$.

SATZ 3. *Es gilt* $L = \inf_{v\in V} \|f-v\|_B$, *und das durch* (2.5) *definierte* $v$ *ist eine Minimallösung zu* $f$.

BEWEIS. Nehmen wir an, $v$ wäre keine Minimallösung zu $f$. Dann ist die Aufgabe $A(v, M(v))$ lösbar mit $\mu(v, M(v)) > 0$. Da nun $M \subset M(v)$ ist, gilt natürlich für die Aufgabe $A(v, M)$:

$$\mu(v, M) \geqslant \mu(v, M(v)) > 0.$$

Wenden wir nun Satz 2 an, so muß andererseits gelten:

$$\mu(v_{k_i}, M_{k_i}) \overset{i\to\infty}{\to} \mu(v, M).$$

Dies ist aber ein Widerspruch zu (2.7). Also ist $v$ Minimallösung zu $f$ und $L = \inf_{v \in V} \|f-v\|_B$.

Es stellt sich die Frage, wie man im konkreten Fall die Mengen $M_k$ in Abhängigkeit von $f$, $V$ und $v_k$ wählen soll, damit die geforderten Eigenschaften (2.5)–(2.7) erfüllt sind. Dazu betrachten wir einige Beispiele.

(i) Auf $B$ sei eine Folge von diskreten Teilmengen $\{B_k\}$ gegeben mit der Eigenschaft $\lim_{k\to\infty} \operatorname{dist}(B, B_k) = 0$ und ein $\tau$ mit $0 < \tau < 1$.

Im $k$-ten Schritt definieren wir zu vorgegebenen $v_k \in V$ und $\delta_k > 0$ die Menge

$$M_k := M(v_k) \cup \{x \in B_k \mid |f(x)-v_k(x)| \geqslant \Delta(v_k)-\delta_k\}. \tag{2.8}$$

Dann definieren wir

$$\delta_{k+1} := \begin{cases} \tau\delta_k, & \text{falls } \mu(v_k, M_k) \leqslant \delta_k, \\ \delta_k, & \text{falls } \mu(v_k, M_k) > \delta_k \end{cases} \tag{2.9}$$

und haben so durch die Vorschrift unter (a) und (b) das Verfahren vollständig beschrieben. Zu untersuchen bleiben die Eigenschaften (2.5)–(2.7) der Folge $\{M_k\}$.

Dazu behaupten wir, daß

$$\delta := \lim_{k\to\infty} \delta_k = 0 \tag{2.10}$$

ist, falls das Verfahren nicht abbricht. Daraus folgt nämlich die Existenz einer Teilfolge $\{M_{k_i}\}$ mit den Eigenschaften (2.5)–(2.7).

Falls $B$ eine diskrete Menge ist und $B_k = B$ $(k = 1, 2, \ldots)$ gesetzt wird, dann ist dieses Verfahren, abgesehen von der Bestimmung von $t_k$, mit demjenigen von Krabs [4] identisch.

(ii) Wir setzen voraus, daß jede Fehlerfunktion $f-v$ nur endlich viele relative Extrema im Innern von $B$ ($\operatorname{Int} B$) habe. Auf dem Rand von $B$ seien diskrete Punktmengen $R_k$ gegeben mit $\lim_{k\to\infty} \operatorname{dist}(B-\operatorname{Int} B, R_k) = 0$.

Im $k$-ten Schritt des Verfahrens definieren wir zu gegebenen $v_k \in V$ und $\delta_k > 0$ die Menge

$$(2.11) \qquad M_k := M(v_k) \cup \{x \in R_k |\, |f(x)-v_k(x)| \geqslant \Delta(v_k)-\delta_k\} \cup \\ \cup \{x \in \operatorname{Int} B |\, |f(x)-v_k(x)| \geqslant \Delta(v_k)-\delta_k \text{ und } \\ x \text{ relatives Extremum von } f-v\}.$$

Dann wird $\delta_{k+1}$ nach (2.9) bestimmt und wir behaupten wie unter (i) die Beziehung (2.10). Daraus folgen wieder die gewünschten Eigenschaften (2.5)–(2.7) von $\{M_k\}$. Als Beispiel für eine solche Vorgehensweise betrachten wir das

BEISPIEL 4. $B$ sei ein kompakter Bereich der reellen $(x, y)$-Ebene und $f(x, y) = f_1(x)+f_2(y)$. Dabei ist $f_1$ (bzw. $f_2$) holomorph in einem Gebiet $G_1$ (bzw. $G_2$), das die Projektion von $B$ auf die $x$-Achse (bzw. $y$-Achse) umfaßt. $V$ werde von $1, v^1, \ldots, v^{n-1}$ aufgespannt, wobei jedes $v^i$ wie $f$ die Form $v^i(x, y) = v_1^i(x)+v_2^i(y)$ mit in $G_1$ holomorphem $v_1^i$ und in $G_2$ holomorphem $v_2^i$ $(i = 1, \ldots, n-1)$ hat.

(iii) Wir setzen jetzt voraus, daß jede Fehlerfunktion $f-v$ nur endlich viele relative Extrema in $B$ habe. Dann definieren wir im $k$-ten Schritt zu $v_k \in V$ und $\delta_k > 0$

$$(2.12) \qquad M_k := M(v_k) \cup \{x \in B |\, |f(x)-v_k(x)| \geqslant \Delta(v_k)-\delta_k \\ \text{und } x \text{ relatives Extremum von } f-v_k\}$$

und bestimmen wie unter (2.9) $\delta_{k+1}$. Falls auch die Relation (2.10) richtig ist, so folgen wie unter (i) auch die Beziehungen (2.5)–(2.7). Als Beispiel für ein solches Vorgehen betrachten wir Beispiel 4, jedoch mit

$$B = \{(x, y) \in \boldsymbol{R}^2 |\, a \leqslant x \leqslant b,\ c \leqslant y \leqslant d\}$$

als Rechteck im $\boldsymbol{R}^2 (a < b,\ c < d)$.

Zu beweisen bleibt demnach noch (2.10) für die Verfahren (i)–(iii).

BEWEIS VON (2.10) FÜR VERFAHREN (i). Nehmen wir $\delta > 0$ an. Dann gibt es nach (2.9) ein $K_0 \in N$, so daß für $k \geqslant K_0$ nur noch der Fall

$$\mu(v_k, M_k) > \delta_k$$

möglich ist. Also gilt für $k \geqslant K_0$:

$$0 < \delta = \delta_k < \mu(v_k, M_k). \tag{2.13}$$

Da die Zahlen $\|v_k\|_B$ und $\|h_k\|_B$ beschränkt sind und $\{M_k\}$ eine Folge von Teilmengen in einem kompakten metrischen Raum ist, so existiert eine Teilfolge $\{k_i\} \subset N$ mit den Eigenschaften:

$$\lim_{i\to\infty} v_{k_i} = v \in V, \tag{2.14}$$

$$\lim_{i\to\infty} h_{k_i} = h \in V, \tag{2.15}$$

$$\lim_{i\to\infty} M_{k_i} = M \subset B. \tag{2.16}$$

Wir behaupten, daß

$$M(v) \subset M. \tag{2.17}$$

Denn ist $x \in M(v)$, dann gibt es eine Folge $\{x_k\} \subset B$ mit $x_k \in B_k$ und $\lim_{k\to\infty} x_k = x$.

Nun ist

$$|f(x_k)-v_k(x_k)|$$
$$\geqslant |f(x)-v(x)|-|f(x_k)-f(x)|-|v(x)-v(x_k)|-|v(x_k)-v_k(x_k)|,$$

also existiert ein $n_0 \in N$, so daß für $i \geqslant n_0$ gilt:

$$|f(x_{k_i})-v_{k_i}(x_{k_i})| \geqslant \Delta(v)-\delta/2.$$

Wegen $\Delta(v_k) \to \Delta(v)$ für $k \to \infty$ können wir $n_1$ so wählen, daß für $i \geqslant n_1$

$$|f(x_{k_i})-v_{k_i}(x_{k_i})| \geqslant \Delta(v_{k_i})-\delta \quad \text{gilt.}$$

Für $i \geqslant n_1$ ist damit $x_{k_i} \in M_{k_i}$ und deshalb $x \in M$. Aus Satz 2 und (2.16) folgt für die Aufgabe $A(v, M)$:

$$\mu(v, M) = \lim_{i\to\infty} \mu_{k_i} \geqslant \delta.$$

Nun ist wieder nach Satz 2 $(h, \mu(v, M))$ eine Lösung von $A(v, M)$ und es gilt wegen (2.17):

$$\Delta(v_{k_i+1}) = G(v_{k_i}, h_{k_i}; t_{k_i}) \overset{i\to\infty}{\to} \min_{t\geqslant 0} G(v, h; t) < \Delta(v).$$

Dies ist aber ein Widerspruch zu $\Delta(v_{k_i+1}) \to \Delta(v)$ für $i \to \infty$.

Beweis von (2.10) für das Verfahren (ii). Nehmen wir wieder $\delta > 0$ an, dann gibt es ein $K_0 \in N$, so daß (2.13) erfüllt ist. Außerdem existiert eine Teilfolge $\{k_i\} \subset N$ mit den Eigenschaften (2.14)–(2.16). Wir behaupten wieder $M(v) \subset M$:

Ist $x \in M(v) \cap (B - \operatorname{Int} B)$, dann zeigt man wie im vorigen Beweis, daß $x \in M$ ist. Ist dagegen $x \in M(v) \cap \operatorname{Int} B$, dann gibt es eine $\varepsilon$-Umgebung $U_\varepsilon(x)$ in $B$, so daß für alle $z \in U_\varepsilon(x)$, $z \neq x$ gilt:

$$\Delta(v) - \frac{\delta}{4} \leqslant |f(z) - v(z)| < \Delta(v).$$

Nun sei $n_0$ so gewählt, daß $\|v_{k_i} - v\| \leqslant \delta/4$ für $i \geqslant n_0$.
Dann gilt für $i \geqslant n_0$ und $z \in U_\varepsilon(x)$:

$$\Delta(v) - \frac{\delta}{2} \leqslant |f(z) - v_{k_i}(z)| \leqslant \Delta(v) + \frac{\delta}{4}. \tag{2.18}$$

Sei nun $\varepsilon' < \varepsilon$ und

$$N(\varepsilon') := \max_{|z-x|=\varepsilon'} |f(z) - v(z)|,$$

dann ist $N(\varepsilon') < \Delta(v)$ und wegen (2.14) gibt es ein $n_1(\varepsilon') \geqslant n_0$, so daß für $i \geqslant n_1(\varepsilon')$

$$\max_{|z-x|=\varepsilon'} |(f - v_{k_i})(z)| \leqslant \tfrac{1}{2}\big(N(\varepsilon') + \Delta(v)\big)$$

und

$$|(f - v_{k_i})(x)| \geqslant \tfrac{1}{2}\big(N(\varepsilon') + \Delta(v)\big)$$

gilt. Also existiert für $i \geqslant n_1(\varepsilon')$ in $U_{\varepsilon'}(x)$ ein relatives Extremum und wegen (2.18) und $\Delta(v_{k_i}) \leqslant \Delta(v) + \delta/4$ gehört dieses zu $M_{k_i}$. Daraus ergibt sich $x \in M$.

Insgesamt haben wir also $M(v) \subset M$ und wie im vorigen Beweis ergibt sich daraus ein Widerspruch.

BEWEIS VON (2.10) FÜR DAS VERFAHREN (iii). Dieser Beweis verläuft analog zu den beiden vorhergehenden. Insbesondere beweist man die wichtige Relation (2.17) wie für das Verfahren (ii).

## LITERATUR

[1] G. B. Dantzig, J. Folkman and N. Shapiro, 'On the Continuity of the Minimum Set of a Continuous Function', *J. Math. Anal. Appl.* **17** (1967) 519–548.

[2] H. Hahn, *Reelle Funktionen*, Akademische Verlagsgesellschaft, Leipzig 1932.

[3] W. Krabs, 'Ein Verfahren zur Lösung gewisser nichtlinearer diskreter Approximationsprobleme', *ZAMM* **50** (1970) 359–368.

[4] W. Krabs, 'Zur stetigen Abhängigkeit des Extremalwertes eines konvexen Optimierungsproblems von einer stetigen Änderung des Problems', *ZAMM* **52** (1972) 359–368.

[5] O. L. Mangasarian, *Non-Linear Programming*, McGraw-Hill Book Company, New York 1969.

[6] G. Meinardus, *Approximation von Funktionen und ihre numerische Behandlung*, Springer-Verlag, Berlin 1964.

[7] F. A. Valentine, *Konvexe Mengen*, Bibliographisches Institut, Mannheim 1968.

[8] L. Wittmeyer-Koch, 'A method of Descent for Chebyshev Approximation', *BIT* **8** (1968) 328–342.

# SPLINE BASES IN FUNCTION SPACES

Z. CIESIELSKI

*Sopot*

**1. Introduction.** The aim of this note is to announce some results on systems of splines which were investigated earlier in the papers [2], [3] and [4].

For each positive integer $n$ a partition $\pi_n = \{s_{n,i}: i = 0, \pm 1, \ldots\}$ is defined as follows: for $n = 1$ $s_{n,i} = i$ and for $n = 2^\mu + \nu$ with $\mu \geqslant 0$, $1 \leqslant \nu \leqslant 2^\mu$,

$$s_{n,i} = \begin{cases} \dfrac{i}{2^{\mu+1}} & \text{for } i \leqslant 2\nu, \\[2ex] \dfrac{i-\nu}{2^\mu} & \text{for } i \geqslant 2\nu. \end{cases}$$

Thus, for each $n \geqslant 1$ we have $s_{n,i} < s_{n,i+1}$ and $s_{n,0} = 0$, $s_{n,n} = 1$. Now, let $I_{n,i}$ be equal to $\langle s_{n,i-1}, s_{n,i})$ for $i < n$, to $(s_{n,i-1}, s_{n,i}\rangle$ for $i > n$ and $I_{n,n} = \langle s_{n,n-1}, s_{n,n}\rangle$. Clearly $\{I_{n,i}, i = 0, \pm 1, \ldots\}$ is a partition of the real line. For each $\pi_n$ the partition of unity of splines of order $m = -1$ is defined as follows

$$N_{n,i}^{(m)}(t) = \begin{cases} 1 & \text{for } t \in I_{n,i}, \\ 0 & \text{for } t \notin I_{n,i}. \end{cases}$$

It is well known that for each $m \geqslant 0$ the following splines of order $m$

$$N_{n,i}^{(m)}(t) = (s_{n,i+m+1} - s_{n,i-1})[s_{n,i-1}, \ldots, s_{n,i+m+1}; (s-t)_+^{m+1}]$$

are non-negative and form a partition of unity corresponding to $\pi_n$, the square bracket denotes the divided difference of $(s-t)_+^{m+1}$ as a function of $s$ taken at the points $s_{n,i-1}, \ldots, s_{n,i+m+1}$.

To fixed $m \geqslant -1$ and $n \geqslant -m$ there corresponds finite dimensional space $S_n^m(I)$ of spline functions of order $m$ defined over $I = \langle 0, 1\rangle$. If $-m \leqslant n \leqslant 0$, then $S_n^m(I)$ is defined as the linear span over $1, t, \ldots, t^{m+n}$, and if $n > 0$, then it is generated by the linearly independent functions

$N_{n,i}^{(m)}$, $i = -m, \ldots, n$. Clearly, $S_n^m(I) \subset S_{n+1}^m(I)$ and $\dim S_n^m = m+n+1$.
Now, let

$$S^m(I) = \bigcup_{n=-m}^{\infty} S_n^m(I).$$

It should be clear that $S^m(I)$ is dense in $C(I)$ for $m \geqslant 0$, and it is dense in $L_p(I)$ for $1 \leqslant p < \infty$, $m \geqslant -1$. Each function $f \in S^m(I)$ has absolutely continuous derivative of order $m$. In what follows it is always required that $D^{m+1}f \in S^{-1}(I)$ for each $f \in S^m$, $m \geqslant 0$, and consequently $D^{m+1}f$ is assumed to be defined everywhere in $I$.

Now let us consider $S_n^m$, $n \geqslant -m$, as subspaces of $L_2(I)$ with the scalar product

$$(f,g) = \int_I f(t)g(t)\,dt. \tag{1.1}$$

There exists orthonormal, with respect to (1.1), system of splines of order $m$, $m \geqslant -1$, $f_n^{(m)}$, $n \geqslant -m$ such that $f_{-m}^{(m)} = 1$, $f_n^{(m)} \in S_n^m \setminus S_{n-1}^m$ for $n > -m$ and $(f_n^{(m)}, f_n^{(m)}) = 1$. Moreover, we define

$$f_n^{(m,k)} = D^k f_n^{(m)} \quad \text{for } n \geqslant k-m,\ 0 \leqslant k \leqslant m+1,$$

and

$$g_n^{(m,k)} = H^k f_n^{(m)} \quad \text{for } n \geqslant k-m,$$

where $0 \leqslant k \leqslant m+2$ and $Hf(t) = \int_t^1 f(u)\,du$.

We know that the system $\{g_i^{(m,k)}, f_j^{(m,k)}, i,j \geqslant k-m\}$, $0 \leqslant k \leqslant m+1$, is biorthogonal with respect to (1.1) Moreover, it was shown in [2] that

$$f = \sum_{j=k-m}^{\infty} (g_j^{(m,k)}, f) f_j^{(m,k)} \tag{1.2}$$

holds in $C(I)$ for $0 \leqslant k \leqslant m$ and in $L_p(I)$ with $1 \leqslant p < \infty$ for $0 \leqslant k \leqslant m+1$.

Our purpose is to state a series of results concerning (1.2): the behaviour of partial sums, the order of approximation and the convergence almost everywhere. Moreover, some more properties of the biorthogonal system $\{g_i^{(m,k)}, f_j^{(m,k)}, i,j \geqslant k-m\}$ will be stated. In particular it turns out that the system $\{1, g_i^{(m,k+1)}, i \geqslant k-m\}$ is a basis in

$C(I)$ for each $k, 0 \leqslant k \leqslant m+1, m \geqslant -1$, and for $k = m+1$ it is an interpolating basis of splines of type $I'$.

**2. Estimates for partial sums and convergence almost everywhere.** We are going to employ the notation

$$(P_n^{(m,k)}f)(t) = \sum_{j=k-m}^{n} (f, g_j^{(m,k)})f_j(t).$$

THEOREM 2.1. *Let $m \geqslant -1$, $0 \leqslant k \leqslant m+1$ and $f \in L_1(I)$. Then there exists constant $M_m$ such that*

$$\sup_{n \geqslant k-m} |(P_n^{(m,k)}f)(t)| \leqslant M_m \sup\left\{\left|\frac{1}{t-s}\int_s^t f(u)du\right| : \; s \neq t, \; s \in I\right\}$$

*holds for each $t \in I$.*

This result and the application of a theorem of Hardy–Littlewood give

THEOREM 2.2. *Let $p > 1, m \geqslant -1$ and $0 \leqslant k \leqslant m+1$. Then there exist constants $M_m$ and $L_m$ such that*

$$\| \sup_{n \geqslant k-m} P_n^{(m,k)}f\|_p \leqslant 2M_m\left(\frac{p}{p-1}\right)^{1/p} \|f\|_p$$

*holds for $f \in L_p(I)$, and*

$$\| \sup_{n \geqslant k-m} P_n^{(m,k)}f\|_1 \leqslant M_m \int_I |f(t)| \log^+ |f(t)|\, dt + L_m$$

*holds for $f \in L_1(I)$.*

Next result in the case of $m = 0$ and $k = 0$ was established in [1].

THEOREM 2.3. *Let $m \geqslant -1, 0 \leqslant k \leqslant m+1$ and $f \in L_1(I)$. Then for $t \in I$ such that*

$$f(t) = \lim_{h \to 0} \frac{1}{h} \int_t^{t+h} f(u)\, du$$

*we have*

$$f(t) = \sum_{j=k-m}^{\infty} (f, g_j^{(m,k)}) f_j^{(m,k)}(t).$$

**3. Order of approximation.** For $1 \leqslant p < \infty$ the modulus of continuity of order $r$ for $f \in L_p(I)$ is defined as follows

$$\omega_r^{(p)}(f; \delta) = \sup_{0<h<\delta} \Big( \int_0^{1-rh} |\Delta_h^r f(t)|^p dt \Big)^{1/p},$$

and similarly in the case of $p = \infty$ for $f \in C(I)$

$$\omega_r^{(\infty)}(f; \delta) = \sup_{0<h<\delta} \sup_{0<t<t+rh<1} |\Delta_h^r f(t)|.$$

THEOREM 3.1. *Assume that* $1 \leqslant p \leqslant \infty, m \geqslant -1, 0 \leqslant k \leqslant m+1$ *and that* $f \in L_p(I)$ *in the case of finite* $p$ *and* $f \in C(I)$ *in the case of* $p = \infty$. *Then there esists a constant* $C_m$ *independent of* $n, p$ *and* $f$ *such that for* $n > 0$

$$\|f - P_n^{(m,k)} f\|_p \leqslant C_m \omega_{m+2-k}^{(p)}\left(f; \frac{1}{n}\right).$$

This is a consequence of known results (cf. the paper by K. Scherer in this volume).

**4. The $L_p(I)$ norms of the biorthogonal functions.** In what follows the symbol $a_n \sim b_n$ will be used whenever $a_n = 0(b_n)$ and $b_n = 0(a_n)$ for large $n$.

THEOREM 4.1. *Let* $m$, $k$ *and* $p$ *be given such that* $m \geqslant -1, 0 \leqslant k \leqslant m+1$ *and* $1 \leqslant p \leqslant \infty$. *Then for large* $n$

$$\|f_n^{(m,k)}\|_p \sim n^{\frac{1}{2}+k-1/p},$$

$$\|g_n^{(m,k)}\|_p \sim n^{\frac{1}{2}-k-1/p},$$

*and the estimates are independent of* $p$.

In the case of Franklin system cf. [1].

**5. Local estimates for the biorthogonal functions.** For given $n > 1$ let $t_n = (2\nu - 1)/2^{\mu+1}$, where $n = 2^\mu + \nu$, $\mu \geqslant 0$ and $1 \leqslant \nu \leqslant 2^\mu$. Moreover, let $t_n^{m,k}$, $s_n^{m,k} \in I$ be such that

$$\|f_n^{(m,k)}\|_\infty = |f_n^{(m,k)}(t_n^{m,k})|,$$

$$\|g_n^{(m,k)}\|_\infty = |g_n^{(m,k)}(s_n^{m,k})|.$$

THEOREM 5.1. *Let* $m \geqslant -1, 0 \leqslant k \leqslant m+1$. *Then there exist constants* $q_m$, $0 < q_m < 1$, *and* $C_m$ *such that*

$$|f_n^{(m,k)}(t)| \leqslant C_m n^{\frac{1}{2}+k} q_m^{n|t-u|}, \qquad u = t_n,\ t_n^{m,k},$$

$$|g_n^{(m,k)}(t)| \leqslant C_m n^{\frac{1}{2}-k} q_m^{n|t-u|}, \qquad u = t_n,\ s_n^{m,k},$$

*hold for all* $t \in I$.

Particular cases of this result were proved in [1] and [3].

**6. The fundamental inequalities.** The theorem below is an extension of a result established earlier in the case of the Franklin system in [1].

THEOREM 6.1. *Let* $m \geqslant -1$, $0 \leqslant k \leqslant m+1$ *and* $1 \leqslant p \leqslant \infty$. *Then there are constants* $C_m$ *and* $M_m$ *such that for any real sequence* $\{a_n\}$ *the following inequalities hold for* $\mu \geqslant 0$,

$$\Big\| \sum_{2^\mu+1}^{2^{\mu+1}} |a_n f_n^{(m,k)}| \Big\|_p \leqslant C_m 2^{\mu(\frac{1}{2}+k-1/p)} A_{p,\mu},$$

$$\Big\| \sum_{2^\mu+1}^{2^{\mu+1}} |a_n g_n^{(m,k)}| \Big\|_p \leqslant C_m 2^{\mu(\frac{1}{2}-k-1/p)} A_{p,\mu},$$

$$2^{\mu(\frac{1}{2}+k-1/p)} A_{p,\mu} \leqslant M_m \Big\| \sum_{2^\mu+1}^{2^{\mu+1}} a_n f_n^{(m,k)} \Big\|_p,$$

$$2^{\mu(\frac{1}{2}-k-1/p)} A_{p,\mu} \leqslant M_m \Big\| \sum_{2^\mu+1}^{2^{\mu+1}} a_n g_n^{(m,k)} \Big\|_p,$$

*where*

$$A_{p,\mu} = \Big( \sum_{2^\mu+1}^{2^{\mu+1}} |a_n|^p \Big)^{1/p}.$$

**7. New spline bases in** $C(I)$ **and** $L_p(I)$. It appears that not only the systems $\{f_n^{(m,k)}, n \geqslant k-m\}$ do form nice bases in various function spaces (see e.g. [2]). The systems of splines $\{g_n^{(m,k)}, n \geqslant k-m\}$, $0 \leqslant k \leqslant m+2$, $m \geqslant -1$, form bases in the same spaces.

THEOREM 7.1. *Let $m \geqslant -1$, $0 \leqslant k \leqslant m+1$ and let $1 \leqslant p < \infty$. Then $\{g_n^{(m,k)}, n \geqslant k-m\}$ is a basis in $L_p(I)$ and for each $f \in L_p(I)$ we have*

$$f = \sum_{n=k-m}^{\infty} (f, f_n^{(m,k)}) g_n^{(m,k)}.$$

THEOREM 7.2. *Let $m \geqslant -1$ and $0 \leqslant k \leqslant m+1$. Then the set of functions $\{1, g_n^{(m,k+1)}, n \geqslant k-m\}$ is a basis in $C(I)$ and for $f \in C(I)$ and $t \in I$ we have*

$$f(t) = f(1) - \sum_{n=k-m}^{\infty} \left(\int_I f_n^{(m,k)}(u)\, df(u)\right) g_n^{(m,k+1)}(t).$$

*Moreover, for $k = m+1$ this system is an interpolating basis of splines.*

The last part of the statement is related to the results presented in [5].

The discussion and complete proofs of the results presented in this note will be published in Studia Mathematica.

## REFERENCES

[1] Z. Ciesielski, 'Properties of the Orthonormal Franklin System, II', *Studia Math.* **27** (1966) 289–323.

[2] Z. Ciesielski and J. Domsta, 'Construction of an Orthonormal Basis in $C^m(I^d)$ and $W_p^m(I^d)$', *ibidem* **41** (1972) 211–224.

[3] Z. Ciesielski and J. Domsta, 'Estimates for the Spline Orthonormal Functions and for Their Derivatives', *ibidem* **44** (1972) 315–320.

[4] J. Domsta, 'A Theorem on *B*-splines', *ibidem* **41** (1972) 291–314.

[5] S. Schonefeld, 'Schauder Bases in the Banach Spaces $C^k(T^q)$', *Trans. Amer. Math. Soc.* **165** (1972) 309–318.

# SATURATIONSTHEORIE AUF KOMPAKTEN TOPOLOGISCHEN GRUPPEN

BERND DRESELER

*Siegen*

*Abstract.* Let $G$ be an arbitrary compact not necessarily abelian topological group. It is the purpose of the present paper to establish some saturation theorems for approximation processes which are generated by families of complex Radon measures on $G$ and which operate on the left Banach module $L^p(G)$, $p \in [\![1, +\infty[\![$, or $C(G)$, over the convolution algebra $M(G)$. A basic tool is the projection method based on the Peter–Weyl spectral decomposition theorem of the space $L^2(G)$. The theory admits a natural extension to the case of compact homogeneous spaces $G/K$, where $K$ is an arbitrary compact subgroup of $G$.

**1. Einleitung.** Die ersten saturationstheoretischen Untersuchungen bei Approximationsverfahren vom Faltungstyp auf beliebigen lokalkompakten abelschen topologischen Gruppen gehen auf H. Buchwalter [2] zurück. Seiner kurzen Darstellung folgen die Arbeiten [7], [8], [13], die eine ausführliche und erweiterte Behandlung des lokalkompakten abelschen Falles enthalten. Es ist das Ziel dieser Arbeit, einen allgemeinen Saturationssatz für Approximationsverfahren vom Faltungstyp auf kompakten nicht notwendig abelschen topologischen Gruppen anzugeben. Die dazu erforderlichen Hilfsmittel, die wir im nächsten Abschnitt kurz zusammenstellen wollen, bleiben in diesem allgemeinen Rahmen überwiegend im algebraischen Bereich. Mit dem für die Anwendung sehr wichtigen Spezialfall, daß die zugrundeliegende Gruppe sogar eine Lie-Gruppe ist, beschäftigen wir uns dann in [9].

Die folgenden Ergebnisse findet man in detaillierter Darstellung in [6].

**2. Hilfsmittel.** Im folgenden bezeichne $G$ stets eine kompakte topologische Gruppe mit dem neutralen Element $1 \in G$ und dem Haar-Maß $\nu$, das durch $\nu(G) = 1$ normalisiert ist. $C(G)$ sei der komplexe Vektorraum der stetigen komplexwertigen Funktionen auf $G$ versehen mit der Topologie der gleichmäßigen Konvergenz auf $G$. Als topologischen Dual von $C(G)$ erhalten wir den Vektorraum $M(G)$ aller komplexen Radon-Maße $\mu$ auf $G$. Hinsichtlich der Norm, die die starke Dualtopologie $\beta(M(G), C(G))$ auf $M(G)$ induziert und der üblichen Faltungsmultiplikation

$$(\mu, \varrho) \rightsquigarrow \mu * \varrho$$

wird $M(G)$ zu einer involutiven komplexen Banachalgebra. Die Involution von $M(G)$ sei gerade die Transponierte der stetigen linearen Abbildung

$$C(G) \ni f \rightsquigarrow (f^* : G \ni x \rightsquigarrow \bar{f}(x^{-1}) \in C) \in C(G).$$

Schließlich sei $A(G)$ im folgenden jeweils eine der komplexen involutiven Banachalgebren $L^p(G) = L^p_C(G;\nu)$, $p \in [\![1, +\infty[\![$, und $C(G)$. Die Injektion

$$A(G) \to M(G)$$

ist ein Monomorphismus hinsichtlich der Kategorie der involutiven topologischen $C$-Algebren. Zur Vereinfachung bezeichnen wir die üblichen Normen von $A(G)$ und $M(G)$ mit demselben Symbol $\|\cdot\|$.

Bekannterweise (Dieudonné [5]) ist $L^2(G)$ eine komplexe (vollständige) Hilbertalgebra hinsichtlich des gewöhnlichen Skalarproduktes. Nach dem Spektralsatz von Ambrose–Gurevič ist die Hilbertsumme

$$L^2(G) = \bigoplus_{\lambda \in \Lambda(G)} \mathfrak{a}_\lambda \tag{1}$$

die Spektralzerlegung des $L^2(G)$. Die minimalen selbstadjungierten zweiseitigen Ideale $\mathfrak{a}_\lambda$ $(\lambda \in \Lambda(G))$ — nach Dieudonné [4] auch Füße genannt — enthalten (selbstadjungierte) Einheiten $u_\lambda$ und für die (endliche) Dimension von $\mathfrak{a}_\lambda$ gilt

$$\dim_C \mathfrak{a}_\lambda = u_\lambda(1) = n_\lambda^2 \qquad (\lambda \in \Lambda(G)).$$

Mit $\mathfrak{a}_{\lambda_0}$ bezeichnen wir den trivialen Fuß der Dimension $n_{\lambda_0}^2 = 1$, der aus den konstanten komplexwertigen Funktionen auf $G$ besteht. Man beachte schließlich, daß der Sockel des Rings $L^2(G)$, d.h. die algebraische direkte Summe

$$\mathfrak{S}_{L^2(G)} = \coprod_{\lambda \in \Lambda(G)} \mathfrak{a}_\lambda,$$

ein Untervektorraum von $C(G)$ und ein überall dichtes zweiseitiges Ideal von $A(G)$ ist. Darüberhinaus existiert eine Bijektion $\eta$ von der

Menge der Füße $\{\mathfrak{a}_\lambda | \lambda \in \Lambda(G)\}$ auf das duale Objekt $\hat{G}$ von $G$ derart, daß die Funktion

$$\chi_\lambda = \frac{1}{n_\lambda} u_\lambda$$

der zum Fuß $\mathfrak{a}_\lambda$ und zur Äquivalenzklasse stetiger irreduzibler unitärer Darstellungen $\eta(\mathfrak{a}) \in \hat{G}$ gehörige Gruppencharakter ist. Es gilt demnach

$$\operatorname{Tr}\big(\eta(\mathfrak{a}_\lambda)\big) = \bar{\chi}_\lambda$$

für alle Parameter $\lambda \in \Lambda(G)$. Damit haben wir auch den Zusammenhang des Spektraltheorems von Ambrose–Gurevič mit dem bekannten Zerlegungstheorem von Peter–Weyl [3], [11] hergestellt.

**3. Ein allgemeiner Saturationssatz für Approximationsverfahren auf $A(G)$.** Es sei $T$ eine nichtleere gerichtete Menge von Parametern. Eine Familie $(I_t)_{t\in T}$ von stetigen $C$-linearen Abbildungen $I_t: A(G) \to A(G)$ wird ein Approximationsverfahren auf dem komplexen Banachraum $A(G)$ genannt, wenn sie hinsichtlich des Abschnittsfilters auf der gerichteten Menge $T$ gegen den identischen Automorphismus von $A(G)$ konvergiert, d.h. wenn

$$\lim_{t\in T} \|I_t(f) - f\| = 0$$

für alle $f \in A(G)$ gilt.

DEFINITION 1. Ein Approximationsverfahren $(I_t)_{t\in T}$ auf $A(G)$ hat die *Saturationsstruktur* $(\varphi; A(G); V)$, wenn folgende Bedingungen erfüllt sind:

(I) Es existiert eine Abbildung $\varphi: T \to \boldsymbol{R}_+^* = ]\!]0, +\infty[\![$, mit $\lim_{t\in T} \varphi(t) = 0$, so daß aus $f \in A(G)$ und

$$\|I_t(f) - f\| = o\big(\varphi(t)\big) \qquad (t \in T) \tag{2}$$

$f \in \mathfrak{a}_{\lambda_0}$ folgt.

(II) Es existiert ein Untervektorraum $V \neq \mathfrak{a}_{\lambda_0}$ von $A(G)$, so daß aus $f \in A(G)$ und der Bedingung

$$\|I_t(f) - f\| = O\big(\varphi(t)\big) \qquad (t \in T) \tag{3}$$

$f \in V$ folgt.

(III) Gilt $f \in V$, so erfüllt $f$ die Bedingung (3).

In diesem Fall wird $\varphi$ die Saturationsordnung und $V$ die Saturationsklasse des Verfahrens $(I_t)_{t\in T}$ genannt.

Wir wollen hier Approximationsverfahren vom Faltungstyp untersuchen. Dazu definieren wir

DEFINITION 2. Eine Familie $(\mu_t)_{t\in T}$ von Maßen aus $M(G)$ wird *approximative Einheit* von $A(G)$ genannt, wenn die Bedingungen

$$\mu_t(G) = 1 \qquad (t \in T)$$

und

$$\lim_{t\in T} \|\mu_t * f - f\| = 0 \qquad (f \in A(G))$$

gelten.

Da $A(G)$ ein linker Banachmodul über $M(G)$ ist, wird durch

$$I_t\colon A(G) \ni f \rightsquigarrow \mu_t * f \in A(G) \qquad (t \in T)$$

ein Approximationsverfahren auf $A(G)$ definiert, das wir das durch die approximative Einheit $(\mu_t)_{t\in T}$ auf $A(G)$ erzeugte Approximationsverfahren nennen.

BEMERKUNG. Aufgrund der Faktorisierung

$$L^1(G) * A(G) = A(G)$$

(siehe Hewitt–Ross [11], Chapter VIII) ist jede linke approximative Einheit von $L^1(G)$ auch linke approximative Einheit von $A(G)$.

DEFINITION 3. Eine linke approximative Einheit $(\mu_t)_{t\in T}$ von $L^1(G)$ wird *linkes Prosaturationsmaß vom Typ* $(\varphi;\psi)$ auf $G$ genannt, wenn folgende Bedingungen gelten:

Es existiert eine Funktion $\varphi\colon T \to \boldsymbol{R}_+^*$ mit $\lim_{t\in T} \varphi(t) = 0$ und eine Funktion $\psi\colon \Lambda(G) \to \boldsymbol{C}$ mit $\psi(\lambda) \neq 0$ für $\lambda \in \Lambda(G) \setminus \{\lambda_0\}$, so daß

$$\lim_{t\in T} \frac{u_\lambda * \mu_t - u_\lambda}{\varphi(t)} = \psi(\lambda)\, u_\lambda \qquad (\lambda \in \Lambda(G))$$

gleichmäßig auf $G$ gilt.

DEFINITION 4. Ein linkes Prosaturationsmaß vom Typ $(\varphi;\psi)$ auf $G$ wird *linkes Saturationsmaß vom Typ* $(\varphi, \psi)$ auf $G$ genannt, wenn folgende Bedingungen erfüllt sind:

Es existiert eine Familie $(\nu_t)_{t\in T}$ von Maßen aus $M(G)$ mit gemeinsamer Normschranke

$$M_0 = \sup_{t>0} \|\nu_t\| < \infty$$

derart, daß die Identität

$$u_\lambda * \nu_t = \begin{cases} \dfrac{u_\lambda * \mu_t - u_\lambda}{\varphi(t)\cdot\psi(\lambda)} & \text{für} \quad \lambda \neq \lambda_0 \\ 1_G & \text{für} \quad \lambda = \lambda_0 \end{cases} \qquad (\lambda \in \Lambda(G);\ t \in T)$$

in $C(G)$ erfüllt ist.

Einen für die konkrete Anwendung wichtigen Spezialfall bilden die zentralen Prosaturationsmaße (Saturationsmaße) auf $G$. Eine ausführliche Behandlung dieses Falles findet man in [9].

Wir sind nun in der Lage, das allgemeine Saturationstheorem zu formulieren. Zuvor vereinbaren wir noch einige Bezeichnungsweisen. Dem Raum $A(G)$ ordnen wir mit folgender Regel den Raum $B(G)$ zu:

$$B(G) = \begin{cases} C(G) & \text{für} \quad A(G) = L^1(G); \\ L^1(G) & \text{für} \quad A(G) = C(G); \\ L^{p'}(G) & \text{für} \quad A(G) = L^p(G) \text{ und } p \in \,]\!]1, +\infty[\![. \end{cases}$$

Wie üblich bezeichne $p'$ den dualen Exponenten von $p$. Wir nennen dann $\big(A(G), B(G)\big)$ ein zulässiges Paar. Unter Berücksichtigung dieser Sprechweisen erhalten wir folgendes allgemeine Saturationstheorem.

Theorem. *Es sei $(I_t)_{t\in T}$ ein Approximationsverfahren auf $A(G)$, das durch das linke Prosaturationsmaß $(\mu_t)_{t\in T}$ vom Typ $(\varphi;\psi)$ auf $G$ erzeugt wird. $\big(A(G), B(G)\big)$ sei ein zulässiges Paar und es sei $V_\psi\big(A(G)\big) = \{f \in A(G) \wr \psi(\lambda) u_\lambda * f = u_\lambda * \varrho,\ \varrho \in B'(G),\ \lambda \in \Lambda(G)\}$. Dann gelten folgende Aussagen:*

1) *$(I_t)_{t\in T}$ hat die Saturationsstruktur $\big(\varphi; A(G); V\big)$ mit*

$$V \subset V_\psi\big(A(G)\big).$$

2) *Ist $(\mu_t)_{t\in T}$ zusätzlich ein linkes Saturationsmaß auf $G$, so gilt sogar*

$$V = V_\psi\big(A(G)\big).$$

Den Beweis dieses Theorems findet man in [6]. Er verläuft mit den in Abschnitt 2 zusammengestellten Hilfsmitteln analog zum klassischen Fall $G = \boldsymbol{T} = \boldsymbol{R}/2\pi\boldsymbol{Z}$. In Teil 1 führen im wesentlichen Stetigkeitsargumentationen und eine Ausnutzung des Theorems von Alaoglu zum

Ziel. Teil 2 erhält man aus der Totalität der Familie von stetigen Orthogonalprojektoren $(P^\lambda)_{\lambda\in\Lambda(G)}$ auf $A(G)$, die von der Familie $(u_\lambda)_{\lambda\in\Lambda(G)}$ der Einheiten durch

$$A(G)\ni f \leadsto P^\lambda f = u_\lambda * f\in A(G) \qquad (\lambda\in\Lambda(G))$$

erzeugt wird. Der Beweis von Teil 1 ist für zentrale Approximationsverfahren auf $A(G)$ auch in [9] enthalten.

**4. Schlußbemerkungen.** Die Untersuchung konkreter Approximationsverfahren auf ihre Saturationsstruktur unter Verwendung des allgemeinen Theorems in Abschnitt 3 findet man in [6] und [9]. Unter anderem werden dort einige Verallgemeinerungen klassischer Verfahren wie die singulären Integrale von Weierstraß und Abel–Poisson für die Gruppen $T^n$, $SU(2)$ und für beliebige kompakte Lie-Gruppen untersucht. Darüberhinaus liegt die Bedeutung einer Saturationstheorie auf beliebigen kompakten topologischen Gruppen $G$ darin, daß sie eine natürliche Vorstufe ist zu einer Theorie auf kompakten homogenen Räumen $G/K$, wobei $K$ eine beliebige abgeschlossene Untergruppe von $G$ sein darf. Eine solche Erweiterung ist ebenfalls in [6] enthalten. Der für die Anwendung wichtige Spezialfall, daß $(G, K)$ sogar ein Gelfand-Paar ist, wird in [10] ausführlich behandelt. Die Saturationssätze auf der $(n-1)$-dimensionalen Sphäre $S^{n-1}$ des $\boldsymbol{R}^n$ in [1] und [12] ergeben sich dann im allgemeineren Rahmen der kompakten homogenen Räume, indem man für $G$ die spezielle orthogonale Gruppe $SO(n)$ des $\boldsymbol{R}^n$ und für $K$ die Gruppe $SO(n-1)$ wählt.

## LITERATUR

[1] H. Berens, P. L. Butzer und S. Pawelke, 'Limitierungsverfahren von Reihen mehrdimensionaler Kugelfunktionen und deren Saturationsverhalten', *Publ. Res. Inst. Math. Sci.* Ser. A. **4** (1968) 201–268.

[2] H. Buchwalter, 'Saturation sur un groupe abélien localement compact', *C.R. Acad. Sci. Paris* **250** (1960) 808–810.

[3] R. R. Coifman et G. Weiss, 'Analyse harmonique noncommutative sur certains espaces homogènes', *Lecture Notes in Mathematics*, Vol. **242** Springer, Berlin–Heidelberg–New York 1971.

[4] J. Dieudonné, 'Sur le socle d'un anneau et les anneaux simples infinis', *Bull. Soc. Math. France* **70** (1942) 46–75.

[5] J. Dieudonné, 'Representaciones de grupos compactos y funciones esfericas', *Cursos y seminarios de matemática*, Fasc. **14**, Universidad de Buenos Aires 1964.

[6] B. Dreseler, *Saturation auf kompakten topologischen Gruppen und homogenen Räumen*, Dissertation, Universität Mannheim 1972.

[7] B. Dreseler and W. Schempp, 'Saturation on Locally Compact Abelian Groups', *Manuscripta Math.* **7** (1972) 141–174.

[8] B. Dreseler and W. Schempp, 'Saturation on Locally Compact Abelian Groups: An Extended Theorem', *Manuscripta Math.* **8** (1973) 271–286.

[9] B. Dreseler and W. Schempp, Central Approximation Processes, *Tôhoku Math. J.* **27** (1975).

[10] B. Dreseler and W. Schempp, Zonal Approximation Processes, *Math. Z.* **133** (1973) 81–92.

[11] E. Hewitt and K. A. Ross, *Abstract Harmonic Analysis*, Vol. II, *Structure and Analysis for Compact Groups*, Springer, Berlin–Heidelberg–New York 1970.

[12] S. Pawelke, *Saturation und Approximation bei Reihen mehrdimensionaler Kugelfunktionen*, Dissertation, Technische Hochschule Aachen 1969.

[13] W. Schempp, 'Zur Theorie der saturierten Approximationsverfahren auf lokalkompakten abelschen Gruppen' (this volume) 181–187.

# UNIFORM APPROXIMATION ON CLOSED SETS BY FUNCTIONS ANALYTIC ON A RIEMANN SURFACE

P. M. GAUTHIER* and W. HENGARTNER**

*College Park, Md.*

Let $E$ be a closed subset of an open Riemann surface $S$; and denote by $C_A(E)$ the set of functions continuous on $E$ and holomorphic in the interior $E^0$ of $E$, by $A_S$ the holomorphic functions on $S$, and by $A_S(E)$ the uniform limits on $E$ of functions holomorphic on $S$. Similarly we denote by $A_E$ the set of germs of functions holomorphic on $E$ and by $A_E(E)$ the uniform limits on $E$ of germs in $A_E$. Clearly

$$C_A(E) \supset A_S(E) \tag{1}$$

and

$$C_A(E) \supset A_E(E). \tag{2}$$

We are interested in knowing when equality holds in (1) and (2) respectively. Should $S$ be the finite plane $\boldsymbol{C}$ and $E$ compact, (1) corresponds to the problem of polynomial approximation and (2) to the problem of rational approximation.

In Section 1 we shall consider (1) and (2) for the case where $E$ is compact and in Section 2 we shall discuss (1) for the general case when $E$ is merely closed.

We employ the following notations. $S^*$ denotes the one-point compactification of $S$; for $E \subset S$, $\partial E$ denotes the boundary of $E$ in $S$.

We extend our thanks to Professor Lawrence Zalcman for suggesting Example 1 of Section 2.

**1. Compact sets.** The following theorem extends Mergelian's theorem on polynomial approximation to arbitrary open Riemann surfaces.

* Supported by NRC of Canada, Grant A-5597 and a grant from the Gouvernement du Québec.

** Supported by NRC of Canada, Grant A-7339 and a grant from the Gouvernement du Québec.

THEOREM A (Bishop [3], Corollary 2, p. 48). *Let $K$ be compact in $S$. If $S^*\setminus K$ is connected, then $C_A(K) = A_S(K)$.*

Actually it is well known that this condition is necessary as well as sufficient. (The necessity we will show in Section 2.) The proof of Theorem A is very deep, but for the special case that $S$ is a plane domain, Arakélian has given an alternate proof (see Theorem C).

Using Bishop's techniques, Ketchum-Kodama has obtained the following theorem on approximation by functions holomorphic on $K$.

THEOREM B (Ketchum-Kodama [4], Corollary 4, p. 1276). *Let $K$ be compact in $S$. If $S^*\setminus K$ has finitely many components*(1), *then $C_A(K) = A_K(K)$.*

Note that in this case uniform approximation by functions holomorphic in a neighborhood of $K$, by meromorphic functions, or by meromorphic functions with finitely many poles are all equivalent.

Our statement of Theorem B is weaker than the original. Again if $S$ is the finite plane $\boldsymbol{C}$, Theorem B is a consequence of a theorem of Mergelian on rational approximation (see [6], p. 317).

In the following we show that Theorem A and Theorem B are equivalent by using elementary facts and well-known theorems.

First we prove that Theorem A implies Theorem B. Suppose then that $S^*\setminus K$ has finitely many components. Let $f$ be continuous on $K$ and holomorphic on the interior of $K$ and let $\varepsilon > 0$. We must find a meromorphic function $g$ on $S$ such that $\|f-g\|_K < \varepsilon$. Let $U_1, U_2, \dots, U_n$ be the components of $S^*\setminus K$, where $U_1$ contains the ideal point and $U_2, \dots, U_n$ are precompact in $S$. Choose $P_j \in U_j, j = 2, 3, \dots, n$, and set

$$S_1 = S\setminus \{P_2, P_3, \dots, P_n\}.$$

By Theorem A, there is a holomorphic function $g_1$ on $S_1$ such that $\|f-g_1\|_K < \varepsilon/2$. Now, by the Runge–Behnke–Stein theorem on approximation by meromorphic functions [2] (Theorem 13, p. 456), there is a meromorphic function $g$ on $S$ such that $\|g_1-g\|_K < \varepsilon/2$. Thus Theorem A implies Theorem B.

(1) A component is a maximal connected set.

Conversely, suppose that Theorem B holds and $S^*\setminus K$ is connected. Let $f$ be continuous on $K$ and holomorphic on $K^0$, and let $\varepsilon > 0$. Then, by Theorem B, there is a meromorphic function $g_2$ on $S$ such that $\|f-g_2\|_K < \varepsilon/2$.

Let $\{P_n\}$, $n = 1, 2, \ldots$, be the poles of $g_2$. Since no component of $S\setminus K$ is precompact, we can construct disjoint simple paths $\gamma_n$, $n = 1, 2, \ldots$, such that each path $\gamma_n$ connects $P_n$ to the ideal point and $\gamma_n$ does not meet $K$. Set

$$S_1 = S\setminus \bigcup_{j=1}^{\infty} \gamma_j.$$

Then $g_2$ is holomorphic on $S_1$ and $S_1$ satisfies the hypotheses for the Runge–Behnke–Stein theorem on approximation by holomorphic functions ([2], Theorem 6, p. 445). Hence there is a $g$ holomorphic on $S$ such that $\|g_2-g\|_K < \varepsilon/2$. This completes the proof that Theorem B implies Theorem A.

**2. Closed sets.** N. U. Arakélian has extended Mergelian's theorem on polynomial approximation as follows:

THEOREM C (Arakélian [1], Theorem 1). *Let $E$ be a (relatively) closed subset of a plane domain $S$. Then $C_A(E) = A_S(E)$ if and only if $S^*\setminus E$ is connected and locally connected.*

We shall see that these conditions remain necessary on arbitrary open Riemann surfaces. However, surprisingly (at least to us), they are not sufficient. Thus Arakélian's theorem, in the form in which we stated it, fails on arbitrary Riemann surfaces. This raises the two following problems:

(a) What are the necessary and sufficient conditions, for uniform approximations on closed sets by holomorphic functions on $S$,

(b) For which surfaces does Arakélian's theorem carry over without modifications.

*Necessity.* Suppose $E$ is a closed subset of an open Riemann surface $S$ and suppose $S^*\setminus E$ is not connected. Then $S\setminus E$ has a precompact component $U$. There exists a meromorphic function $f$ on $S$ having a pole at a point $P_0 \in U$ and nowhere else and which is different from zero on $\bar{U}$. Suppose that for each $n \in N$, there is a function $g_n$, holomorphic on $S$, such that

$$\|f-g_n\|_E < 1/n.$$

Then

$$\|1-g_n/f\|_{\partial U} < \|f^{-1}\|_{\partial U} n^{-1}.$$

Hence, by the maximum principle,

$$\|1-g_n/f\|_U < \|f^{-1}\|_{\partial U} n^{-1}.$$

In particular, at $P_0$ we have

$$1 \leqslant \|f\|_{\partial U}^{-1} n^{-1},$$

which is a contradiction for large $n$. Thus the condition "$S^* \setminus E$ connected" is necessary.

Now suppose $S^* \setminus E$ is not locally connected. We may assume $S^* \setminus E$ is connected. It follows that there exists a compact domain $K \subset S$ such that the precompact components of

$$(S \setminus E) \cap (S \setminus K) \tag{3}$$

are unbounded in the following sense. We may choose a sequence $P_n$, $n \in N$, of points from distinct precompact components $D_n$ of (3) such that $P_n$ converges to the ideal point at "infinity". For each $n \in N$, let $\psi_n$ be a holomorphic function on $S$ with a simple zero at $P_n$ and bounded by $1/n$ on $K \cup \partial D_n$. Now let $f$ be a meromorphic function on $S$ having simple poles at the $P_n$'s and whose principal parts have the property

$$(f \cdot \psi_n)(P_n) = 1.$$

Therefore $f \in C_A(E)$; suppose there is an analytic function $g$, on $S$, such that

$$\|f-g\|_E < 1.$$

Set

$$M = \sup_K \{1+|f|+|g|\}$$

and

$$h_n = \varphi_n(f-g).$$

Then $h_n$ is holomorphic in $D_n$ and continuous on $\bar{D}_n$. On $\partial D_n$, we have

$$|h_n(P)| < n^{-1}M.$$

Thus by the maximum principle, the same is true inside $D_n$. However, at $P_n$, we have $h_n(P_n) = 1$, and so

$$1 < n^{-1}M$$

which, for large $n$, gives the desired contradiction. We have shown that the conditions of Theorem C remain necessary on Riemann surfaces.

We shall show that these conditions are not sufficient.

THEOREM D. *There exists a Riemann surface $S$ with the property that $S \setminus K \in O_{AB}$ for each compact $K \subset S$ such that $S \setminus K$ is connected.*

This theorem is due to Kuramochi [5], Theorem 5.1, who showed that any surface in $O_{HB} \setminus O_G$ has the required property. The strict inclusion $O_G \subset O_{HB}$ was shown by Tôki [8] and Sario [7].

Lawrence Zalcman has provided us with the following explicit example of a Riemann surface having the property of Theorem D.

EXAMPLE 1. For $n = 0, \pm 1, \pm 2, \ldots$, let $S_n$ be a copy of the finite plane slit along

$$\left[n+\frac{1}{2k+1}, n+\frac{1}{2k}\right] \quad \text{and} \quad \left[n+1+\frac{1}{2k+1}, n+1+\frac{1}{2k}\right] \tag{4}$$

for $k = 1, 2, \ldots$ The surface $S$ is constructed by welding the sheets $S_n$ together in the obvious way so as to leave no free edges.

Suppose now $K$ is a compact set in $S$ and $f$ is a bounded holomorphic function defined on $S \setminus K$. Choose $n$ so large that

$$K \cap \overline{(S_{n-1} \cup S_n \cup S_{n+1})} = \emptyset,$$

where the closure is taken in $S$.

Let $\pi$ be the projection of $S_n$ into the complex plane. Then $f \circ \pi^{-1} = F$ bounded and holomorphic in the plane slit along (4). We shall show that $F$ extends to an entire function.

Let $S'$ be the two sheeted covering of $(|z-n| < 1/2)$ formed by joining the portions of $S_{n-1}$ and $S_n$ which lie over $(0 < |z-n| < 1/2)$. Let $\varphi$ be the conformal interchange of sheets on $S'$. Then $(f-f \circ \varphi)^2$ yields a bounded (single-valued) holomorphic function on $(|z-n)| < 1/2$, having zeros at $n+1/2k$, $k = 1, 2, \ldots$ Hence this function is identically zero. Thus $f$ takes the same value at any two points of $S_n$ and $S_{n-1}$

having the same projection. Thus $F$ extends holomorphically to the slits $\left[n+\frac{1}{2k+1}, n+\frac{1}{2k}\right]$, $k = 1, 2, \ldots$, and to the point $n$ since the extended $F$ is bounded and holomorphic in a deleted neighbourhood of $n$. An analogous argument for the sheets $S_n$ and $S_{n+1}$ shows that $F$ extends to the remaining cuts in (4) and to the point $n+1$. Thus $F$ extends to a bounded entire function. Therefore $f$ is a constant function, and we have shown that the surface $S$ has the property in Theorem D.

We may now present an example to show that the conditions in Theorem C are not sufficient on Riemann surfaces.

EXAMPLE 2. Let $S$ be a surface having the property in Theorem $D$, and let $D$ be a parametric disc in $S$ with centre $P_0$. Choose $P_1 \neq P_0$ in $D$. Now set $S_0 = S \setminus \{P_0\}$ and $E = (S \setminus D) \cup \{P_1\}$. Let $f$ be zero on $S \setminus D$ and one at $P_1$. Then $S_0^* \setminus E$ is connected and locally connected, and $f$ is continuous on $E$ and holomorphic on $E^0$. Suppose there is a function $g$ holomorphic on $S_0$ such that

$$\|f-g\|_E < 1/2. \tag{5}$$

Then $g$ is holomorphic and bounded on $S \setminus \bar{D}$. By our choice of $S$, $g$ is constant on $S \setminus \bar{D}$. But then $g$ must be constant on its whole domain of definition $S_0$. However, this contradicts (5), and so Theorem C fails on $S_0$.

## REFERENCES


[1] N. U. Arakélian, 'Approximation complexe et propriétés des fonctions analytiques', *Actes, Congrès intern. Math.* **2** (1970) 595–600.

[2] H. Behnke and K. Stein, 'Entwicklung analytischer Funktionen auf Riemannschen Flächen', *Math. Ann.* **120** (1949) 430–461. MR10-696.

[3] E. Bishop, 'Subalgebras of Functions on a Riemann Surface', *Pacific J. Math.* **8** (1958) 29–50. MR20-3300.

[4] L. Ketchum-Kodama, 'Boundary Measures of Analytic Differentials and Uniform Approximation on a Riemann Surface', *Pacific J. Math.* **15** (1965) 1261–1277. MR32-7740.

[5] Z. Kuramochi, 'On Covering Surfaces', *Osaka Math. J.* **5** (1953), 155–201; errata **6** (1954) 167, MR15-518.

[6] S. N. Mergelyan, 'Uniform Approximation to Functions of a Complex Variable, *Amer. Math. Soc. Trans.* **101**.

[7] L. Sario, 'Functionals on Riemann Surfaces', Lectures on functions of a complex variable, 245–256. The University of Michigan Press, Ann Arbor, 1955, MR19-736.
[8] Y. Tôki, 'On the Examples in the Classification of Open Riemann Surfaces I', *Osaka Math. J.* **5** (1953) 267–280. MR15-519.

# DIE ABLEITUNGEN DER ALGEBRAISCHEN POLYNOME BESTER APPROXIMATION

M. v. GOLITSCHEK*

*Würzburg*

*Abstract.* Let $P_n$ denote the algebraic polynomial of degree $n$ of weighted best approximation to $f \in C[-1, 1]$, if $p = \infty$, or to $f \in L_p(-1, 1)$, if $1 \leqslant p < \infty$, relative to

(I) $$E_n^{(\lambda)}(f)_p := \inf_{P \in \pi_n} \left\| (\max\{1/n; \sqrt{1-x^2}\})^{-\lambda}(f(x) - P(x)) \right\|_p.$$

The following question is considered: Under which conditions are the statements

(II) $$E_n^{(\lambda)}(f)_p = O(n^{-\beta}),$$

(III) $$\left\| (\max\{1/n; \sqrt{1-x^2}\})^{r-\lambda} P_n^{(r)}(x) \right\|_p = O(n^{r-\beta}), \qquad n \to \infty,$$

equivalent, where $r$ is a positive integer and $\beta$ a real number, $0 < \beta < r$.

The answer is different for $\lambda \leqslant 0$ and $\lambda > 0$. If $\lambda \leqslant 0$, then Theorems 2–4 state that (II) and (III) are equivalent. For $\lambda > 0$ the situation is more complicated: From (II) follows (III) for all integers $r > \max\{\beta; (\lambda+\beta)/2\}$; but conversely, the statement (II) follows from (III) only under the additional assumption

(IV) $$\lim_{m \to \infty} E_m^{(\lambda)}(f)_p = 0.$$

In the second part of this paper two classes $K$ of functions are defined for which (II) and (III) are equivalent:

(a) $p = \infty, 0 < \lambda < r, k := [\lambda/3]$;

$$K := \{f \in C^k[-1, 1], \quad \lim_{h \to 0} \omega(f^{(k)}; h) h^{k-\lambda/3} = 0\};$$

(b) $p = \infty, 0 < \lambda < 1, K := \{f \in C[-1, 1], \operatorname{Var} f < \infty\}$.

In a counterexample functions $f \in \operatorname{Lip} \lambda/3, 0 < \lambda \leqslant 3$, are constructed for which

$$E_n^{(\lambda)}(f)_\infty = 1 \quad \text{and} \quad P_n \equiv 0 \quad \text{for} \quad n = 2, 3, 4, \ldots$$

Therefore condition (a) is best possible in a certain sense.

* Diese Arbeit enthält zum Teil Ergebnisse der Habilitationsschrift des Autors, die im November 1971 an der Universität Würzburg eingereicht wurde.

Ausgangspunkt meiner Betrachtungen ist das folgende wohlbekannte Ergebnis:

SATZ 1. *Es sei g eine* $2\pi$*-periodische Funktion, die zur Klasse* $L_p(-\pi, \pi)$, $1 \leqslant p < \infty$, *oder zur Klasse C gehört. Ferner sei* $T_n(g)$ *das trigonometrische Polynom bester Approximation aus* $\pi_n^*$ *von g bezüglich der* $L_p$*-bzw. sup-Norm. Dann ist*

$$\|T_n^{(r)}(g)\|_p = O(n^{r-\beta}), \quad r > \beta > 0, \tag{1}$$

*wobei r eine positive ganze Zahl ist, äquivalent zu*

$$\|g - T_n(g)\|_p = O(n^{-\beta}), \quad n \to \infty. \tag{2}$$

Daß aus (2) die Gleichung (1) folgt, und zwar für jedes $r \in N$, $r > \beta$, bewies M. Zamansky [8] schon im Jahre 1949. Daß man aber auch umgekehrt aus dem Verhalten (1) der Polynomableitungen auf die Größenordnung der besten Approximation schließen kann, wurde erstmals von P. L. Butzer und S. Pawelke [1], S. 182, vermutet und in derselben Arbeit für die $L_2$-Norm bewiesen. Den Beweis für die sup-Norm und die $L_p$-Normen, $1 \leqslant p < \infty$, führte G. Sunouchi [6] im Jahre 1968.

Naheliegend ist nun die Frage, ob es möglich ist, dem Satz 1 entsprechende Ergebnisse auch für die algebraischen Polynome bester Approximation anzugeben. Eine allgemeine Antwort hierauf wird in dieser Arbeit gegeben.

Wir vereinbaren folgende Bezeichnungen: Es sei für $n \in N$

$$\delta_n(x) := \max\{1/n; \sqrt{1-x^2}\},$$

und für $F \in C[-1, 1]$, bzw. $F \in L_p(-1, 1)$, $1 \leqslant p < \infty$,

$$\|F\|_\infty := \max_{-1 \leqslant x \leqslant 1} |F(x)|, \qquad \|F\|_p := \left(\int_{-1}^{1} |F(x)|^p \, dx\right)^{1/p}.$$

Wertvolle Hilfe leistet uns die wohlbekannte Verallgemeinerung des Jacksonkernes: Für die nichtnegative ganze Zahl $r$ definieren wir

$$\gamma_m = \int_{-\pi}^{\pi} \left(\frac{\sin(mt/2)}{m \sin(t/2)}\right)^{2r+4} dt.$$

Es ist bekannt, daß

$$K(t) = \frac{1}{\gamma_m}\left(\frac{\sin(mt/2)}{m\sin(t/2)}\right)^{2r+4} \tag{3}$$

ein gerades trigonometrisches Polynom der Ordnung $(r+2)(m-1)$ ist, das den Bedingungen

$$\int_{-\pi}^{\pi} K(t)\,dt = 1, \qquad \int_{-\pi}^{\pi} |t|^\beta K(t)\,dt \leqslant \frac{C}{m^\beta}, \qquad 0 < \beta < 2r+3, \tag{4}$$

genügt; $C$ ist unabhängig von $m$.

Ist $f$ auf dem Intervall $[-1, 1]$ integrierbar, dann ist $Q^*(x)$,

$$Q^*(\cos\theta) = \int_{-\pi}^{\pi} f(\cos(\theta+t))K(t)\,dt, \tag{5}$$

ein algebraisches Polynom vom Grade $(r+2)(m-1)$.

Mit Hilfe dieses Polynoms $Q^*$ können wir die folgende Ungleichung vom Jackson-Typ herleiten:

LEMMA 1([1]). *Die Funktion $F$ möge eine Ableitung $F'$ besitzen, die der Ungleichung*

$$\left\|\frac{F'(x)}{(\sqrt{1-x^2}+1/n)^s}\right\|_p \leqslant M < \infty \tag{6}$$

*genügt; hierbei sei $s$ eine beliebige reelle Zahl; $1 \leqslant p \leqslant \infty$. Dann existiert ein Polynom $Q(x) \in \pi_n$ derart, daß*

$$I = \left\|\frac{F(x)-Q(x)}{(\sqrt{1-x^2}+1/n)^{s+1}}\right\|_p \leqslant \frac{\tilde{C}M}{n}, \tag{7}$$

*wobei $\tilde{C}$ von $s$, nicht aber von $n$, $M$ und $F$ abhängt ($n \geqslant |s|+2$).*

BEWEIS. Es sei $m$ die größte natürliche Zahl, die der Ungleichung $(r+2)(m-1) \leqslant n$ genügt, wobei $r := [|s|]$. Es sei $Q$ das Polynom in (5). Wir setzen in (7) $x = \cos\theta$ und erhalten

(a) *für* $1 \leqslant p < \infty$:

$$I = \left(\frac{1}{2}\int_{-\pi}^{\pi}\left|\int_{-\pi}^{\pi} K(t)\frac{F(\cos\theta)-F(\cos(\theta+t))}{(|\sin\theta|+1/n)^{s+1}}\,dt\right|^p |\sin\theta|\,d\theta\right)^{1/p}.$$

([1]) Wir folgen einer Beweisidee von M. K. Potapov [3].

Verwenden wir die verallgemeinerte Minkowski-Ungleichung, so folgt hieraus

$$I \leqslant \int_{-\pi}^{\pi} K(t)\, I_1(t)\, dt$$

$$= \int_{-\pi}^{\pi} K(t) \left\{ \frac{1}{2} \int_{-\pi}^{\pi} \left| \frac{F(\cos\theta) - F(\cos(\theta+t))}{(|\sin\theta| + 1/n)^{s+1}} \right|^p |\sin\theta|\, d\theta \right\}^{1/p} dt.$$

Wir berechnen das innere Integral $I_1(t)$ und beachten, daß $F$ eine Ableitung besitzt. Wir formen $I_1$ um und wenden erneut die Minkowski-Ungleichung an; wir erhalten für $t \geqslant 0$

$$I_1(t) = \left\{ \frac{1}{2} \int_{-\pi}^{\pi} \left| \int_0^t \frac{F'(\cos(\theta+u)) \sin(\theta+u)}{(|\sin\theta| + 1/n)^{s+1}}\, du \right|^p |\sin\theta|\, d\theta \right\}^{1/p}$$

$$\leqslant \int_0^t \left\{ \frac{1}{2} \int_{-\pi}^{\pi} \left| \frac{F'(\cos(\theta+u)) \sin(\theta+u)}{(|\sin\theta| + 1/n)^{s+1}} \right|^p |\sin\theta|\, d\theta \right\}^{1/p} du.$$

Wir wenden nun die leicht zu beweisende Ungleichung

$$\frac{(|\sin(\theta+u)| + 1/n)^{sp} |\sin(\theta+u)|^p |\sin\theta|}{(|\sin\theta| + 1/n)^{(s+1)p}} \leqslant C_1^p (1 + m|u|)^{(|s|+1)p} |\sin(\theta+u)|$$

an. Für $0 \leqslant u \leqslant t$ erhalten wir dann

$$I_1(t) \leqslant C_1 M t (1 + mt)^{|s|+1}.$$

Analog verfahren wir für $t < 0$; wir ersetzen nur $t$ durch $|t|$. Insgesamt ist dann

$$I \leqslant C_1 M \int_{-\pi}^{\pi} |t|\, (1 + m|t|)^{|s|+1} K(t)\, dt.$$

Mit Hilfe von (4) folgt schließlich

$$I \leqslant \frac{\tilde{C} M}{n}.$$

(b) Analog verläuft der Beweis für $p = \infty$: Es ist

$$I = \sup_{\theta \in R} \left| \int_{-\pi}^{\pi} \frac{K(t) \{F(\cos\theta) - F(\cos(\theta+t))\}}{(|\sin\theta| + 1/n)^{s+1}}\, dt \right|$$

$$= \sup_{\theta \in R} \left| \int_{-\pi}^{\pi} K(t) \left\{ \int_0^t \frac{F'(\cos(\theta+u)) \sin(\theta+u)}{(|\sin\theta| + 1/n)^{s+1}}\, du \right\} dt \right|.$$

Nun ist aber

$$\frac{(|\sin(\theta+u)|+1/n)^s|\sin(\theta+u)|}{(|\sin\theta|+1/n)^{s+1}} \leqslant C_1'(1+m|u|)^{|s|+1}$$

und daher für $t \geqslant 0$

$$\int_0^t \left|\frac{F'(\cos(\theta+u))\sin(\theta+u)}{(|\sin\theta|+1/n)^{s+1}}\right| du \leqslant C_1' \int_0^t \frac{|F'(\cos(\theta+u))|(1+mu)^{|s|+1}}{(|\sin(\theta+u)|+1/n)^s} du$$

$$\leqslant C_1' Mt(1+mt)^{|s|+1}.$$

Ebenso verfahren wir für $t < 0$. Es folgt nun, daß

$$I \leqslant C_1' M \int_{-\pi}^{\pi} |t|(1+m|t|)^{|s|+1} K(t)\, dt.$$

Mit Hilfe von (4) erhalten wir hieraus die Ungleichung (7) für $p = \infty$. Somit ist Lemma 1 vollständig bewiesen.

Aus Lemma 1 gewinnen wir sofort

Lemma 2. *Es sei s eine reelle und r eine natürliche Zahl. Sei* $1 \leqslant p \leqslant \infty$. *Die Funktion F möge der Ungleichung*

$$\|(\delta_n(x))^{-s} F^{(r)}(x)\|_p \leqslant M^* < \infty \tag{8}$$

*genügen. Dann existiert ein Polynom* $P \in \pi_n$, $n \geqslant |s|+r+1$, *mit*

$$\|(\delta_n(x))^{-s-r}(F(x)-P(x))\|_p \leqslant M^* C^* n^{-r}. \tag{9}$$

*Hierbei hängt* $C^*$ *von s und r, nicht von* $M^*$, *F und n ab.*

Beweis. Nach Lemma 1 gibt es ein $Q_1 \in \pi_{n-r+1}$ mit

$$\|(\delta_n(x))^{-s-1}(F^{(r-1)}(x)-Q_1(x))\|_p \leqslant M^* C_1 n^{-1};$$

$C_1$ unabhängig von $M^*$, $F$ und $n$. Für $r \geqslant 2$ wenden wir Lemma 1 nun auf

$$F^{(r-2)}(x) - \int_0^x Q_1(t)\, dt$$

an. Rekursiv erhalten wir so nach $r$ Schritten die Aussage (9).

In den folgenden Sätzen soll nun der Zusammenhang zwischen der besten Approximation und den Ableitungen der algebraischen Polynome bester Approximation wiedergegeben werden. Die Sätze vom Jackson–Timan Typ lassen es wünschenswert erscheinen, die gewichtete beste

Approximation

$$E_n^{(\lambda)}(f)_p := \inf_{P \in \pi_n} \| (\delta_n(x))^{-\lambda} (f(x) - P(x)) \|_p \tag{10}$$

zu betrachten, wobei $\lambda$ eine vorgegebene reelle Zahl ist.

Ist $f \in C[-1, 1]$, bzw. $f \in L_p(-1, 1)$, $1 \leqslant p < \infty$, und bezeichnet $P_n = P_{np\lambda}$ das Polynom bester Approximation von $f$ bezüglich (10), so erhalten wir für $1 \leqslant p \leqslant \infty$:

SATZ 2. *Es seien $\lambda$ und $\beta$ reelle Zahlen, $\beta > 0$. Erfüllt die beste Approximation die Bedingung*

$$E_n^{(\lambda)}(f)_p = O(n^{-\beta}) \quad \text{für } n \to \infty, \tag{11}$$

*dann gilt für alle $r \in \boldsymbol{N}$ mit $r > \max\{\beta; (\lambda+\beta)/2\}$*

$$\| (\delta_n(x))^{r-\lambda} P_n^{(r)}(x) \|_p = O(n^{r-\beta}), \quad n \to \infty. \tag{12}$$

BEWEIS. Nach der Voraussetzung (11) existiert eine Konstante $A > 0$, so daß für alle $n \geqslant n_0 = 2^{s_0}$, $s_0 \in \boldsymbol{N}$, gilt:

$$\| \delta_n^{-\lambda}(f - P_n) \|_p \leqslant A n^{-\beta}.$$

Hieraus folgt für alle $q_1, q_2 \in \boldsymbol{N}$ mit $n_0 \leqslant q_1 < q_2 \leqslant 2q_1$

$$\| \delta_{q_2}^{-\lambda}(P_{q_2} - P_{q_1}) \|_p \leqslant A_0 q_2^{-\beta}.$$

Nach einem Satz von G. K. Lebed' [2] folgt hieraus für $r = 1, 2, \ldots$

$$\| \delta_{q_2}^{r-\lambda}(P_{q_2}^{(r)} - P_{q_1}^{(r)}) \|_p \leqslant A_r q_2^{r-\beta}, \tag{13}$$

wobei die Zahlen $A_r$ unabhängig von $q_1$ und $q_2$ sind.

Sei nun $s \in \boldsymbol{N}$ bestimmt durch $2^s < n \leqslant 2^{s+1}$. Dann ist

$$\begin{aligned} \| \delta_n^{r-\lambda} P_n^{(r)} \|_p \leqslant \| \delta_n^{r-\lambda}(P_n^{(r)} - P_{2^s}^{(r)}) \|_p + \| \delta_n^{r-\lambda} P_{2^{s_0}}^{(r)} \|_p + \\ + \sum_{k=s_0+1}^{s} \| \delta_n^{r-\lambda}(P_{2^k}^{(r)} - P_{2^{k-1}}^{(r)}) \|_p. \end{aligned} \tag{14}$$

Ist $\lambda \leqslant r$, so folgt aus (13) und (14) wegen $\delta_n(x) \leqslant \delta_{2^k}(x)$, $k \leqslant s$, sofort die Aussage (12). Ist $\lambda > r$ und zusätzlich $\lambda + \beta < 2r$, dann ist

$$\delta_{2^k}(x) \leqslant 2^{s+1-k} \delta_n(x);$$

und aus (13) und (14) folgt ebenfalls sofort die Aussage (12). Damit ist Satz 2 bewiesen.

Während der Beweis von Satz 2 nur wenig Schwierigkeiten bietet und sich im wesentlichen auf das Ergebnis von G. K. Lebed' über Ungleichungen der Polynomableitungen vom Bernstein-Typ stützt, ist bei der Umkehrung des Satzes 2 die Ungleichung vom Jackson-Typ in Lemma 2 wichtigstes Hilfsmittel.

Um das wesentliche Ergebnis vorwegzunehmen: Wir werden sehen, daß im Falle $\lambda \leqslant 0$ die Aussagen (11) und (12) äquivalent sind, daß aber im Falle $\lambda > 0$ diese Äquivalenz nicht immer gewährleistet ist.

SATZ 3(²). *Es sei* $1 \leqslant p \leqslant \infty$ *und* $f \in C[-1, 1]$ *bzw.* $f \in L_p(-1, 1)$. *Es seien* $\lambda$ *und* $\beta$ *reelle Zahlen,* $r$ *eine natürliche Zahl, mit* $\lambda \geqslant 0$ *und* $0 < \beta < r$. *Erfüllen die Polynome bester Approximation* $P_n$ *bezüglich* (10) *für dieses feste* $r$ *die Bedingung* (12) *und ist zusätzlich*

$$\lim_{m \to \infty} E_m^{(\lambda)}(f)_p = 0, \tag{15}$$

*dann genügt* $E_n^{(\lambda)}(f)_p$ *der Gleichung* (11).

Für $\lambda = 0$ ist die Bedingung (15) für alle $f \in C[-1, 1]$ bzw. $f \in L_p(-1, 1)$ erfüllt. Für $\lambda > 0$ hingegen ist (15) im allgemeinen nicht richtig, wie ein späteres Beispiel zeigen wird. Eine genaue Untersuchung hierüber im Falle $p = \infty$ wird im letzten Teil dieser Arbeit durchgeführt. Im Falle $\lambda < 0$ sind die Bedingungen (11) und (12) äquivalent:

SATZ 4. *Es sei* $1 \leqslant p \leqslant \infty$ *und* $f \in C[-1, 1]$ *bzw.* $f \in L_p(-1, 1)$. *Es seien* $\lambda$ *und* $\beta$ *reelle Zahlen,* $r$ *eine natürliche Zahl, mit* $\lambda < 0$ *und* $0 < \beta < r$. *Erfüllen die Polynome* $P_n$ *bester Approximation bezüglich* (10) *für dieses feste* $r$ *die Bedingung* (12), *dann genügt die beste Approximation* $E_n^{(\lambda)}(f)_p$ *der Gleichung* (11).

BEWEIS ZU SATZ 3. Sei $\lambda \in \boldsymbol{R}$. Nach Voraussetzung (12) gibt es eine Konstante $A_0$, so daß

$$\|\delta_n^{r-\lambda} P_{2n}^{(r)}\|_p \leqslant A_0 n^{r-\beta} \qquad \text{für alle } n \geqslant n_0.$$

Nach Lemma 2 existiert dann ein Polynom $Q \in \pi_n$ mit

$$\|\delta_n^{-\lambda}(P_{2n} - Q)\|_p \leqslant A n^{-\beta}$$

mit von $n$ unabhängiger Konstanten $A$. Daher ist

$$E_n(f) \leqslant \|\delta_n^{-\lambda}(f - Q)\|_p \leqslant A n^{-\beta} + \|\delta_n^{-\lambda}(f - P_{2n})\|_p. \tag{16}$$

(²) Unabhängig von mir bewiesen im Falle $\lambda = \beta, p = \infty$ K. Scherer und H. J. Wagner [4] das Ergebnis des Satzes 3.

Sei $\lambda \geqslant 0$. Wegen $\delta_{2n} \leqslant \delta_n$ folgt aus (16)

$$E_n(f) \leqslant An^{-\beta} + E_{2n}(f)$$

und hieraus rekursiv für jedes $k \in N$

(17) $$E_n(f) \leqslant An^{-\beta}\big(1 + 2^{-\beta} + \ldots + (2^{k-1})^{-\beta}\big) + E_{n2^k}(f).$$

Ist zusätzlich (15) erfüllt, so folgt aus (17) für $k \to \infty$ die Aussage (11). Satz 3 ist daher vollständig bewiesen.

Beweis zu Satz 4. Für $1 \leqslant p < \infty$ ist der Beweis elementar, aber lang. Aus diesem Grunde führen wir hier nur den Beweis für den wichtigen Fall $p = \infty$, $\lambda < 0$, $f \in C[-1, 1]$:

Nach P. L. Tschebyscheff (siehe z.B. bei H. Werner [7]) ist das Polynom $P_n$ bester Approximation bezüglich (10) für $p = \infty$ eindeutig charakterisiert durch die Existenz einer Alternante der Länge $n+2$,

(18) $$-1 \leqslant x_1 < x_2 < \ldots < x_{n+2} \leqslant 1, \qquad x_j = x_j^{(n)},$$

so daß

(19) $$f(x_j) - P_n(x_j) = \eta(-1)^j E_n(f)\,\{\delta_n(x_j)\}^\lambda, \qquad \eta := \pm 1,$$

für $j = 1, 2, \ldots, n+2$ erfüllt ist.

Annahme 1. *In* $[-1, -1+1/n^2] =: I_1$ *gebe es mindestens* $r+1$ *Alternantenpunkte* $x_j$, $j = 1, \ldots, r+1$.

Dann folgt aus (19) wegen $\delta_n(x_j) < \sqrt{2}/n$

(20) $$|P_n(x_{j+1}) - P_n(x_j)| \geqslant 2^{1+\lambda/2} n^{-\lambda} E_n(f) - \omega(f; 1/n^2)$$

für $j = 1, \ldots, r$, wobei $\omega(f;\ )$ der Stetigkeitsmodul von $f$ ist.

Ist

(21) $$\omega(f; 1/n^2) \leqslant 2^{\lambda/2} n^{-\lambda} E_n(f),$$

so folgt aus (19), daß

$$\operatorname{sgn}\big(P_n(x_{j+1}) - P_n(x_j)\big) = \eta(-1)^j, \qquad j = 1, \ldots, r.$$

Daher besitzt $P_n^{(\nu)}$, $\nu = 1, \ldots, r-1$, in $I_1$ mindestens $r-\nu$ Nullstellen. Dann ist für $j = 1, \ldots, r$ unter Verwendung von (12)

(22) $$|P_n(x_{j+1}) - P_n(x_j)| \leqslant (x_{r+1} - x_1)^r \max_{x \in I_1} |P_n^{(r)}(x)| \leqslant C_0 n^{-\lambda-\beta}.$$

Aus (20), (21) und (22) erhalten wir schließlich

$$E_n(f) \leqslant 2^{-\lambda/2} C_0 n^{-\beta}.$$

Berücksichtigen wir auch den Fall, daß (21) nicht erfüllt ist, so gilt stets unter der Annahme 1

$$E_n(f) \leqslant C_1 \max\{n^{-\beta}; n^{\lambda}\omega(f; 1/n^2)\}, \tag{23}$$

wobei die Konstante $C_1$ nicht von $n$ abhängt.

ANNAHME 2. *In* $[-1+1/n^2, 0] =: I_2$ *gebe es* $N(n)$ *Alternantenpunkte* $x_j$; *und es sei* $N(n) \geqslant n/4$.

Für jedes $x_j \in I_2$ ist

$$\begin{aligned} E_n(f) &= (1-x_j^2)^{-\lambda/2}|f(x_j)-P_n(x_j)| \\ &\leqslant E_{2n}(f)+(1-x_j^2)^{-\lambda/2}|P_{2n}(x_j)-P_n(x_j)|. \end{aligned} \tag{24}$$

Unter Verwendung von (19) erhalten wir

$$P_{2n}(x_j)-P_n(x_j) = (1-x_j^2)^{\lambda/2}\{\eta(-1)^j E_n(f)+\varepsilon_j E_{2n}(f)\} \tag{25}$$

mit $|\varepsilon_j| \leqslant 1$. Ist $E_{2n}(f) < E_n(f)$, so folgt aus (25)

$$\operatorname{sgn}\big(P_{2n}(x_j)-P_n(x_j)\big) = \eta(-1)^j$$

für alle $x_j \in I_2$. Dann ist aber

$$|P_{2n}(x_j)-P_n(x_j)| \leqslant (x_{j+r}-x_j)^r \max_{x\in[x_j, x_{j+r}]} |P_{2n}^{(r)}(x)-P_n^{(r)}(x)|$$

und bei Verwendung von (12)

$$|P_{2n}(x_j)-P_n(x_j)| \leqslant C_2 n^{r-\beta}(1-x_j^2)^{\lambda/2-r/2}(x_{j+r}-x_j)^r. \tag{26}$$

Da im Falle $\lambda < 0$ immer $E_{2n}(f) \leqslant E_n(f)$ ist, erhalten wir aus (24) und (26)

$$E_n(f) \leqslant E_{2n}(f)+C_2 n^{r-\beta}(1-x_j^2)^{-r/2}(x_{j+r}-x_j)^r. \tag{27}$$

Wenden wir nun den Hilfssatz 2 des Anhangs an, so existiert wegen $N(n) \geqslant n/4$ eine von $n$ unabhängige positive Zahl $K$ und eine natürliche Zahl $j$ mit $x_j, x_{j+r} \in I_2$ und

$$(1-x_j^2)^{-1/2}(x_{j+r}-x_j) \leqslant K/n.$$

Für dieses $x_j$ folgt aus (27)

$$E_n(f) \leqslant E_{2n}(f)+C_3 n^{-\beta}, \tag{28}$$

wobei auch $C_3$ unabhängig von $n$ ist.

Für die Intervalle $[0, 1-1/n^2]$ und $[1-1/n^2, 1]$ können wir natürlich ebenso verfahren wie für $I_2$ und $I_1$. Da es zu jedem $n \in N$ stets $n+2$ Alternantenpunkte gibt, ist Annahme 1 oder Annahme 2 für jedes genügend große $n$ zumindest für eines der vier Intervalle richtig. Daher erhalten wir aus (23) bzw. (28)

(29) $$E_n(f) \leqslant E_{2n}(f) + C \max\{n^{-\beta}, n^{\lambda}\omega(f; 1/n^2)\},$$

wobei die Zahl $C$ positiv und unabhängig von $n$ ist.

Wenden wir nun (29) rekursiv auf $E_{2n}, E_{4n}, \ldots$ an, so erhalten wir wegen

$$\lim_{m\to\infty} E_m(f) = 0$$

die Ungleichung

(30) $$E_n(f) \leqslant C^* \max\{n^{-}\ , n^{\lambda}\omega(f; 1/n^2)\}.$$

Im Falle $\lambda \leqslant -\beta$ ist aus (30) unmittelbar die Aussage des Satzes 4 für $p = \infty$ abzulesen. Im Falle $-\beta < \lambda < 0$ folgt aus (30)

(31) $$\lim_{m\to\infty} m^{-\lambda} E_m(f) = 0.$$

Da die Ungleichung (16) auch für $\lambda < 0$ gilt, folgt aus ihr wegen $\delta_n(x) \leqslant 2\delta_{2n}(x)$

(32) $$E_n(f) \leqslant An^{-\beta} + 2^{-\lambda} E_{2n}(f).$$

Wenden wir nun (32) rekursiv auf $E_{2n}, E_{4n}, E_{8n}, \ldots$ an, so erhalten wir unter Berücksichtigung der Eigenschaft (31) sofort die Aussage des Satzes 4 über das asymptotische Verhalten von $E_n(f)$. Damit ist Satz 4 für $p = \infty$ vollständig bewiesen.

Den zweiten Teil dieser Arbeit beginnen wir mit einem Beispiel:

Es sei $p = \infty$ und $0 < \lambda \leqslant 3$. Für $m = 2, 4, 8, \ldots, 2^k, \ldots$ definieren wir die Punkte

$$x_{jm} := -1 + 2/m^2 - j/(2m^3) \quad \text{für } j = 0, 1, 2, \ldots, 2m.$$

Dann ist für $0 \leqslant j \leqslant 2m-2$, $m = 2^k$,

$$\ldots > x_{jm} > x_{j+1,m} > x_{2m,m} = -1 + 1/m^2 > x_{0,2m} > \ldots$$

und für $0 \leqslant j \leqslant 2m-1$

$$x_{jm} - x_{j+1,m} = 1/(2m^3), \qquad x_{2m,m} - x_{0,2m} = 1/(2m^2).$$

Die Funktion $f \in C[-1, 1]$ sei dann der lineare Streckenzug, der die Punkte $(x_{jm}, f(x_{jm}))$ linear verbindet, wobei

$$f(x_{jm}) := (-1)^j (1 - x_{jm}^2)^{\lambda/2} \quad \text{für } j = 0, 1, \ldots, 2m,\ m = 0, 2, 4, 8, \ldots;$$
$$f(-1) := 0; \quad x_{00} := 0; \quad f(x) := 0 \text{ für } 0 \leqslant x \leqslant 1.$$

Es ist $m^{-\lambda} < |f(x_{jm})| < 2^{\lambda} m^{-\lambda}$ und daher

$$|f(x_{j+1,m}) - f(x_{jm})| < 2^{1+\lambda} m^{-\lambda} = 2^{1+4\lambda/3} |x_{j+1,m} - x_{jm}|^{\lambda/3}.$$

Es läßt sich hieraus sofort folgern, daß die Funktion $f$ in $[-1, 1]$ der Bedingung $f \in \text{Lip}\, \lambda/3$ genügt.

Sei nun $n$ eine natürliche Zahl; sei $m$ bestimmt durch $m = 2^k \leqslant n < 2m$. Dann gilt für alle $x \in [-1, 1]$

$$\{\delta_n(x)\}^{-\lambda} |f(x)| \leqslant 1,$$

aber

$$\{\delta_n(x_{jm})\}^{-\lambda} f(x_{jm}) = (-1)^j \quad \text{für } j = 0, 1, \ldots, 2m.$$

Dann ist aber $P_n \equiv 0$ das Polynom bester Approximation von $f$ bezüglich (10), denn es existiert eine Tschebyscheff-Alternante $x_{jm}$, $j = 0, 1, \ldots, n+1$, der Länge $n+2$ (vergleiche (19)); daher ist für alle $n \geqslant 2$

$$E_n^{(\lambda)}(f)_\infty = 1.$$

Ergebnis. Obwohl die Funktion $f \in \text{Lip}\, \lambda/3$ und obwohl die Bedingung (12) für das Polynom $P_n \equiv 0$ bester Approximation von $f$ bezüglich (10) für alle $n \geqslant 2$ und alle $r$ erfüllt ist, ist (15) und natürlich auch (11) falsch. Für die Funktion $f$ unseres Beispiels ist außerdem $\text{Var}\, f = \infty$ im Falle $0 < \lambda \leqslant 1$ und $\text{Var}\, f < \infty$ im Falle $\lambda > 1$.

In dem nun folgenden Satz wird gezeigt, daß die Aussagen (11) und (12) äquivalent sind, falls die Bedingung $f \in \text{Lip}\, \lambda/3$ unseres Beispiels etwas verschärft wird oder falls im Falle $0 < \lambda < 1$, $\text{Var}\, f < \infty$, erfüllt ist:

Satz 5. *Sei $p = \infty$ und $f \in C[-1, 1]$. Es seien $\lambda$ und $\beta$ reelle Zahlen, $r$ eine natürliche Zahl, mit $0 < \beta < r$. Die Polynome $P_n$ bester Approximation von $f$ bezüglich* (10) *mögen für dieses feste $r$ der Bedingung* (12) *genügen. Dann gilt für die beste Approximation $E_n(f)$ die Aussage* (11), *wenn nur eine der beiden Zusatzbedingungen erfüllt ist:*

(a) *Sei $0 < \lambda < r$. Die nichtnegative ganze Zahl $k$ sei definiert durch $k \leqslant \lambda/3 < k+1$; außerdem sei*

$$f \in C^k[-1, 1], \quad \lim_{h \to 0} \omega(f^{(k)}; h) h^{k-\lambda/3} = 0.$$

(b) *Sei $0 < \lambda < 1$ und $\text{Var}\, f < \infty$ in $[-1, 1]$.*

Beweis. Es sei $n$, $n \geqslant 8r$ fest gewählt. Sei weiter $x_j$, $j = 1, \ldots, n+2$, die Alternante (18) von $P_n$ mit der Eigenschaft (19). Wir bilden die

Intervalle $I_\nu$ durch

$$I_\nu := [x_{(\nu-1)r+1}, x_{\nu r+1}], \qquad \nu = 1, 2, \ldots, A(n),$$

$A(n) := [(n+1)/r]$, die jeweils $r+1$ Alternantenpunkte $x_j$ enthalten. Für die Anzahl $A(n)$ dieser Intervalle $I_\nu$ gilt

$$A(n) > 7n/(8r). \tag{33}$$

Wir wählen nun ein festes, aber beliebiges Intervall $I_\nu$ und setzen

$$y_s := x_{(\nu-1)r+s}, \qquad s = 1, \ldots, r+1. \tag{34}$$

Die Funktion $f$ erfülle nun die Voraussetzung (a) des Satzes. Wir definieren das Polynom $p$ $k$-ten Grades

$$p(x) := \sum_{\nu=0}^{k} f^{(\nu)}(y_1)\,(x-y_1)^\nu/\nu!.$$

Für die Funktion $f^* := f-p$ ist dann

$$|f^*(x)| \leqslant (y_{r+1}-y_1)^k \omega(f^{(k)}; y_{r+1}-y_1), \qquad x \in I_\nu. \tag{35}$$

Setzen wir $P^* := P_n - p$, so genügt $P^*$ wegen $k < r$ der Bedingung (12); außerdem erfüllt $f^* - P^* = f - P_n$ die Alternantenbedingung (19).

Erfüllt die Funktion $f$ die Voraussetzung (b) des Satzes, so setzen wir $f^* := f - f(y_1)$, $P^* := P_n - f(y_1)$. Dann genügen $P^*$ und $f^* - P^*$ ebenfalls den Bedingungen (12) und (19).

Für ein Intervall $I_\nu \subset [-1, 0]$ substituieren wir gemäß (34). Wir müssen dann die beiden folgenden Möglichkeiten unterscheiden:

*Fall* 1. Es sei

$$|f^*(y_s)| \leqslant \tfrac{1}{2} E_n(f)\,\{\delta_n(y_s)\}^\lambda \quad \text{für alle } s = 1, \ldots, r+1. \tag{36}$$

Dann folgt aus (19)

$$\operatorname{sgn} P^*(y_s) = -\operatorname{sgn} P^*(y_{s+1}).$$

Daher besitzt $P^*$ in $I_\nu$ $r$ Nullstellen; und es ist

$$|P^*(y_s)| \leqslant \max_{x\in I_\nu} |P^{*(r)}(x)|\,(y_{r+1}-y_1)^r.$$

Bei Verwendung von (12) folgt hieraus

$$|P^*(y_s)| \leqslant C n^{r-\beta} \{\delta_n(y_1)\}^{\lambda-r} (y_{r+1}-y_1)^r. \tag{37}$$

Aus (19), (36) und (37) folgt schließlich

(38) $$E_n(f) \leqslant 2Cn^{r-\beta}\{\delta_n(y_1)\}^{-r}(y_{r+1}-y_1)^r.$$

*Fall* 2. Es gebe ein $s \in \{1, 2, \ldots, r+1\}$ mit

(39) $$|f^*(y_s)| > \tfrac{1}{2}E_n(f)\ \{\delta_n(y_s)\}^\lambda.$$

Wir diskutieren nun alle möglichen Verteilungen der Alternantenpunkte $x_j$ des Polynoms bester Approximation $P_n$. Zur Abkürzung führen wir noch die Bezeichnungen

$$l_\nu := (y_{r+1}-y_1), \qquad e_\nu := (y_{r+1}-y_1)\,(1-y_1^2)^{-1/2}$$

für jedes Intervall $I_\nu$ ein.

ANNAHME 1. *Fall* 1 *trete für mindestens* $n/(16r)$ *Intervalle* $I_\mu \subset [-1, -1+1/n^2]$ *auf.*

Für jedes dieser Intervalle folgt aus (38)

(40) $$E_n(f) \leqslant 2Cn^{-\beta}.$$

ANNAHME 2. *Fall* 1 *trete für mindestens* $n/(16r)$ *Intervalle* $I_\mu \subset [-1+1/n^2, 0]$ *auf.*

Nach Hilfssatz 2 des Anhangs ist dann für mindestens eines dieser Intervalle, $I_\nu$, $e_\nu \leqslant 16r\,K/n$ und daher wegen (38)

(41) $$E_n(f) \leqslant C_1 n^{-\beta}, \qquad C_1 := 2C(16rK)^r.$$

ANNAHME 3. *Es gelte die Voraussetzung* (b) *des Satzes. Fall* 2 *trete für mindestens* $n/(8r)$ *Intervalle* $I_\mu \subset [-1, 0]$ *auf.*

Bei Verwendung von (39) ist dann

$$\operatorname{Var} f \geqslant \sum_\mu |f^*(y_{s_\mu})| \geqslant n^{1-\lambda}E_n(f)/(16r)$$

und daher

(42) $$E_n(f) \leqslant 16rn^{\lambda-1}\operatorname{Var} f.$$

ANNAHME 4. *Es gelte die Voraussetzung* (a). *Fall* 2 *trete für mindestens* $n/(16r)$ *Intervalle* $I_\mu \subset [-1, -1+1/n^2]$ *auf.*

Dann gibt es unter diesen sicher ein Intervall, $I_\nu$, der Länge $l_\nu \leqslant 16r/n^3$. Für dieses Intervall folgt aus (39) und (35)

(43) $$E_n(f) \leqslant C_2\,\omega(f^{(k)}; 1/n^3)n^{\lambda-3k}; \qquad C_2 := 2(16r)^{k+1}.$$

ANNAHME 5. *Es gelte die Voraussetzung* (a). *Fall 2 trete für mindestens* $n/(16r)$ *Intervalle* $I_\mu \subset [-1+1/n^2, 0]$ *auf.*

Unter diesen gibt es mindestens $n/(32r)$ Intervalle der Intervallänge $l_\mu \leqslant 32r/n$ und nach Hilfssatz 2 darunter sicher eines, $I_\nu$, mit

(44) $$g_\nu := (y_{r+1}-y_1)/(1-y_1^2)^{3/2} \leqslant m(3/2; n, n/(32r)) \leqslant \bar{K},$$

wobei die Konstante $\bar{K}$ nicht von $n$ abhängt. Für dieses Intervall $I_\nu$ folgt aus (39) und (35)

(45) $$E_n(f) < 2(g_\nu)^{\lambda/3}\, \omega(f^{(k)}; l_\nu)\, (l_\nu)^{s-\lambda/3}$$

und wegen (44)

(46) $$E_n(f) \leqslant C_3 \omega(f^{(k)}; l_\nu)\, (l_\nu)^{k-\lambda/3},$$

wobei $C_3$ nicht von $n$ abhängt und wobei $l_\nu \leqslant 32r/n$ ist.

Die eben durchgeführte Diskussion können wir natürlich in gleicher Weise für die Intervalle $I_\mu \subset [0, 1]$ durchführen. Da es aber nach (33) mehr als $7n/(8r)$ Intervalle $I_\mu$ gibt, ist für jedes genügend große $n$ zumindest eine der Annahmen richtig. Aus (40)–(43) und (46) folgt dann unter der Voraussetzung (a) oder (b), daß die beste Approximation $E_n(f)$ der Bedingung (15) genügt. Nach Satz 3 ist dann aber sogar (11) richtig. Somit ist Satz 5 vollständig bewiesen.

Die Bedingungen (a) und (b) des Satzes 5 sind hinreichend für die Äquivalenz der Aussagen (11) und (12), aber nicht notwendig. Im wesentlichen entscheidet das lokale Verhalten der Funktion $f \in C[-1, 1]$ an den Stellen $x = -1$ und $x = +1$, ob (11) und (12) äquivalent sind. Aus dem Beweis zu Satz 5 ist in einfacher Weise zu ersehen, daß Satz 5 auch richtig ist, wenn die Voraussetzungen (a) oder (b) ersetzt werden durch die folgenden:

(a)* *Sei* $0 < \lambda < r$. *Sei* $\varepsilon > 0$ *eine feste reelle Zahl und* $D := [-1, -1+\varepsilon] \cup [1-\varepsilon, 1]$. *Sei wieder* $k$ *die ganze Zahl mit* $k \leqslant \lambda/3 < k+1$; $f$ *sei in* $D$ *k-mal stetig differenzierbar, und für den Stetigkeitsmodul bezüglich* $D$ *gelte*

$$\lim_{h\to 0} \omega(f^{(k)}; h)_D h^{k-\lambda/3} = 0.$$

(b)* *Sei* $0 < \lambda < 1$ *und* $\operatorname*{Var}_D f < \infty$ *bezüglich* $D$.

Eine genauere Untersuchung über weitere hinreichende oder gar notwendige Bedingungen für die Äquivalenz von (11) und (12) soll hier nicht durchgeführt werden, obwohl der Beweis zu Satz 5 hierzu zahlreiche Ansatzmöglichkeiten zur Verschärfung enthält.

## ANHANG

Seien $n$ und $M$ positive ganze Zahlen. Wir betrachten die Menge aller $M$-Tupel $Z = \{z_j\}, j = 1, \ldots, M$, mit der Eigenschaft

$$1/n^2 \leqslant z_1 < z_2 < \ldots < z_M \leqslant 1. \tag{1}$$

Für $\gamma = 1/2, 1, 3/2$ untersuchen wir die Größen

$$m(\gamma; n, M) := \sup_{Z} \min_{1 \leqslant k \leqslant M-1} (z_{k+1} - z_k)(z_k)^{-\gamma}. \tag{2}$$

Hilfssatz 1. *Für $M \geqslant 1 + 2\log n$ ist*

$$m(1; n, M) < 8M^{-1}\log n. \tag{3}$$

Beweis. Es ist sofort einzusehen, daß für jedes $\gamma$ das folgende $M$-Tupel $Z$ den maximalen Wert in (2) liefert:

$$z_1 := 1/n^2, \quad z_M := 1, \quad z_{k+1} := z_k + m(z_k)^\gamma \tag{4}$$

für $k = 1, 2, \ldots, M-1$, wobei $m := m(\gamma; n, M)$ die gesuchte Zahl ist. Für $\gamma = 1$ ist dann nach (4)

$$1 = z_M = (1+m)z_{M-1} = (1+m)^{M-1}z_1 = (1+m)^{M-1}n^{-2};$$

und wegen $M - 1 \geqslant 2\log n$ folgt dann

$$m = n^{2/(M-1)} - 1 \leqslant 2(e-1)(M-1)^{-1}\log n,$$

womit (3) bewiesen ist.

Hilfssatz 2. *Für $M \geqslant 2 + 2\log n/\log 2$ ist*

$$m(1/2; n, M) < 12/M, \tag{5}$$

$$m(3/2; n, M) < 48n/M. \tag{6}$$

Beweis. Sei die positive ganze Zahl $q$ bestimmt durch $2^{q-1} < n \leqslant 2^q$. Dann ist $q < M/2$. Für $i = 1, \ldots, q$ sei $M_i$ die Anzahl der $z_k$ des $M$-Tupels (4) mit $z_k \in J_i := [4^{-i}, 4^{-i+1}]$, wobei dieses maximale $M$-Tupel $Z$ in (4) natürlich von $\gamma$ abhängt. Es gibt Indizes $\mu$ und $\nu \in \{1, \ldots, q\}$ mit

$$M_\mu \geqslant 1 + M2^{-\mu-1}, \tag{7}$$

$$M_\nu \geqslant 1 + M2^{\nu-3}/n. \tag{8}$$

Denn wäre (7) falsch für alle $i = 1, \ldots, q$, so wäre

$$M \leqslant \sum_{i=1}^{q} M_i < q + M \sum_{i=1}^{q} 2^{-i-1} < M;$$

und wäre (8) falsch für alle $i = 1, \ldots, q$, so wäre wegen $n > 2^{q-1}$

$$M \leqslant \sum_{i=1}^{q} M_i < q + M \sum_{i=1}^{q} 2^{i-q-2} < M.$$

Da $M_\mu$ und $M_\nu$ ganzzahlig sind, folgt aus (7) und (8), daß $M_\mu \geqslant 2$ und $M_\nu \geqslant 2$. Für $j := \mu$ und $j := \nu$ gibt es dann zwei Elemente des entsprechenden $M$-Tupels $Z$ in (4), $z_k$ und $z_{k+1} \in J_j$, mit

$$z_{k+1} - z_k \leqslant |J_j|/(M_j - 1) \leqslant 6 \cdot 4^{-j}/M_j, \tag{9}$$

wobei $|J_j| = 3 \cdot 4^{-j}$ die Länge des Intervalls $J_j$ ist. Dann folgt aus (4), (7) und (9)

$$m(1/2; n, M) = (z_{k+1} - z_k)\,(z_k)^{-1/2} \leqslant 6 \cdot 4^{-\mu} 2^{\mu}/M_\mu < 12/M;$$

und aus (4), (8) und (9)

$$m(3/2; n, M) = (z_{k+1} - z_k)\,(z_k)^{-3/2} \leqslant 6 \cdot 4^{-\nu} 4^{3\nu/2}/M_\nu < 48n/M.$$

Damit ist der Hilfssatz 2 bewiesen.

## LITERATUR

[1] P. L. Butzer und S. Pawelke, 'Ableitungen von trigonometrischen Approximationsprozessen', *Acta Sci. Math. (Szeged)* **28** (1967) 173–183.

[2] G. K. Lebed', 'Ungleichungen für Polynome und ihre Ableitungen', *Dokl. Akad. Nauk* **117** (1957) 570–572.

[3] M. K. Potapov, 'Über die Sätze vom Jackson-Typ in der Metrik $L_p$', *Dokl. Akad. Nauk* **111** (1956) 1185–1188.

[4] K. Scherer und H.-J. Wagner, 'An Equivalence Theorem on Best Approximation of Continuous Functions by Algebraic Polynomials', *Applicable Analysis* **1** (1972).

[5] S. B. Stečkin, 'Über die beste Approximation stetiger Funktionen', *Izv. Akad. Nauk* **15** (1951) 219–242.

[6] G. Sunouchi, 'Derivatives of a Polynomial of Best Approximation', *Jber. Deutsch. Math. Verein.* **70** (1968) 165–166.

[7] H. Werner, *Vorlesung über Approximationstheorie*, Springer-Verlag 1966.

[8] M. Zamansky, 'Classes de saturation de certains procédés d'approximation des séries de Fourier des fonctions continues et applications a quelques problèmes d'appıoximation', *Ann. Sci. Ecole Norm. Sup.* **66** (1949) 19–93.

# ZUR APPROXIMATIONSTHEORIE FÜR SUMMATIONSPROZESSE VON FOURIER-ENTWICKLUNGEN IN BANACH-RÄUMEN: VERGLEICHSSÄTZE UND UNGLEICHUNGEN VOM BERNSTEIN-TYP*

E. GÖRLICH, R. J. NESSEL, W. TREBELS

*Aachen*

*Abstract.* In this paper certain aspects of an approximation theory for summation processes of Fourier expansions in Banach spaces $X$ are discussed. Let $\{P_k\}$ be a total sequence of mutually orthogonal projections and consider summation processes $\sum \tau_\varrho(k) P_k f$, $\varrho \to \infty$, of the expansion $f \sim \sum P_k f$. As an immediate consequence of the underlying multiplier structure, e.g. the proof of a comparison theorem for two different processes or of a Bernstein-type inequality may be translated into the verification of uniform multiplier conditions. In case $X$ admits a Cesàro decomposition, i.e. the $(C, j)$-means of the expansion $\sum P_k f$ are bounded for some $j$, a multiplier criterion is given in terms of $\int_0^\infty t^j |d\tau_\varrho^{(j)}(t)| < \infty$. The results are applied, e.g. to the typical and Abel–Cartwright means in the particular instance of Hermite series.

In dieser Arbeit sollen exemplarisch einige Aspekte einer Approximationstheorie für Summationsprozesse von Fourier-Entwicklungen in Banach-Räumen diskutiert werden. Zu diesem Zweck sei zunächst der allgemeine Rahmen terminologisch abgesteckt.

Sei $X$ ein beliebiger Banach-Raum und $[X]$ die Banach-Algebra aller beschränkten, linearen Operatoren von $X$ in sich. Mit $\boldsymbol{P}$, der Menge aller nichtnegativen, ganzen Zahlen, sei $\{P_k\}_{k \in \boldsymbol{P}} \subset [X]$ ein vorgegebenes System von stetigen, linearen Projektoren, die paarweise orthogonal und total sind, d.h.: (i) $P_j P_k = \delta_{jk} P_k$, wobei $\delta_{jk}$ das Kronecker-Symbol ist, (ii) aus $P_k f = 0$ für alle $k \in \boldsymbol{P}$ folgt $f = 0$. Dann kann jedem Element

* Dies ist eine Zusammenfassung der beiden Vorträge, die von den beiden erstgenannten Autoren gehalten wurden.

$f \in X$ seine Fourier-Entwicklung

$$f \sim \sum_{k=0}^{\infty} P_k f \qquad (f \in X) \tag{1}$$

in ein-eindeutiger Weise zugeordnet werden.

Sei $s$ die Menge aller Folgen von Skalaren. Dann heißt $\tau \in s$ ein Multiplikator (auf $X$ bezüglich des Systems $\{P_k\}$), falls zu jedem $f \in X$ ein Element $f^\tau \in X$ existiert, so daß $\tau_k P_k f = P_k f^\tau$ für alle $k \in \boldsymbol{P}$ gilt. Offenbar wird dann durch $Tf = f^\tau$ ein beschränkter, linearer Operator $T \in [X]$ erzeugt, dem die Entwicklung

$$Tf \sim \sum_{k=0}^{\infty} \tau_k P_k f \qquad (f \in X) \tag{2}$$

mit (Entwicklungs-) Koeffizienten $\{\tau_k\}$ entspricht. Sei $M = M(X; \{P_k\})$ die Menge aller Multiplikatoren $\tau$ und $[X]_M$ die entsprechende Menge der Multiplikatorenoperatoren $T$. Mit den üblichen Vektoroperationen, koordinatenweiser Multiplikation und der Norm

$$\|\tau\|_M = \sup_{\|f\| \leqslant 1} \|f^\tau\| = \|T\|_{[X]} \tag{3}$$

ist $M$ eine kommutative Banach-Algebra mit Einselement, die isometrisch isomorph zu $[X]_M$ ist: $M \cong [X]_M \subset [X]$.

Für eine beliebige Folge $\alpha \in s$ sei $X^\alpha$ die Menge aller $f \in X$, für die ein Element $f^\alpha \in X$ existiert, so daß $\alpha_k P_k f = P_k f^\alpha$ für alle $k \in \boldsymbol{P}$ gilt. Ist dann $B^\alpha$ der Operator mit Definitionsbereich $X^\alpha_{\mathfrak{l}} \subset X$ und Wertebereich in $X$, der durch $B^\alpha f = f^\alpha$ definiert ist, so ist $B^\alpha$ ein abgeschlossener, linearer Operator für jedes $\alpha \in s$.

Im folgenden wird nun grundsätzlich davon ausgegangen, daß die betrachteten Operatoren vom Multiplikatoren-Typ (im obigen Sinne) sind, mit anderen Worten, eine gewisse Faltungsstruktur besitzen. In diesem Rahmen werden (starke) Approximationsprozesse betrachtet, also Familien $\{T(\varrho)\}_{\varrho > 0} \subset [X]_M$ von Operatoren, für die (mit Konstante $D > 0$)

$$\|T(\varrho)f\| \leqslant D\|f\|, \qquad \lim_{\varrho \to \infty} \|T(\varrho)f - f\| = 0 \qquad (f \in X) \tag{4}$$

gilt. Im Falle, daß das System $\{P_k\}$ fundamental ist, d.h., die lineare Hülle von $\bigcup_{k=0}^{\infty} P_k(X)$ ist dicht in $X$, kann Bedingung (4) nach dem

Satz von Banach–Steinhaus äquivalent über die entsprechende Familie $\{\tau(\varrho)\} \subset M$ von Multiplikatoren ausgedrückt werden durch

(4*) $$\|\tau(\varrho)\|_M \leqslant D, \quad \lim_{\varrho\to\infty} \tau_k(\varrho) = 1 \qquad (k \in \boldsymbol{P}).$$

Diese Möglichkeit des Abwälzens auf die zugehörigen Koeffizientenfolgen $\{\tau(\varrho)\}$ soll nun an zwei weiterführenden approximationstheoretischen Fragestellungen verfolgt werden.

Ein Aspekt einer genaueren Diskussion des Approximationsverhaltens in (4) ist nach Favard [2] (für bibliographische Details vgl. [1]) das Vergleichsproblem: Ist $\{S(\varrho)\} \subset [X]_M$ ein weiterer Approximationsprozess, so stellt sich die Frage nach direkten Abschätzungen zwischen den Größen $\|T(\varrho)f-f\|$ und $\|S(\varrho)f-f\|$, also etwa die Frage der Existenz einer Konstanten $A > 0$, so daß

(5) $$\|T(\varrho)f-f\| \leqslant A\|S(\varrho)f-f\| \qquad (f \in X, \varrho > 0)$$

gilt. In diesem Fall sagt man, daß das Verfahren $\{T(\varrho)\}$ besser als $\{S(\varrho)\}$ ist. Gilt auch die Umkehrung, so nennt man die Verfahren äquivalent und kennzeichnet dies durch $T(\varrho)-I \approx S(\varrho)-I$ (mit $I$, dem Identitätsoperator).

Wegen der Multiplikatorenstruktur läßt sich auch hier das Problem auf die entsprechenden Koeffizientenfolgen abwälzen.

Satz 1. *Zu zwei Familien* $\{T(\varrho)\}, \{S(\varrho)\} \subset [X]_M$ *von Operatoren, die entsprechend durch die Multiplikatoren* $\{\tau(\varrho)\}, \{\sigma(\varrho)\} \subset M$ *erzeugt werden, existiere eine Familie* $\{\eta(\varrho)\} \subset M$ *von gleichmäßig beschränkten Multiplikatoren*: $\|\eta(\varrho)\|_M \leqslant A$, *so daß*

(5*) $$\tau_k(\varrho)-1 = \eta_k(\varrho)\,[\sigma_k(\varrho)-1] \qquad (k \in \boldsymbol{P}, \varrho > 0)$$

*gilt. Dann ist* $\{T(\varrho)\}$ *besser als* $\{S(\varrho)\}$.

Zum Beweis ist nur zu bemerken, daß für jedes $f \in X, k \in \boldsymbol{P}, \varrho > 0$

$$P_k\big(T(\varrho)f-f\big) = \eta_k(\varrho)\,[\sigma_k(\varrho)-1]P_kf = P_k\big(E(\varrho)\,[S(\varrho)f-f]\big)$$

ist, wobei der von $\eta(\varrho) \in M$ erzeugte Operator mit $E(\varrho) \in [X]_M$ bezeichnet wurde. Da das System $\{P_k\}$ total ist, folgt

$$T(\varrho)f-f = E(\varrho)\,[S(\varrho)f-f]$$

und hieraus (vgl. (3)) die Behauptung.

Neben dem Vergleichsproblem ist vor allem auch das Aufstellen von Ungleichungen vom Bernstein-Typ für einen vorgegebenen Approximationsprozess $\{T(\varrho)\}$ außerordentlich wichtig. Wegen der vorausgesetzten Multiplikatorenstruktur läßt sich auch dieses Problem unmittelbar auf die Koeffizienten übertragen.

SATZ 2. *Sei* $\alpha \in s$ *beliebig und* $\{T(\varrho)\} \subset [X]_M$ *erzeugt durch* $\{\tau(\varrho)\} \subset M$. *Existiert eine auf* $(0, \infty)$ *nichtnegative, monoton wachsende Funktion* $\Omega(\varrho)$ *mit* $\lim_{\varrho\to\infty} \Omega(\varrho) = \infty$ *und eine Familie* $\{\eta(\varrho)\} \subset M$ *von gleichmäßig beschränkten Multiplikatoren*: $\|\eta(\varrho)\|_M \leqslant A$ *mit*

$$\alpha_k \tau_k(\varrho) = \Omega(\varrho)\, \eta_k(\varrho) \qquad (k \in \boldsymbol{P}, \varrho > 0), \tag{6*}$$

*so gilt die Ungleichung vom Bernstein-Typ*

$$\|B^\alpha T(\varrho) f\| \leqslant A\Omega(\varrho) \|f\| \qquad (f \in X, \varrho > 0). \tag{6}$$

Der Beweis verläuft analog zu dem von Satz 1, wobei zu beachten ist, daß Bedingung (6*) insbesondere $T(\varrho)(X) \subset X^\alpha$ impliziert.

Im vorliegenden Rahmen, d.h., falls die betrachteten Operatoren obige Multiplikatorenstruktur haben, lassen sich also die beiden Probleme (5), (6) wie auch weitere, klassische Approximationsprobleme (z.B. das Saturationsproblem, siehe [1]) über die zugehörigen Koeffizienten in Form der (gleichmäßigen) Multiplikatorenbedingungen (5*), (6*) ausdrücken. Danach bestehen die Probleme jetzt darin, Bedingungen (5*), (6*) zu diskutieren, insbesondere möglichst handliche, hinreichende Kriterien für gleichmäßig beschränkte Multiplikatoren herzuleiten. Diese Kriterien sollten auf strukturellen Eigenschaften der Folgen beruhen, etwa darauf, daß eine Folge $\{\eta_k\}_{k \in \boldsymbol{P}}$ zu einer Funktion $e(x)$ auf $(0, \infty)$ erweitert werden kann, die gewisse Differenzierbarkeitseigenschaften hat.

Ohne weitere Information über das System $\{P_k\}$ bzw. den Raum $X$ scheint dieses Programm schwerlich durchführbar zu sein. Deshalb wird hier die zusätzliche Forderung gestellt, daß das System $\{P_k\}$ für den Raum $X$ im wesentlichen eine Cesàro-Zerlegung definiert.

Dazu seien für $j \in \boldsymbol{P}$ die $(C, j)$ Mittel der Fourier-Reihe (1) von $f$ definiert durch

$$(C, j)_n f = (A_n^j)^{-1} \sum_{k=0}^{n} A_{n-k}^j P_k f, \qquad A_n^j = \binom{n+j}{n}.$$

Die Annahme ist nun, daß für ein gewisses, festes $j \in \boldsymbol{P}$ die Operatoren $(C,j)_n \subset [X]$ gleichmäßig in $n$ beschränkt sind, daß also eine Konstante $C_j$ existiert, so daß

$$\|(C,j)_n f\| \leqslant C_j \|f\| \qquad (f \in X, n \in \boldsymbol{P}). \tag{7}$$

Um die Bedingung etwas zu erläutern und um einen Spezialfall in (1) sofort mitzuerledigen, sei vermerkt, daß man es in den Anwendungen häufig mit totalen Biorthogonalsystemen $\{f_k; f_k^*\}$ zu tun hat mit $\{f_k\} \subset X$, $\{f_k^*\} \subset X^*$, dem dualen Raum von $X$. Die (1), (2) entsprechenden Fourier-Entwicklungen schreiben sich dann

$$f \sim \sum_{k=0}^{\infty} f_k^*(f) f_k, \qquad Tf \sim \sum_{k=0}^{\infty} \tau_k f_k^*(f) f_k,$$

und $P_k(X)$ ist gerade der eindimensionale, durch $f_k$ aufgespannte, lineare Raum. Ist weiterhin die Folge $\{f_k\}$ fundamental in $X$, so ist es auf Grund des Satzes von Banach–Steinhaus klar, daß Bedingung (7) für $j = 0$ äquivalent zu der Annahme ist, daß die Folge $\{f_k\}$ eine Schauder-Basis ist, d.h.:

$$\lim_{n \to \infty} \left\| f - \sum_{k=0}^{n} f_k^*(f) f_k \right\| = 0 \qquad (f \in X),$$

während für $j = 1$ die Bedingung (7) dann äquivalent zu der Aussage ist, daß die Folge $\{f_k\}$ eine Cesàro-Basis bildet, d.h.:

$$\lim_{\to \infty} \left\| f - \sum_{k=0}^{n} \left(1 - \frac{k}{n+1}\right) f_k^*(f) f_k \right\| = 0 \qquad (f \in X).$$

Hier soll aber weiter mit einem System $\{P_k\}$ von Projektoren, d.h. statt mit Basen mit Zerlegungen, gearbeitet werden, da es insbesondere von den Anwendungen her (z.B. Entwicklungen nach Kugelfunktionen) wünschenswert ist, den Bildraum $P_k(X)$ mehrdimensional zuzulassen. Für Grundlagen aus der Theorie der Basen, bzw. Zerlegungen von Banach-Räumen sei auf die Bücher von J. T. Marti bzw. I. Singer hingewiesen.

Um nun Multiplikatoren in Verbindung mit Projektoren $P_k$, die der Bedingung (7) für ein $j \in \boldsymbol{P}$ genügen, zu studieren, führt man die folgenden Folgenräume ein:

$$bv_{j+1} = \left\{ \tau \in l^\infty \,\middle|\, \|\tau\|_{bv_{j+1}} = \sum_{k=0}^{\infty} \binom{k+j}{j} |\Delta^{j+1} \tau_k| + \lim_{m \to \infty} |\tau_m| < \infty \right\}. \tag{8}$$

Hierbei ist $\Delta\tau_k = \tau_k - \tau_{k+1}$, $\Delta^{j+1} = \Delta(\Delta^j)$, und $l^\infty$ die Menge aller beschränkten Zahlenfolgen. Es sei vermerkt, daß aus $\tau \in l^\infty$ und der Konvergenz der Reihe in (8) die Existenz des Grenzwertes $\lim \tau_m = \tau_\infty$ folgt. Weiterhin ist $bv_{j+1} \subset bv_j$ im Sinne stetiger Einbettung, und es ist klar, daß $bv_1$ nichts anderes als den Raum der Folgen von beschränkter Variation und $bv_2$ den Raum der quasi-konvexen Folgen bedeutet. Dann gilt (vgl. Orlicz [4a])

SATZ 3. *Erfüllt das System* $\{P_k\}$ *Bedingung* (7) *für ein* $j \in \boldsymbol{P}$, *so ist* $bv_{j+1} \subset M$, *und für jedes* $\tau \in bv_{j+1}$ *gilt* $\|\tau\|_M \leqslant C_j \|\tau\|_{bv_{j+1}}$.

Der Beweis folgt unmittelbar aus dem Ansatz

$$f^\tau = \sum_{k=0}^{\infty} \binom{k+j}{j} \Delta^{j+1}\tau_k \cdot (C, j)_k f + \tau_\infty f.$$

Um also die entscheidenden Bedingungen (5*), (6*) aus Satz 1, 2 zu verifizieren, muß man unter den gegenwärtigen Bedingungen gleichmäßige Schranken für die $bv_{j+1}$-Normen der entsprechenden Folgen $\eta(\varrho)$ bestimmen. Ein direktes Vorgehen ist dabei wegen der komplizierten Struktur der Reihe in (8) zumeist noch recht unhandlich. Anders wird die Situation dagegen, wenn die Folgen zu Funktionen auf $(0, \infty)$ mit entsprechenden Differenzierbarkeitseigenschaften erweitert werden können; denn dann gilt mit $BV_{j+1}$, der Menge aller auf $[0, \infty]$ stetigen Funktionen $t(x)$, für die $t^{(j-1)}$ lokal auf $(0, \infty)$ absolut stetig(1) und für die $t^{(j)}$ lokal auf $(0, \infty)$ von beschränkter Variation mit $\int_0^\infty x^j |dt^{(j)}(x)| < \infty$ ist:

SATZ 4. *Für* $\tau \in s$ *existiere eine Funktion* $t(x) \in BV_{j+1}$, *so daß* $\tau_k = t(k)$ *für alle* $k \in \boldsymbol{P}$ *gilt. Dann ist* $\tau \in bv_{j+1}$ *und*

$$\|\tau\|_{bv_{j+1}} \leqslant \frac{1}{j!} \int_0^\infty x^j |dt^{(j)}(x)| + \lim_{x\to\infty} |t(x)|.$$

Der Beweis folgt durch einige direkte, elementare Abschätzungen; für die (bestmögliche) Konstante $(1/j!)$ vgl. [6].

Hieraus folgt unmittelbar ein handliches Kriterium bezüglich gleichmäßig beschränkter Multiplikatoren:

(1) d.h., auf jedem kompakten Teilintervall von $(0, \infty)$ absolut stetig.

SATZ 5. *Das System* $\{P_k\}$ *erfülle Bedingung* (7) *für ein* $j \in \mathbf{P}$. *Zu der Familie von Folgen* $\{\tau(\varrho)\} \subset s$ *existiere eine Funktion* $t(x) \in BV_{j+1}$, *so daß* $\tau_k(\varrho) = t(k/\varrho)$ *für alle* $k \in \mathbf{P}$, $\varrho > 0$ *gilt, d.h., die Familie* $\{\tau(\varrho)\}$ *ist in ihrer Abhängigkeit vom Parameter* $\varrho > 0$ *vom Fejér'schen Typ. Dann ist* $\{\tau(\varrho)\} \in M$ *und*

$$\|\tau(\varrho)\|_M \leqslant C_j \left[ \frac{1}{j!} \int_0^\infty x^j |dt^{(j)}(x)| + \lim_{x\to\infty} |t(x)| \right]$$

*gleichmäßig für* $\varrho > 0$.

Es sollen nun einige Anwendungen dieser Theorie betrachtet werden, wobei in dieser Arbeit Bedingung (7) auf den Fall $j = 1$ beschränkt bleiben soll. Zunächst sei der Raum $X$ und das System $\{P_k\}$ noch beliebig. Als Beispiele von Verfahren sollen die typischen Mittel $\{R_\varkappa(\varrho)\}$ bzw. die Abel–Cartwright-Mittel $\{W_\varkappa(\varrho)\}$ der Ordnung $\varkappa > 0$ betrachtet werden:

$$R_\varkappa(\varrho) f = \sum_{k<\varrho} \left(1 - \left(\frac{k}{\varrho}\right)^\varkappa\right) P_k f, \qquad W_\varkappa(\varrho) f = \sum_{k=0}^\infty e^{-(k/\varrho)^\varkappa} P_k f,$$

wobei zu beachten ist, daß $\|P_k\|_{[X]} = O(k)$ aus (7) für $j = 1$ folgt. Beide Verfahren sind vom Fejér'schen Typ mit

$$t(x) = r_\varkappa(x) = \begin{cases} 1 - x^\varkappa, & 0 \leqslant x \leqslant 1, \\ 0, & x > 1, \end{cases} \qquad t(x) = w_\varkappa(x) = e^{-x^\varkappa}.$$

Da mit $q_\varkappa(x) = [r_\varkappa(x) - 1][w_\varkappa(x) - 1]^{-1}$ die Bedingungen $q_\varkappa, q_\varkappa^{-1} \in BV_2$ erfüllt sind, folgt aus Satz 1, 5 bezüglich des Vergleichsproblems:

KOROLLAR 1. *Sei* $X$ *ein Banach-Raum und* $\{P_k\}$ *ein System von Projektoren, das Bedingung* (7) *für* $j = 1$ *erfüllt. Dann gilt* $R_\varkappa(\varrho) - I \approx W_\varkappa(\varrho) - I$ *für jedes* $\varkappa > 0$.

Bezüglich einer Ungleichung vom Bernstein-Typ sei $\{G(n)\} \subset [X]_M$ die Folge von Operatoren, die durch $\{\gamma(n)\} \subset M$ mit $\gamma_k(n) = 1$ für $0 \leqslant k \leqslant n$, $= 2 - (k/n)$ für $n < k \leqslant 2n$, $= 0$ für $k > 2n$ erzeugt wird. Wegen (7) für $j = 1$ ist dann $\gamma(n) \in bv_2$ gleichmäßig für alle $n \in \mathbf{P}$, und die Restriktion von $G(n)$ auf die direkte Summe $\Pi_n = \bigoplus_{k=0}^n P_k(X)$ ist der Identitätsoperator. Sei nun $\alpha \in s$ nichtnegativ, d.h. $\alpha_k \geqslant 0$. Da $\alpha\gamma(n) = \eta(n)\gamma(n)$ mit

(9) $\qquad \eta_k(n) = \alpha_k \quad$ für $0 \leqslant k \leqslant 2n, \quad = \alpha_{2n} \quad$ für $k > 2n$,

genügt es für eine Anwendung von Satz 2, 3, etwas über das Verhalten von $\|\eta(n)\|_{bv_2}$ zu wissen. Deshalb folgt für die Restriktion von $B^\alpha G(n)$ auf $\Pi_n$:

KOROLLAR 2. *Das System* $\{P_k\}$ *erfülle* (7) *für* $j = 1$. *Für nichtnegatives* $\alpha \in s$ *sei* $\eta(n) \in s$ *durch* (9) *definiret und erfülle* $\|\eta(n)\|_{bv_2} \leqslant A\alpha_{2n}$ *für alle* $n \in \boldsymbol{P}$. *Dann gilt die Bernstein-Typ-Ungleichung* $\|B^\alpha f\| \leqslant A\alpha_{2n}\|f\|$ *für alle* $f \in \Pi_n$.

Dies kann insbesondere auf alle konkaven Folgen $\alpha$ angewendet werden, da dann $\|\eta(n)\|_{bv_2} = \alpha_{2n} - \alpha_0$ ist.

Ist nun speziell $\alpha = \{k^\omega\}$, $\omega > 0$, so ergibt sich

KOROLLAR 3. *Das System* $\{P_k\}$ *erfülle* (7) *für* $j = 1$. *Dann gelten für jedes* $\omega > 0$ (*und* $\varkappa > 0$) *die Bernstein-Typ-Ungleichungen*

$$\Big\|\sum_{k=0}^{n} k^\omega P_k f\Big\| \leqslant A n^\omega \Big\|\sum_{k=0}^{n} P_k f\Big\| \qquad (f \in X,\ n \in \boldsymbol{P}),$$

$$\Big\|\sum_{k=0}^{\infty} k^\omega e^{-(k/\varrho)^\varkappa} P_k f\Big\| \leqslant A\varrho^\omega \|f\| \qquad (f \in X,\ \varrho > 0).$$

Bisher wurden die allgemeinen Resultate von Satz 1–5 in Verbindung mit konkreten Beispielen von Approximationsverfahren $\{T(\varrho)\}$ bzw. Folgen $\alpha$ diskutiert. Der Raum $X$ und das System $\{P_k\}$ waren dabei noch beliebig, sieht man einmal davon ab, daß (7) für $j = 1$ erfüllt sein soll. Durch entsprechende Wahl von $X$, $\{P_k\}$ gelangt man jetzt zu einer Vielzahl von konkreten Anwendungen auf klassische Orthogonalentwicklungen.

Sei etwa $X = L^p(-\infty, \infty)$, $1 \leqslant p \leqslant \infty$, mit üblicher Norm und $\varphi_k(x) = \big(2^k k!\sqrt{\pi}\big)^{-1/2} e^{-x^2/2} H_k(x)$, $H_k(x) = (-1)^k e^{x^2} (d/dx)^k e^{-x^2}$ die Hermite-Funktionen bzw. -Polynome. Dann erfüllt das System von Projektionen

$$P_k f(x) = \Big[\int_{-\infty}^{\infty} f(u)\,\varphi_k(u)\,du\Big]\varphi_k(x) \tag{10}$$

Bedingung (7) für $j = 1$ (siehe [3a], [5]). Da außerdem $\varphi_k(x)$ Eigenfunktion des Differentialoperators $(d^2/dx^2) + (1 - x^2)$ zum Eigenwert $-2k$ ist, folgt aus Korollar 1, 3:

KOROLLAR 4. *Sei* $X = L^p(-\infty, \infty)$, $1 \leqslant p \leqslant \infty$, *und das System* $\{P_k\}$ *durch* (10) *gegeben. Dann existieren zu jedem* $\varkappa > 0$ *Konstanten*

*$A, A_1, A_2$, so daß*

$$A_1 \left\| f - \sum_{k=0}^{\infty} e^{-(k/\varrho)^\varkappa} P_\varkappa f \right\|_p \leqslant \left\| f - \sum_{k<\varrho} (1-(k/\varrho)^\varkappa) P_k f \right\|_p$$

$$\leqslant A_2 \left\| f - \sum_{k=0}^{\infty} e^{-(k/\varrho)^\varkappa} P_k f \right\|_p$$

*für alle $f \in L^p$, $\varrho > 0$ ist. Weiterhin gilt*

$$\left\| \left[ \frac{d^2}{dx^2} + (1-x^2) \right] \sum_{k=0}^{n} c_k \varphi_k(x) \right\|_p \leqslant An \left\| \sum_{k=0}^{n} c_k \varphi_k(x) \right\|_p, \tag{11}$$

*wobei $c_1, \ldots, c_n$ beliebige Koeffizienten sind.*

Es sei vermerkt, daß die konkrete Bernstein-Typ-Ungleichung (11) schon in [3] enthalten ist.

In entsprechender Weise lassen sich Anwendungen auf Entwicklungen nach Jacobi-Polynomen bzw. Bessel-, Laguerre-, Haar-, Walsh-Funktionen, nach Kugelfunktionen usw. geben (vgl. [1], [4]). Entscheidend für die Anwendbarkeit dieser Theorie auf konkrete Orthogonalsysteme ist dabei natürlich die Frage, für welche Werte von $j$ Bedingung (7) gültig ist. Hier sind insbesondere in letzter Zeit in einer zumeist sehr subtilen Analysis große Fortschritte erzielt worden, die mit Namen wie H. Pollard, G. M. Wing, E. M. Stein, R. Askey, I. I. Hirschman, Jr., S. Wainger, B. Muckenhoupt, A. Benedek, R. Panzone, G. Gasper, um nur einige zu nennen, verbunden sind, so daß eine auf (7) basierende Theorie ein breites Anwendungsspektrum vorfindet. In diesen Arbeiten wurden die $(C, \alpha)$-Mittel nicht nur für ganzzahlige Werte $\alpha = j \in \boldsymbol{P}$ betrachtet, sondern insbesondere auch für beliebige, fraktionierte Werte $\alpha \geqslant 0$. So ist Bedingung (7) für viele konkrete Orthogonalentwicklungen in optimalen $\alpha$-Intervallen nachgewiesen. Dies macht es wünschenswert, nicht nur, wie in dieser Arbeit, Multiplikatorenkriterien für ganzzahlige Werte $\alpha = j \in \boldsymbol{P}$ herzuleiten, sondern beliebige, fraktionierte Werte $\alpha \geqslant 0$ zuzulassen, um somit auch in diesem Rahmen auf der Basis der entsprechenden Bedingung (7) optimale Multiplikatorensätze zu bekommen. Hierfür wie auch für weitere approximationstheoretische Probleme und Anwendungen sei auf [6] verwiesen, wo insbesondere auch die hier diskutierten Kriterien in die klassische Limitierungstheorie divergenter, numerischer Reihen eingebettet werden.

## LITERATUR


[1] P. L. Butzer, R. J. Nessel and W. Trebels, 'On Summation Processes of Fourier Expansions in Banach Spaces, I: Comparison Theorems, II: Saturation Theorems', *Tôhoku Math. J.* **24** (1972) 127–140, 551–559.

[2] J. Favard, 'On the Comparison of the Processes of Summation', *SIAM J. Numer. Anal. Ser. B* **1** (1964) 38–52.

[3] G. Freud, 'On an Inequality of Markov Type', *Soviet Math. Dokl.* **12** (1971) 570–573.

[3a] G. Freud and S. Knapowski, 'On Linear Processes of Approximation III', *Studia Math.* **25** (1965) 374–383.

[4] E. Görlich, R. J. Nessel and W. Trebels, 'Bernstein-Type Inequalities for Families of Multiplier Operators in Banach Spaces with Cesàro Decompositions, I: General Theory, II: Applications', *Acta Sci. Math.* (*Szeged*) **34** (1973) 121–130, **36** (1974) 39–48.

[4a] W. Orlicz, 'Über $k$-fach monotone Folgen', *Studia Math.* **6** (1936) 149–159.

[5] E. L. Poiani, 'Mean Cesàro summability of Laguerre and Hermite series' *Trans. Amer. Math. Soc.* **173** (1972) 1–31.

[6] W. Trebels, 'Multipliers for $(C, \alpha)$-bounded Fourier Expansions in Banach Space and Approximation Theory', Lecture Notes in Mathematics 329, Springer 1973.

# APPROXIMATION OF FUNCTIONS OF SEVERAL VARIABLES BY THE SUMS OF TWO FUNCTIONS OF A SMALLER NUMBER OF VARIABLES

I. I. IBRAGIMOV and M.-B. A. BABAEV

*Baku*

In recent years many papers (see e.g. [1]–[6], [8]–[10], [13], [16]–[19]) are devoted to the development of the theory of the best approximation of functions of several variables $f(x_1, \ldots, x_n)$ by the sum of some functions of a smaller number of variables. Apart of their theoretical value, those topics find an application in the solution of various problems (see e.g. [7], [13]–[16]). One of the principal problems of this theory, also of a great importance for applications, consists in working out an effective method for finding a best approximation to a given function $f(x_1, \ldots, x_n)$ from a given class (see [5], [6], [11], [13]). On the other hand, methods of finding and exactly estimating the distance to a function from a class, also are of an unquestionable interest.

In the sequel we shall denote, for brevity, by $x_{\bar{k}}$ the system of the first $k$ variables $(x_1, \ldots, x_k)$ and by $y_{\bar{q}}$ the system of the last $q$ variables $(x_{k+1}, \ldots, x_n)$, where $k < n$ and $k+q = n$; the inequality $x''_{\bar{k}} \geqslant x'_{\bar{k}}$ means that $x''_1 \geqslant x'_1, \ldots, x''_k \geqslant x'_k$.

We denote by $\Pi_{k,q}$ the class(¹) of functions $f(x_1, \ldots, x_n)$ defined in an $n$-dimensional parallelepiped $Q = [a_1, b_1, \ldots, a_n, b_n]$ and satisfying the condition

(1) $$f(x''_{\bar{k}}, y''_{\bar{q}})+f(x'_{\bar{k}}, y'_{\bar{q}}) \geqslant f(x''_{\bar{k}}, y'_{\bar{q}})+f(x'_{\bar{k}}, y''_{\bar{q}})$$

for any $x''_{\bar{k}} \geqslant x'_{\bar{k}}$ and $y''_{\bar{q}} \geqslant y'_{\bar{q}}$. The class of functions $f(x_1, \ldots, x_n)$ with $(-f) \in \Pi_{k,q}$ is denoted by $\Pi^*_{k,q}$. The quantity

$$E_f = E[f; \varphi(x_{\bar{k}})+\psi(y_{\bar{q}})]$$
$$= \inf_{\varphi+\psi\in C[Q]} \|f-\varphi(x_{\bar{k}})-\psi(y_{\bar{q}})\|_C = \|f-\varphi_0-\psi_0\|_C$$

---

(¹) Classes $\Pi_{k,q}$ and $\Pi^*_{k,q}$ were introduced by one of the authors of this paper M.-B. A. Babaev in [6], [7] and, for the case $n = 2$, in [5].

is called the *distance* to the function $f(x_1, \ldots, x_n)$ from the class of continuous functions of the form $\varphi(x_{\bar{k}})+\psi(y_{\bar{q}})$; the function $\varphi_0(x_{\bar{k}})+$ $+\psi_0(y_{\bar{q}})$ is called a *best approximation* to $f$ from that class.

In Section 1 of this paper we give two methods of construction a best approximation to a function $f \in \Pi_{k,q}(Q)$ or $\Pi^*_{k,q}(Q)$ by functions of the form $\varphi(x_{\bar{k}})+\psi(y_{\bar{q}})$.

In Section 2 we introduce the notion of the derivative with respect to a system of variables, an extension of the directional derivative, and we give methods of calculation of such derivatives. Further on we introduce the class $\Pi^{(\lambda)}_{k,q}$ generalizing the class $\Pi_{k,q}$ and, applying the extended notion of the derivative, we find sufficient conditions for a function to belong to this class. Finally, we give a method for computing the distance to a function $f \in \Pi^{(\lambda)}_{k,q}$ from functions $\varphi(x_{\bar{k}})+\psi(y_{\bar{q}})$ and we find exact two-sided estimates for this distance for functions which possess at least the mixed "second order" derivative with respect to the systems of variables $x_{\bar{k}}$ and $y_{\bar{q}}$.

**Section 1.** The first method (the $A$ method) of finding a best approximation of the form $\varphi(x_{\bar{k}})+\psi(y_{\bar{q}})$ applies in the case when the function $f(x_1, \ldots, x_n)$ is a generalized polynomial of the form:

(2)
$$Z = \sum_{p_1=0}^{m_1} \cdots \sum_{p_n=0}^{m_n} A_{p_1 \ldots p_n} U_{p_1 \ldots p_k}(x_{\bar{k}}) V_{p_{k+1} \ldots p_n}(y_{\bar{q}})$$
$$= \sum_{p_1=0}^{m_1} \cdots \sum_{p_n=0}^{m_n} A_{p_{\bar{n}}} U_{p_{\bar{k}}}(x_{\bar{k}}) V_{p_{\bar{q}}}(y_{\bar{q}})$$
$$= \sum_{p_1} \cdots \sum_{p_n} A_{p_{\bar{n}}} U(x_{\bar{k}}) V(y_{\bar{q}}),$$

where $U$ and $V$ are continuous and monotonic (with respect to each variable) functions.

The other method (the $B$ method) applies for any continuous function $f \in \Pi_{k,q}$ or $f \in \Pi^*_{k,q}$. The $A$ method is not a consequence of the $B$ method.

Our further considerations are heavily based on the following theorem due to Babaev [7]:

THEOREM A. *If a function $f(x_1, \ldots, x_n)$ belongs to $\Pi_{k,q}(Q)$ and is continuous in the parallelepiped $Q$, then the distance* $E_f = E[f; \varphi(x_{\bar{k}})+$

$+\psi(y_{\bar{q}})]_C$ *is expressed by the formula*

$$E_f = \tfrac{1}{4}[f(b_{\bar{k}}, b_{\bar{q}})+f(a_{\bar{k}}, a_{\bar{q}})-f(b_{\bar{k}}, a_{\bar{q}})-f(a_{\bar{k}}, b_{\bar{q}})] = L.$$

COROLLARY. *If* $f \in \Pi^*_{k,q}(Q)$ *is continuous in* $Q$, *then* $E_f = -L$.

One of the main results of this paper is the following

THEOREM 1. *Let* $Z$ *be a generalized polynomial* (2) *with non-negative coefficients* $A_{p_{\bar{n}}} \geqslant 0$ *and such that* $U(x_{\bar{k}}) \geqslant 0$, $V(y_{\bar{q}}) \geqslant 0$ *are continuous monotonic functions; let*

$$g_0 = g_0(x_{\bar{k}}) = \tfrac{1}{2}\Big( \widehat{\max_{\substack{x_j \in [a_j, b_j] \\ j = \overline{k+1, n}}}} Z + \widecheck{\min_{\substack{x_j \in [a_j, b_j] \\ j = \overline{k+1, n}}}} Z\Big),$$

$$h_0 = h_0(y_{\bar{q}}) = \tfrac{1}{2}\Big[ \widehat{\max_{\substack{x_i \in [a_i, b_i] \\ i \in \overline{1, k}}}} (Z-g_0) + \widecheck{\min_{\substack{x_i \in [a_i, b_i] \\ i \in \overline{1, k}}}} (Z-g_0)\Big],$$

*where*

$$\begin{Bmatrix} \widehat{\max} \\ \widecheck{\min} \end{Bmatrix} Z = \sum_{p_1=0}^{m_1} \cdots \sum_{p_n=0}^{m_n} \begin{Bmatrix} \max \\ \min \end{Bmatrix} A_{p_{\bar{n}}} U_{p_{\bar{k}}}(x_{\bar{k}}) V_{p_{\bar{q}}}(y_{\bar{q}}).$$

*Then, if the functions* $U_{p_{\bar{k}}}$ *and* $V_{p_{\bar{q}}}$ *are both increasing or decreasing, then the generalized polynomial*

$$(3) \quad Z_0 = g_0 + h_0 = \tfrac{1}{2} \sum_{p_1=0}^{m_1} \cdots \sum_{p_n=0}^{m_n} A_{p_{\bar{n}}} \{U(x_{\bar{k}})\, [V_{p_{\bar{q}}}(a_{\bar{q}}) + V_{p_{\bar{q}}}(b_{\bar{q}})] + \\ + V_{p_{\bar{q}}}(y_{\bar{q}})\, [U_{p_{\bar{k}}}(a_{\bar{k}}) + U_{p_{\bar{k}}}(b_{\bar{k}})] - \\ - \tfrac{1}{2}[U_{p_{\bar{k}}}(a_{\bar{k}}) + U_{p_{\bar{k}}}(b_{\bar{k}})]\, [V_{p_{\bar{q}}}(a_{\bar{q}}) + V_{p_{\bar{q}}}(b_{\bar{q}})]\}$$

*is a best approximation to* $Z$ *by functions of the form* $\varphi(x_{\bar{k}})+\psi(y_{\bar{q}})$; *and if one of the functions* $U_{p_{\bar{k}}}$ *and* $V_{p_{\bar{q}}}$ *is increasing and the other is decreasing, then* $(-Z_0)$ *is a best approximation to* $Z$.

PROOF. We have

$$g_0(x_{\bar{k}}) = \tfrac{1}{2}[\widehat{\max_{y_{\bar{q}}}} Z + \widecheck{\min_{y_{\bar{q}}}} Z] \\ = \tfrac{1}{2} \sum_{p_1} \cdots \sum_{p_n} A_{p_{\bar{n}}} U_{p_{\bar{k}}}(x_{\bar{k}})\, [V_{p_{\bar{q}}}(b_{\bar{q}}) + V_{p_{\bar{q}}}(a_{\bar{q}})];$$

$$Z - g_0 = \sum_{p_1} \cdots \sum_{p_n} A_{p_{\bar{n}}} U_{p_{\bar{k}}}(x_{\bar{k}})\, \{V_{p_{\bar{q}}}(y_{\bar{q}}) - \tfrac{1}{2}[V_{p_{\bar{q}}}(b_{\bar{q}}) + V_{p_{\bar{q}}}(a_{\bar{q}})]\}.$$

Write

$$N_1 = A_{p_{\bar{n}}} U_{p_{\bar{k}}}(b_{\bar{k}}) \{V_{p_{\bar{q}}}(y_{\bar{q}}) - \tfrac{1}{2}[V_{p_{\bar{q}}}(a_{\bar{q}}) + V_{p_{\bar{q}}}(b_{\bar{q}})]\};$$

$$N_2 = A_{p_{\bar{n}}} U_{p_{\bar{k}}}(a_{\bar{k}}) \{V_{p_{\bar{q}}}(y_{\bar{q}}) - \tfrac{1}{2}[V_{p_{\bar{q}}}(a_{\bar{q}}) + V_{p_{\bar{q}}}(b_{\bar{q}})]\}.$$

Fix arbitrarily the variables $y_{\bar{q}}$. Then the set of all index systems $p_{\bar{q}} = (p_{k+1}, \ldots, p_n)$ with $0 \leqslant p_j \leqslant m_j$, $j = \overline{k+1, n}$, can be decomposed into two subsets $R_1$ and $R_2$ such that for $(p_{\bar{q}}) \in R_1$ we have

$$V_{p_{\bar{q}}}(y_{\bar{q}}) \geqslant \tfrac{1}{2}[V_{p_{\bar{q}}}(a_{\bar{q}}) + V_{p_{\bar{q}}}(b_{\bar{q}})]$$

and for $(p_{\bar{q}}) \in R_2$ the reverse inequality holds.

Applying the introduced notation, we can write:

$$\widehat{\max_{x_{\bar{k}}}}(Z-g_0) = \sum_{p_1} \cdots \sum_{p_k} \Big[\sum \underset{(R_1)}{\cdots} \sum + \sum \underset{(R_2)}{\cdots} \sum\Big] \cdot A_{p_{\bar{n}}} \times$$
$$\times \max U_{p_{\bar{k}}}(x_{\bar{k}}) \{V_{p_{\bar{q}}}(y_{\bar{q}}) - \tfrac{1}{2}[V_{p_{\bar{q}}}(a_{\bar{q}}) + V_{p_{\bar{q}}}(b_{\bar{q}})]\}$$
$$= \sum_{p_1} \cdots \sum_{p_k} \sum \underset{(R_1)}{\cdots} \sum N_1 + \sum_{p_1} \cdots \sum_{p_k} \sum \underset{(R_2)}{\cdots} \sum N_2.$$

Similarly

$$\widecheck{\min_{x_{\bar{k}}}}(Z-g_0) = \sum_{p_1} \cdots \sum_{p_k} \sum \underset{(R_1)}{\cdots} \sum N_2 + \sum_{p_1} \cdots \sum_{p_k} \sum \underset{(R_2)}{\cdots} \sum N_1.$$

Thus

$$h_0(y_{\bar{q}}) = \tfrac{1}{2}[\widehat{\max_{x_{\bar{k}}}}(Z-g_0) + \widecheck{\min_{x_{\bar{k}}}}(Z-g_0)]$$
$$= \tfrac{1}{2}\Big[\sum_{p_1} \cdots \sum_{p_k} \sum \underset{(R_1)}{\cdots} \sum (N_1+N_2) +$$
$$+ \sum_{p_1} \cdots \sum_{p_k} \sum \underset{(R_2)}{\cdots} \sum (N_1+N_2)\Big]$$
$$= \tfrac{1}{2} \sum_{p_1} \cdots \sum_{p_k} (N_1+N_2)$$
$$= \tfrac{1}{2} \sum_{p_1} \cdots \sum_{p_n} A_{p_{\bar{n}}}[U(a_{\bar{k}}) + U(b_{\bar{k}})] \{V(y_{\bar{q}}) - \tfrac{1}{2}[V(a_{\bar{q}}) + V(b_{\bar{q}})]\}.$$

It remains to show that the function $g_0+h_0$ in fact is a best approximation to the polynomial $Z$. Let $U_{p_{\bar{k}}}(x_{\bar{k}})$ and $V_{p_{\bar{q}}}(y_{\bar{q}})$ be both increasing. We first show that $Z \in \Pi_{k,q}$. Indeed, since $A_{p_{\bar{n}}} \geqslant 0$, we have in this case

$$Z(x''_{\bar{k}}, y''_{\bar{q}})+Z(x'_{\bar{k}}, y'_{\bar{q}})-Z(x''_{\bar{k}}, y'_{\bar{q}})-Z(x'_{\bar{k}}, y''_{\bar{q}})$$
$$= \sum_{p_1} \dots \sum_{p_n} A_{p_{\bar{n}}}[U(x''_{\bar{k}})-U(x'_{\bar{k}})]\,[V(y''_{\bar{q}})-V(y'_{\bar{q}})] \geqslant 0$$

for any $x''_{\bar{k}} \geqslant x'_{\bar{k}}$, $y''_{\bar{q}} \geqslant y'_{\bar{q}}$. Then by Theorem A

$$E[Z, \varphi(x_{\bar{k}})+\psi(y_{\bar{q}})] = \tfrac{1}{4}[Z(b_{\bar{k}}, b_{\bar{q}})+Z(a_{\bar{k}}, a_{\bar{q}})-Z(a_{\bar{k}}, b_{\bar{q}})-Z(b_{\bar{k}}, a_{\bar{q}})]$$
$$= \tfrac{1}{4}\sum_{p_1} \dots \sum_{p_n} A_{p_{\bar{n}}}[U(b_{\bar{k}})-U(a_{\bar{k}})]\,[V(b_{\bar{q}})-V(a_{\bar{q}})].$$

Further

$$\varrho[Z, g_0+h_0]_C = \max\{\sup_{p\in Q}[Z-(g_0+h_0)]\,(p),\ -\inf_{p\in Q}[Z-(g_0+h_0)]\,(p)\}.$$

Clearly, the relations

$$U(a_{\bar{k}}) \leqslant U(x_{\bar{k}}) \leqslant U(b_{\bar{k}}),$$
$$V(a_{\bar{q}}) \leqslant V(y_{\bar{q}}) \leqslant V(b_{\bar{q}})$$

imply

$$-\tfrac{1}{2}[U(b_{\bar{k}})-U(a_{\bar{k}})] \leqslant U(x_{\bar{k}})-\tfrac{1}{2}[U(a_{\bar{k}})+U(b_{\bar{k}})] \leqslant \tfrac{1}{2}[U(b_k)-U(a_k)],$$
$$-\tfrac{1}{2}[V(b_{\bar{q}})-V(a_{\bar{q}})] \leqslant V(y_{\bar{q}})-\tfrac{1}{2}[V(a_{\bar{q}})+V(b_{\bar{q}})] \leqslant \tfrac{1}{2}[V(b_{\bar{q}})-V(a_{\bar{q}})].$$

We thus have

$$Z-g_0-h_0$$
$$= \sum_{p_1} \dots \sum_{p_n} A_{p_{\bar{n}}}\{U(x_{\bar{k}})-\tfrac{1}{2}[U(a_{\bar{k}})+U(b_{\bar{k}})]\}\,\{V(y_{\bar{q}})-\tfrac{1}{2}[V(a_{\bar{q}})+V(b_{\bar{q}})]\}$$
$$\leqslant \tfrac{1}{4}\sum_{p_1} \dots \sum_{p_n} A_{p_{\bar{n}}}[U(b_{\bar{k}})-U(a_{\bar{k}})]\,[V(b_{\bar{q}})-V(a_{\bar{q}})] = E(Z).$$

But

$$[Z-g_0-h_0]\,(b_{\bar{k}}, b_{\bar{q}}) = E(Z).$$

Consequently,

$$\sup_{p\in Q}\,[Z-g_0-h_0] = E(Z).$$

Similarly

$$-\inf[Z-g_0-h_0] = E(Z).$$

Hence

$$\varrho[Z, g_0+h_0] = E(Z).$$

The case when both functions $U_{p_{\bar{k}}}(x_{\bar{k}})$ and $V_{p_{\bar{q}}}(y_{\bar{q}})$ are decreasing can be discussed analogously. It is not difficult to see, that if one of the functions $U_{p_{\bar{k}}}(x_{\bar{k}})$, $V_{p_{\bar{q}}}(y_{\bar{q}})$ is increasing and the other one is decreasing, then $f = U_{p_{\bar{k}}}(x_{\bar{k}}) V_{p_{\bar{q}}}(y_{\bar{q}}) \in \Pi^*_{k,q}$. In this case the proof can be accomplished by repeating the argument, just applying the corollary to Theorem A instead of the theorem itself. The proof is thus complete.

The above theorem gives a method of construction of a function $Z_0$ of the form $\varphi(x_{\bar{k}}) + \psi(y_{\bar{q}})$ which realizes a best approximation to a generalized polynomial (2) in $Q$. This is the above mentioned method $A$.

COROLLARY 1. *If* $A_{p_{\bar{n}}} \leqslant 0$, *then, under the remaining assumptions of Theorem* 1, *the function* $(-Z_0)$, *resp.* $Z_0$, *is a best approximation to a generalized polynomial* (2), *depending on whether both functions* $U_{p_{\bar{k}}}$, $V_{p_{\bar{q}}}$ *are increasing or decreasing, resp. one of them is increasing and the other one is decreasing.*

In particular, setting

$$U_{p_{\bar{k}}}(x_{\bar{k}}) = x_1^{p_1} \dots x_k^{p_k}, \qquad V_{p_{\bar{q}}}(y_{\bar{q}}) = x_{k+1}^{p_{k+1}} \dots x_n^{p_n},$$

we obtain a best approximation to an algebraic polynomial of $n$ variables:

$$P = P_{m_1 \dots m_n}(x_1, \dots, x_n) = \sum_{p_1=0}^{m_1} \dots \sum_{p_n=0}^{m_n} A_{p_1 \dots p_n} x_1^{p_1} \dots x_n^{p_n}. \tag{4}$$

COROLLARY 2. *For an algebraic polynomial* (4) *with non-negative coefficients* $A_{p_{\bar{n}}} \geqslant 0$ *the polynomial*

$$\begin{aligned} P_0 = g_0 + h_0 = \tfrac{1}{2} \sum_{p_1=0}^{m_1} \dots \sum_{p_n=0}^{m_n} A_{p_{\bar{n}}} [x_1^{p_1} \dots x_k^{p_k} (a_{k+1}^{p_{k+1}} \dots a_n^{p_n} + \\ + b_{k+1}^{p_{k+1}} \dots b_n^{p_n}) + x_{k+1}^{p_{k+1}} \dots x_n^{p_n} (a_1^{p_1} \dots a_k^{p_k} + b_1^{p_1} \dots b_k^{p_k}) - \\ - \tfrac{1}{2} (a_1^{p_1} \dots a_k^{p_k} + b_1^{p_1} \dots b_k^{p_k}) (a_{k+1}^{p_{k+1}} \dots a_n^{p_n} + b_{k+1}^{p_{k+1}} \dots b_n^{p_n})] \end{aligned} \tag{5}$$

is a best approximation by functions of the form $\varphi(x_{\bar{k}}) + \psi(y_{\bar{q}})$ on the parallelepiped $Q = [a_1, b_1, \dots, a_n, b_n]$ with $a_i \geqslant 0$, $i = \overline{1, n}$. If, conversely, $A_{p_{\bar{n}}} \leqslant 0$, then $(-P_0)$ is a best approximation.

In the case $n = 2$ Corollary 2 yields a result of the paper [5] obtained there by another method.

THEOREM 2. *Let a function* $f(x_{\bar{k}}, y_{\bar{q}})$ *of class* $\Pi_{k,q}$ *be continuous in the domain* $Q$; *put*

$$g(x_{\bar{k}}, y_{\bar{q}}) = f(x_{\bar{k}}, y_{\bar{q}}) - f(x_{\bar{k}}, a_{\bar{q}}) - f(a_{\bar{k}}, y_{\bar{q}}) + f(a_{\bar{k}}, a_{\bar{q}}),$$
$$\varphi_g^0(x_{\bar{k}}) = g(x_{\bar{k}}, y_{\bar{q}}^0),$$

*where* $y_{\bar{q}} = y_{\bar{q}}^0$ *is the solution of the equation*

$$g(b_{\bar{k}}, y_{\bar{q}}) = \tfrac{1}{2} g(b_{\bar{k}}, b_{\bar{q}})$$

*and*

$$\psi_g^0(y_{\bar{q}}) = \tfrac{1}{2}[g(b_k, y_{\bar{q}}) - \tfrac{1}{2} g(b_{\bar{k}}, b_{\bar{q}})].$$

*Then the function* $\varphi_g^0(x_{\bar{k}}) + \psi_g^0(y_{\bar{q}})$ *is a best approximation to the function* $g(x_{\bar{k}}, y_{\bar{q}})$.

The proof of Theorem 2 is omitted here, since it is in fact contained in the proof of Theorem A (though the result has not been clearly stated as a theorem).

Theorem 2 indicates a method of construction of the function $\varphi_g^0(x_{\bar{k}}) + \psi_g^0(y_{\bar{q}})$ which realizes a best approximation to the function $g(x_{\bar{k}}, y_{\bar{q}})$, and consequently also to $f(x_{\bar{k}}, y_{\bar{q}})$. This is precisely the $B$ method referred to above:

COROLLARY 2. *Under the conditions and notation of Theorem 2 the function*

$$\varphi_f^0 + \psi_f^0 = \varphi_g^0 + \psi_g^0 + f(x_{\bar{k}}, a_{\bar{q}}) + f(a_{\bar{k}}, y_{\bar{q}}) - f(a_{\bar{k}}, a_{\bar{q}})$$

*is a best approximation to the function* $f$. *If the condition* $f \in \Pi_{k,q}$ *is replaced by* $f \in \Pi_{k,q}^*$, *then the function* $-(\varphi_f^0 + \psi_f^0)$ *is a best approximation.*

Let now $\tilde{\Pi}_{k,q}$ denote the class of functions $f$ of the form $f(x_{\bar{k}}, y_{\bar{q}}) = U(x_{\bar{k}}) V(y_{\bar{q}}) \in \Pi_{k,q}$. The class $\tilde{\Pi}_{k,q}^*$ is defined analogously. Clearly $\tilde{\Pi}_{k,q} \subset \Pi_{k,q}$, $\tilde{\Pi}_{k,q}^* \subset \Pi_{k,q}^*$. The trivial case when $f$ is of the form $\varphi(x_{\bar{k}}) + \psi(y_q)$ will not be dealt with here. For the classes $\tilde{\Pi}_{k,q}$ and $\tilde{\Pi}_{k,q}^*$ the following lemma holds true:

LEMMA 1. *If* $f$ *belongs to one of the classes* $\tilde{\Pi}_{k,q}$ *or* $\tilde{\Pi}_{k,q}^*$, *then*

$$U(b_{\bar{k}}) \neq U(a_{\bar{k}}), \quad V(b_{\bar{q}}) \neq V(a_{\bar{q}}).$$

PROOF. We have

$$g(x_{\bar{k}}, y_{\bar{q}}) = f(x_{\bar{k}}, y_{\bar{q}}) - f(x_{\bar{k}}, a_{\bar{q}}) - f(a_{\bar{k}}, y_{\bar{q}}) + f(a_{\bar{k}}, a_{\bar{q}}) \tag{6}$$
$$= [U(x_{\bar{k}}) - U(a_{\bar{k}})]\,[V(y_{\bar{q}}) - V(a_{\bar{q}})].$$

Thus $g(x_{\bar{k}}, y_{\bar{q}}) \neq 0$, since otherwise

$$f(x_{\bar{k}}, y_{\bar{q}}) = f(x_{\bar{k}}, a_{\bar{q}}) + f(a_{\bar{k}}, y_{\bar{q}}) - f(a_{\bar{k}}, a_{\bar{q}}),$$

i.e. $f$ has the form $\varphi(x_{\bar{k}}) + \psi(y_{\bar{k}})$. So there exists a point $(x_{\bar{k}}^0, y_{\bar{q}}^0)$ such that

$$g(x_{\bar{k}}^0, y_{\bar{q}}^0) > 0 \tag{7}$$

(as $g(x_{\bar{k}}, a_{\bar{q}}) = 0$ and $g$ increases in $x_{\bar{k}}$ for a fixed $y_{\bar{q}}$).

From (6) and (7) follows

$$[U(x_{\bar{k}}^0) - U(a_{\bar{k}})]\,[V(y_{\bar{q}}^0) - V(a_{\bar{q}})] > 0.$$

Suppose that

$$U(x_{\bar{k}}^0) > U(a_{\bar{k}}). \tag{8}$$

Then

$$V(y_{\bar{q}}^0) > V(a_{\bar{q}}).$$

Further, since $g$ increases in $x_{\bar{q}}$,

$$g(b_{\bar{k}}, y_{\bar{q}}^0) - g(x_{\bar{k}}^0, y_{\bar{q}}^0) = [U(b_{\bar{k}}) - U(x_{\bar{k}}^0)]\,[V(y_{\bar{q}}^0) - V(a_{\bar{q}})] \geqslant 0,$$

whence

$$U(b_{\bar{k}}) \geqslant U(x_{\bar{k}}^0).$$

This, in view of (8), gives $U(b_{\bar{k}}) > U(a_{\bar{k}})$. The case $U(x_{\bar{k}}^0) < U(a_{\bar{k}})$ is fully analogous. The lemma is thus proved.

As can be seen from Theorem 2, in order to find a best approximation by the $B$ method one has to employ the solution of the equation

$$g(b_{\bar{k}}, y_{\bar{q}}) = \tfrac{1}{2} g(b_{\bar{k}}, b_{\bar{q}}).$$

The following theorem shows an application of this method.

THEOREM 3. *For a function $f \in \tilde{\Pi}_{k,q}$, the function*

$$f_0 = \varphi_f^0 + \psi_f^0 = \tfrac{1}{2} U(x_{\bar{k}})\,[V(a_{\bar{q}}) + V(b_{\bar{q}})] + \tfrac{1}{2} V(y_{\bar{q}})\,[U(a_{\bar{k}}) + U(b_{\bar{k}})]$$
$$= \tfrac{1}{4}[U(a_{\bar{k}}) + U(b_{\bar{k}})]\,[V(a_{\bar{q}}) + V(b_{\bar{q}})],$$

*is a best approximation by functions $\varphi(x_{\bar{k}}) + \psi(y_{\bar{q}})$.*

PROOF. For a function $f \in \tilde{\Pi}_{k,q}$ the equation

$$g(b_{\bar{k}}, y_{\bar{k}}) = \tfrac{1}{2}g(b_{\bar{k}}, b_{\bar{q}})$$

has the form

$$[U(b_{\bar{k}})-U(a_{\bar{k}})]\,[V(y_{\bar{q}})-V(a_{\bar{q}})] = \tfrac{1}{2}[U(b_{\bar{k}})-U(a_{\bar{k}})]\,[V(b_{\bar{q}})-V(a_{\bar{q}})]$$

or

$$[U(b_{\bar{k}})-U(a_{\bar{k}})]\,\{V(y_{\bar{q}})-\tfrac{1}{2}[V(a_{\bar{q}})+V(b_{\bar{q}})]\} = 0,$$

whence

$$V(y_{\bar{q}}) = \tfrac{1}{2}[V(a_{\bar{q}})+V(b_{\bar{q}})]. \tag{9}$$

($U(b_{\bar{k}}) \neq U(a_{\bar{k}})$ by Lemma 1.)

The continuous function $V(y_{\bar{q}})$ assumes at the points $y_{\bar{q}} = a_{\bar{q}}$ and $y_{\bar{q}} = b_{\bar{q}}$ the values $V(a_{\bar{q}})$ and $V(b_{\bar{q}})$. From the fact that one of the inequalities

$$V(a_{\bar{q}}) \leqslant \tfrac{1}{2}[V(a_{\bar{q}})+V(b_{\bar{q}})] \leqslant V(b_{\bar{q}})$$

or

$$V(a_{\bar{q}}) \geqslant \tfrac{1}{2}[V(a_{\bar{q}})+V(b_{\bar{q}})] \geqslant V(b_{\bar{q}})$$

necessarily holds, it follows that there exists at least one solution $y_{\bar{q}} = \bar{y}_{\bar{q}}$ of equation (9).

Let us define the functions:

$$\begin{aligned}
\varphi_1(x_{\bar{k}}) &= g(x_{\bar{k}}, \bar{y}_{\bar{q}}) = [U(x_{\bar{k}})-U(a_{\bar{k}})]\,[V(\bar{y}_{\bar{q}})-V(a_{\bar{q}})] \\
&= \tfrac{1}{2}[U(x_{\bar{k}})-U(a_{\bar{k}})]\,[V(b_{\bar{q}})-V(a_{\bar{q}})]; \\
\psi_1(y_{\bar{q}}) &= \tfrac{1}{2}[g(b_{\bar{k}}, y_{\bar{q}})-\tfrac{1}{2}g(b_{\bar{k}}, b_{\bar{q}})] \\
&= \tfrac{1}{2}\{[U(b_{\bar{k}})-U(a_{\bar{k}})]\,[V(y_{\bar{q}})-V(a_{\bar{q}})]- \\
&\quad -\tfrac{1}{2}[U(b_{\bar{k}})-U(a_{\bar{k}})]\,[V(b_{\bar{q}})-V(a_{\bar{q}})]\} \\
&= \tfrac{1}{2}[U(b_{\bar{k}})-U(a_{\bar{k}})]\,\{V(y_{\bar{q}})-\tfrac{1}{2}[V(a_{\bar{q}})+V(b_{\bar{q}})]\}.
\end{aligned}$$

Then the function

$$\begin{aligned}
\varphi_1(x_{\bar{k}})+\psi_1(y_{\bar{q}}) &= \tfrac{1}{2}U(x_{\bar{k}})\,[V(b_{\bar{q}})-V(a_{\bar{q}})]+ \\
&+\tfrac{1}{2}V(y_{\bar{q}})\,[U(b_{\bar{k}})-U(a_{\bar{k}})]-\tfrac{1}{4}[U(b_{\bar{k}})-U(a_{\bar{k}})]\,[V(a_{\bar{q}})+V(b_{\bar{q}})]- \\
&-\tfrac{1}{2}U(a_{\bar{k}})\,[V(b_{\bar{q}})-V(a_{\bar{q}})]
\end{aligned}$$

is a best approximation to the function $g(x_{\bar{k}}, y_{\bar{q}})$. Finally, the function $\varphi_f^0(x_{\bar{k}})+\psi_f^0(y_{\bar{q}})$, defined and calculated below, is a best approximation to the function $f(x_{\bar{k}}, y_{\bar{q}})$:

$$\begin{aligned}\varphi_f^0(x_{\bar{k}})+\psi_f^0(y_{\bar{q}}) &= \varphi_1(x_{\bar{k}})+\psi_1(y_{\bar{q}})+f(x_{\bar{k}}, a_{\bar{q}})+f(a_{\bar{k}}, y_{\bar{q}})-f(a_{\bar{k}}, a_{\bar{q}})\\ &= \varphi_1(x_{\bar{k}})+\psi_1(y_{\bar{q}})+U(x_{\bar{k}})V(a_{\bar{q}})+U(a_{\bar{k}})V(y_{\bar{q}})-U(a_{\bar{k}})V(a_{\bar{q}})\\ &= \tfrac{1}{2}U(x_{\bar{k}})\,[V(a_{\bar{q}})+V(b_{\bar{q}})]+\tfrac{1}{2}V(y_{\bar{q}})\,[U(a_{\bar{k}})+-U(b_{\bar{k}})]-\\ &\quad -\tfrac{1}{4}[U(b_{\bar{k}})-U(a_{\bar{k}})]\,[V(a_{\bar{q}})+V(b_{\bar{q}})]-\tfrac{1}{2}U(a_{\bar{k}})\,[V(a_{\bar{q}})+V(b_{\bar{q}})]\\ &= \tfrac{1}{2}U(x_{\bar{k}})\,[V(a_{\bar{q}})+V(b_{\bar{q}})]-+\tfrac{1}{2}V(y_{\bar{q}})\,[U(a_{\bar{k}})+U(b_{\bar{k}})]-\\ &\quad -\tfrac{1}{4}[U(a_{\bar{k}})+U(b_{\bar{k}})]\,[V(a_{\bar{q}})+V(b_{\bar{q}})].\end{aligned}$$

Theorem 3 is thus proved.

COROLLARY 1. *If a continuous function* $f = U(x_{\bar{k}})V(y_{\bar{q}})$ *is in the class* $\tilde{\Pi}^*_{k,q}$, *then* $(-f_0)$ *is its best approximation.*

COROLLARY 2. *Consider a generalized polynomial* (2) *with functions* $U_{p_{\bar{k}}}(x_{\bar{k}})$ *and* $V_{p_{\bar{q}}}(y_{\bar{q}})$ *continuous. If* $Z \in \Pi_{k,q}$ *and* $A_{p_{\bar{n}}} = B_{p_{\bar{k}}}\cdot C_{p_{\bar{q}}}$, *then the function* (3) *is a best approximation to the polynomial* $Z$; *if* $Z \in \Pi^*_{k,q}$, *then the function* (3) *with the reverse sign is a best approximation to* $Z$.

This follows from the fact that under the condition $A_{p_{\bar{n}}} = B_{p_{\bar{k}}}\cdot C_{p_{\bar{q}}}$ the function $Z$ can be written in the form

$$Z = \Big(\sum_{p_1=0}^{m_1}\cdots\sum_{p_k=0}^{m_k} B_{p_{\bar{k}}}U_{p_{\bar{k}}}(x_{\bar{k}})\Big)\Big(\sum_{p_{k+1}=0}^{m_{k+1}}\cdots\sum_{p_n=0}^{m_n} C_{p_{\bar{q}}}V_{p_{\bar{q}}}(y_{\bar{q}})\Big)$$

and Theorem 3 applies.

COROLLARY 3. *For an algebraic polynomial* (4) *of class* $\Pi_{k,q}$ *with* $A_{p_{\bar{n}}} = B_{p_{\bar{k}}}\cdot C_{p_{\bar{q}}}$ *the polynomial* (5) *is a best approximation from the class of functions* $\varphi(x_{\bar{k}})+\psi(y_{\bar{q}})$; *if the polynomial* (4) *is of class* $\Pi^*_{k,q}$, *then* (5) *with reverse sign is its best approximation.*

Observe that in Corollaries 2 and 3, contrary to Theorem 1, restrictions on the signs of coefficients $A_{p_{\bar{n}}}$ and on the monotonicity of functions $U_{p_{\bar{k}}}(x_{\bar{k}})$ and $V_{p_{\bar{q}}}(y_{\bar{q}})$ do not occur.

**Section 2.** We consider a function $f = f(x_1, \ldots, x_n) = f(p) = f(x_{\bar{\alpha}}, p/x_{\bar{\alpha}})$ (the latter notation explained below) defined in a domain $Q$ in the Euclidean space $R_n$. Fix a system of $\alpha$ variables $x_{\bar{\alpha}} = (x_1, \ldots, x_\alpha)$, where $\alpha \leqslant n$, and an arbitrary vector $\bar{a} = [\nu_1, \ldots, \nu_\alpha]$, where $\nu_i$, $i \in \overline{1, \alpha} = \bar{\alpha}$ are real numbers. The quantity

$$f'_{x_{\bar{\alpha}}, |\bar{a}|} = \lim_{s \to 0} \frac{f(x_{\bar{\alpha}} + s\bar{a}, p/x_{\bar{\alpha}}) - f(p)}{s \sum_{i \in \bar{\alpha}} |\nu_i|}, \qquad s\text{—real},$$

whenever exists, will be called the (first order) *derivative* of the function $f$ with respect to the system of variables $x_{\bar{\alpha}}$ in the direction of the vector $\bar{a}$. A function $f$ will be said to have a $\lambda$-derivative with respect to the system $x_{\bar{\alpha}}$ ($\lambda \in [1, \infty]$) if it has directional derivatives $f'_{x_{\bar{\alpha}}, |\bar{a}|}$ in the direction of any vector $a = [\nu_1, \ldots, \nu_\alpha]$ satisfying the condition

$$(10) \qquad 0 \leqslant \frac{1}{\lambda} \leqslant \left| \frac{\nu_i}{\nu_j} \right| \leqslant \lambda, \quad i, j \in \bar{\alpha}.$$

The symbol $f'_{x_{\bar{\alpha}}, |\bar{a}_\lambda|}$ will mean, that $f$ has such a derivative. When $\lambda$ decreases, the class of functions such that $f'_{x_{\bar{\alpha}}, |\bar{a}_\lambda|}$ exists, increases. The smallest of those classes is the class of functions which have the derivative $f'_{x_{\bar{\alpha}}, |\bar{a}_\infty|} = f'_{x_{\bar{\alpha}}}$, where no restrictions are posed on the increments $s\nu_i$, $i \in \bar{\alpha}$. In the case, when all the coordinates of $\bar{a}$ are positive, we shall also employ the symbols $f'_{x_{\bar{\alpha}}, [\bar{a}]}$, $f'_{x_{\bar{\alpha}}, [\bar{a}_\lambda]}$. If $x_{\bar{\alpha}}$ reduces to a single variable, then $f'_{x_{\bar{\alpha}}, [\bar{a}]}$ is just the usual partial derivative.

The following lemma holds true:

LEMMA 2. *Suppose that the partial derivatives $f'_{x_i}$, $i \in \bar{\alpha}$, exist. Then for any vector $a = [\nu_1, \ldots, \nu_\alpha]$ the directional derivative with respect to the system $x_{\bar{\alpha}}$ in the direction $\bar{a}$ exists and is expressed by the formula*

$$(11) \qquad f'_{x_{\bar{\alpha}}, |\bar{a}|} = \frac{1}{|\nu_1| + \ldots + |\nu_\alpha|} \sum_{i \in \bar{\alpha}} \nu_i f'_{x_i}.$$

PROOF.

$$f'_{x_{\bar{\alpha}}, |\bar{a}|} = \lim_{s \to 0} \frac{f(x_{\bar{\alpha}} + s\bar{a}, p/x_{\bar{\alpha}}) - f(p)}{s \sum_{i \in \alpha} |\nu_i|}$$

$$= \lim_{s \to 0} \sum_{i=1}^{\alpha} \frac{f(x_1 + s\nu_1, \ldots, x_i + s\nu_i, p/x_{\bar{i}}) - f(x_1 + s\nu_1, \ldots, x_{i-1} + s\nu_{i-1}, p/x_{\overline{i-1}})}{s\nu_i} \times$$

$$\times \frac{\nu_i}{\sum_{i \in \bar{\alpha}} |\nu_i|} = \frac{1}{|\nu_1| + \ldots + |\nu_\alpha|} \sum_{i \in \alpha} \nu_i f'_{x_i}.$$

REMARK. The class of functions which have the derivative $f'_{x_{\bar{\alpha}},|\bar{a}|}$ is essentially larger than that of functions which have all partial derivatives $f'_{x_i}$, $i \in \bar{\alpha}$. E.g. the function $f = |xy|$ does not possess the partial derivative $f'_x$ at the point $x = 0, y = 1$; however, for any vector $\bar{a} = [\nu_1, \nu_2]$ the derivative $f'_{(xy),|\bar{a}|}$ exists for all $x$ and $y$.

In particular, if the system $x_{\bar{\alpha}}$ consists of all variables $x_{\bar{\alpha}} = x_{\overline{1,n}}$, then there is a connection between $f'_{x_{\bar{\alpha}},|\bar{a}|}$ and the usual directional derivative. This connection is established by the following

LEMMA 3. *The first order derivative with respect to the system of all variables $f'_{x_{\overline{1,n}},|\bar{a}|}$ in the direction of a vector $\bar{a} = [\nu_1, \ldots, \nu_n]$ exists if and only if the usual directional derivative $\partial f/\partial\bar{a}$ exists. In this case the following formula holds true:*

$$f'_{x_{\overline{1,n}},|\bar{a}|} = \frac{\sqrt{\nu_1^2 + \ldots + \nu_n^2}}{|\nu_1| + \ldots + |\nu_n|} \cdot \frac{\partial f}{\partial \bar{a}}.$$

PROOF. The "only if" part. Suppose that $f'_{x_{\overline{1,n}},|\bar{a}|}$ exists. Then

$$\frac{\partial f}{\partial \bar{a}} = \lim_{\substack{\Delta x_i \to 0 \\ i=\overline{1,n}}} \frac{f(x_1+\Delta x_1, \ldots, x_n+\Delta x_n) - f(x_1, \ldots, x_n)}{\sqrt{\Delta x_1^2 + \ldots + \Delta x_n^2}}$$

$$= \lim_{\substack{\Delta x_i \to 0 \\ i=\overline{1,n}}} \frac{f(x_1+\Delta x_1, \ldots, x_n+\Delta x_n) - f(x_1, \ldots, x_n)}{\sum_{i=\overline{1,n}} |\Delta x_i|} \cdot \frac{\sum_{i=\overline{1,n}} |\Delta x_i|}{\sqrt{\Delta x_1^2 + \ldots + \Delta x_n^2}}$$

$$= f'_{x_{\overline{1,n}},|\bar{a}|} \cdot \lim_{\substack{x_i \to 0 \\ i=\overline{1,n}}} \frac{\dfrac{\sum_{i=\overline{1,n}} |\Delta x_i|}{|\Delta x_1|}}{\sqrt{\dfrac{\sum_{i=\overline{1,n}} |\Delta x_i^2|}{\Delta x_1^2}}}$$

$$= f'_{x_{\overline{1,n}},|\bar{a}|} \cdot \frac{1 + \left|\dfrac{\nu_2}{\nu_1}\right| + \ldots + \left|\dfrac{\nu_n}{\nu_1}\right|}{\sqrt{1 + \dfrac{\nu_2^2}{\nu_1^2} + \ldots + \dfrac{\nu_n^2}{\nu_1^2}}} = \frac{|\nu_1| + \ldots + |\nu_n|}{\sqrt{\nu_1^2 + \ldots + \nu_n^2}} \cdot f'_{x_{\overline{1,n}},|\bar{a}|}.$$

The "if" part is similar.

The $\lambda$-derivative of order $m$ with respect to the system $x_{\bar{\alpha}}$ is defined by induction:

$$f^{(m)}_{x_{\bar{\alpha}}, |\bar{a}_\lambda|} = [f^{(m-1)}_{x_{\bar{\alpha}}, |\bar{a}_\lambda|}]'_{x_{\bar{\alpha}}, |\bar{a}_\lambda|}.$$

The mixed $[\lambda_1, \ldots, \lambda_m]$-derivative of order $(\beta_1 + \ldots + \beta_m)$ with respect to the systems $\eta = (\bar{\alpha}_1, \ldots, \bar{\alpha}_m)$ is defined by:

$$f^{(\beta_1+\ldots+\beta_m)}_{x_{\bar{\alpha}_1}\ldots x_{\bar{\alpha}_m}, |\bar{a}_{\lambda_1}, \ldots, \bar{a}_{\lambda_m}|} = [(f^{(\beta_1)}_{x_{\bar{\alpha}_1}, |\bar{a}_{\lambda_1}|})^{(\beta_2)}_{x_{\bar{\alpha}_2}, |\bar{a}_{\lambda_2}|} \ldots]^{(\beta_m)}_{x_{\bar{\alpha}_m}, |\bar{a}_{\lambda_m}|}.$$

In particular, in the case $\beta_1 = \ldots = \beta_m = 1$ we obtain the mixed $[\lambda_1, \ldots, \lambda_m]$-derivative with respect to $\eta = (\bar{\alpha}_1, \ldots, \bar{\alpha}_m)$:

$$f^{(m)}_{x_{\bar{\alpha}_1}\ldots x_{\bar{\alpha}_m}, |\bar{a}_{\lambda_1}\ldots \bar{a}_{\lambda_m}|},$$

and if, moreover, $\lambda_1 = \ldots = \lambda_m = \lambda$, then we obtain the mixed $\lambda$-derivative with respect to $\eta$:

$$f^{(m)}_{x_{\bar{\alpha}_1}\ldots x_{\bar{\alpha}_m}, |\bar{a}_\lambda|}.$$

LEMMA 4. *Suppose that there exist the usual mixed derivatives* $f^{(m)}_{x_{i_1}\ldots x_{i_m}}$, $i_j \in \bar{\alpha}, j \in \overline{1, m}$. *Then the m-th order derivative with respect to the system* $x_{\bar{\alpha}} = (x_1, \ldots, x_\alpha)$ *in the direction of any vector* $\bar{a} = [\nu_1, \ldots, \nu_\alpha]$ *exists and can be calculated by means of the formula*

$$\text{(12)} \qquad f^{(m)}_{x_{\bar{\alpha}}, |\bar{a}|} = \frac{1}{(|\nu_1| + \ldots + |\nu_\alpha|)^m} \sum_{i_1 \in \bar{\alpha}} \ldots \sum_{i_m \in \bar{\alpha}} \nu_{i_1} \ldots \nu_{i_m} f^{(m)}_{x_{i_1}\ldots x_{i_m}}$$

$$= \frac{1}{(|\nu_1| + \ldots + |\nu_\alpha|)^m} \sum_{i_1+\ldots+i_m=m}^{m\alpha} \nu_{i_1} \ldots \nu_{i_m} f^{(m)}_{x_{i_1}\ldots x_{i_m}}.$$

PROOF. The proof is by induction on $m$. For $m = 1$ formula (12) coincides with (11). Suppose that (12) holds with $m = r$. Then

$$f^{(r+1)}_{x_{\bar{\alpha}}, |\bar{a}|} = [f^{(r)}_{x_{\bar{\alpha}}, |\bar{a}|}]'_{x_{\bar{\alpha}}, |\bar{a}|}$$

$$= \left[\frac{1}{(|\nu_1| + \ldots + |\nu_\alpha|)^r} \sum_{\nu_1+\ldots+\nu_r=r}^{r\alpha} \nu_{i_1} \ldots \nu_{i_r} f^{(r)}_{x_{i_1}\ldots x_{i_r}}\right]'_{x_{\bar{\alpha}}, |\bar{a}|}$$

$$= \frac{1}{(|\nu_1| + \ldots + |\nu_\alpha|)} \sum_{i_{r+1}=1}^{\alpha} \nu_{i_{r+1}} \left[\frac{1}{(|\nu_1| + \ldots + |\nu_\alpha|)^r} \sum_{i_1+\ldots+i_r=r}^{r\alpha} \nu_{i_1} \ldots \nu_{i_r} \times \right.$$

$$\left. \times f^{(r)}_{x_{i_1}\ldots x_{i_r}}\right]'_{x_{i_{r+1}}} = \frac{1}{(|\nu_1| + \ldots + |\nu_\alpha|)^{r+1}} \sum_{i_1+\ldots+i_{r+1}=r+1}^{(r+1)\alpha} \nu_{i_1} \ldots \nu_{i_{r+1}} f^{(r+1)}_{x_{i_1}\ldots x_{i_{r+1}}}.$$

THEOREM 4. *Suppose that there exist the usual mixed derivatives*

$$f^{(\beta_1+\ldots+\beta_m)}_{x_{i_{11}}\ldots x_{i_{m\beta_m}}}, \quad i_{kq_k} \in \bar{\alpha}_k, \; k \in \overline{1, m}, \; q_k \in \overline{1, \beta_k}.$$

*Then there exists the mixed derivative of order* $(\beta_1 + \ldots + \beta_m)$ *with respect to* $\eta = (\bar{\alpha}_1, \ldots, \bar{\alpha}_m)$ *in the direction of any corresponding vectors* $\bar{a}_k = [\nu_{k1}, \ldots, \nu_{k\alpha_k}]$. *This derivative can be calculated by the formula*

$$(13) \qquad f^{(\beta_1+\ldots+\beta_m)}_{x_{\bar{\alpha}_1}\ldots x_{\bar{\alpha}_m}, |\bar{a}_1, \ldots, \bar{a}_m|}$$

$$= \frac{\sum\limits_{i_{11}+\ldots+i_{1\beta_1}=\beta_1}^{\beta_1\alpha_1} \cdots \sum\limits_{i_{m1}+\ldots+i_{m\beta_m}=\beta_m}^{\beta_m\alpha_m} \nu_{1_{i_{11}}} \cdots \nu_{m_{i_{m\beta_m}}}}{\prod\limits_{k=1}^{m} (|\nu_{k_1}| + \ldots |\nu_{k\alpha_k}|)^{\beta_k}} f^{(\beta_1+\ldots+\beta_m)}_{x_{i_{11}}\ldots x_{i_{m\beta_m}}}$$

$$= \frac{\sum\limits_{i_1=1}^{\alpha_1} \cdots \sum\limits_{i_{1\beta_1}=1}^{\alpha_1} \cdots \sum\limits_{i_m=1}^{\alpha_m} \cdots \sum\limits_{i_{m\beta_m}=1}^{\alpha_m} \nu_{1_{i_{11}}} \cdots \nu_{m_{i_{m\beta_m}}} f^{(\beta_1+\ldots+\beta_m)}_{x_{i_{11}}\ldots x_{i_{m\beta_m}}}}{\prod\limits_{k=1}^{m} (|\nu_{k_1}| + \ldots + |\nu_{k\alpha_k}|)^{\beta_k}}.$$

PROOF. For $m = 1$ (13) follows from Lemma 4. Suppose that (13) holds with $m = r$. Applying Lemma 4 we get

$$f^{(\beta_1+\ldots+\beta_{r+1})}_{x_{\bar{\alpha}_1}\ldots x_{\bar{\alpha}_{r+1}}, |\bar{a}_1, \ldots, \bar{a}_{r+1}|} = [f^{(\beta_1+\ldots+\beta_r)}_{x_{\bar{\alpha}_1}\ldots x_{\bar{\alpha}_r}, |\bar{a}_1, \ldots, \bar{a}_r|}]^{(\beta_{r+1})}_{x_{\bar{\alpha}_{r+1}}, |\bar{a}_{r+1}|}$$

$$= \left( \frac{\sum\limits_{i_{11}=1}^{\alpha_1} \sum\limits_{i_{12}=1}^{\alpha_2} \cdots \sum\limits_{i_{r\beta_r}=1}^{\alpha_r} \nu_{1i_{11}} \cdots \nu_{ri_{r\beta_r}} f^{(\beta_1+\ldots+\beta_r)}_{x_{i_{11}}\ldots x_{i_{r\beta_r}}}}{\prod\limits_{j=1}^{r} (|\nu_{j_1}| + \ldots + |\nu_{j\alpha_j}|)^{\beta_j}} \right)^{(\beta_{r+1})}_{x_{\bar{\alpha}_{r-1}}, |\bar{a}_{r+1}|}$$

$$= \frac{\sum\limits_{i_{(r+1)1}=1}^{\alpha_{r+1}} \cdots \sum\limits_{i_{(r+1)\beta_{r+1}}=1}^{\alpha_{r+1}} \nu_{(r+1)i_{(r+1)i}} \cdots \nu_{(r+1)i_{(r+1)\beta_{r+1}}}}{(|\nu_{(r+1)1}| + \ldots + |\nu_{(r+1)\alpha_{r+1}}|)^{\beta_r+1}} \left( \; \right)^{(\beta_{r+1})}_{x_{i_{(r+1)1}}\ldots x_{i_{(r+1)\beta_{r+1}}}}$$

$$= \frac{\sum\limits_{i_{11}=1}^{\alpha_1} \sum\limits_{i_{12}=1}^{\alpha_1} \cdots \sum\limits_{i_{(r+1)\beta_{r+1}}=1}^{\alpha_{r+1}} \nu_{1i_{11}} \nu_{1i_{12}} \cdots \nu_{(r+1)i_{(r+1)\beta_{r+1}}} f^{(\beta_1+\ldots+\beta_{r+1})}_{x_{11}x_{12}\ldots x_{i_{(r+1)\beta_{r+1}}}}}{\prod\limits_{j=1}^{r+1} (|\nu_{j_1}| + \ldots + |\nu_{j\alpha_j}|)^{\beta_j}}.$$

The theorem is thus proved.

Putting in Theorem 4 $\beta_1 = \ldots = \beta_m = 1$ we obtain

COROLLARY. *The mixed derivative with respect to a system* $(x_{\bar{\alpha}_1}, \ldots, x_{\bar{\alpha}_m})$ *in the direction of vectors* $\bar{a}_1, \ldots, \bar{a}_m$ *can be calculated by the formula*

$$f^{(m)}_{x_{\bar{\alpha}_1}\ldots x_{\bar{\alpha}_m}, |\bar{a}_1,\ldots,\bar{a}_m|} = \frac{\sum\limits_{i_1\in\bar{\alpha}_1} \ldots \sum\limits_{i_m\in\bar{\alpha}_m} \nu_{1i_1} \ldots \nu_{mi_m} f_{x_{i_1}\ldots x_{i_m}}}{\prod\limits_{k=1}^{m} (|\nu_{k_1}| + \ldots + |\nu_{k_{\alpha_k}}|)}.$$

A function $f = f(x_{\bar{\alpha}}, p/x_{\bar{\alpha}})$ will be called *increasing* with respect to a system of variables $x_{\bar{\alpha}}$ in the direction of a vector $\bar{a} = [\nu_1, \ldots, \nu_\alpha]$ if for any $x'_{\bar{\alpha}}$, $x''_{\bar{\alpha}}$ with $x''_{\bar{\alpha}} \geqslant x'_{\bar{\alpha}}$ (i.e. $x''_1 \geqslant x'_1, \ldots, x''_\alpha \geqslant x'_\alpha$),

$$\frac{x''_i - x'_i}{x''_j - x'_j} = \frac{\nu_i}{\nu_j}, \qquad i, j \in \bar{\alpha},$$

the following inequality holds:

$$f(x''_{\bar{\alpha}}, p/x_{\bar{\alpha}}) \geqslant f(x'_{\bar{\alpha}}, p/x_{\bar{\alpha}}).$$

A function $f$ will be called *$\lambda$-increasing* with respect to $x_{\bar{\alpha}}$ if it is increasing with respect to $x_{\bar{\alpha}}$ in the direction of any vector $\bar{a}$ from the cone defined by (10).

LEMMA 5. *If* $f'_{x_{\bar{\alpha}}, [\bar{a}_\lambda]} \geqslant 0$, *then $f$ is $\lambda$-increasing with respect to* $x_{\bar{\alpha}}$.

PROOF. We have to show that for any $x''_{\bar{\alpha}} \geqslant x'_{\bar{\alpha}}$ with

$$0 \leqslant \frac{1}{\lambda} \leqslant \frac{x''_i - x'_i}{x''_j - x'_j} \leqslant \lambda$$

(this condition being assumed throughout the sequel) we have

$$f(x''_{\bar{\alpha}}, p/x_{\bar{\alpha}}) \geqslant f(x'_{\bar{\alpha}}, p/x_{\bar{\alpha}}). \tag{14}$$

Without a loss of generality we may put $\alpha = (1, \ldots, k) = \bar{k}$. Suppose $\Delta x_i > 0$, $i \in \overline{1, k}$. From the relation $f'_{x_{\bar{\alpha}}} \geqslant 0$ we infer that there is $\beta$ such that for $\sum\limits_{i=1}^{k} \Delta x_i \leqslant \beta$ we have

$$f((x+\Delta x)_{\bar{k}}, p/x_{\bar{k}}) \geqslant f(x_{\bar{k}}, p/x_{\bar{k}}). \tag{15}$$

Write $\delta_i = x''_i - x'_i$. Two cases may occur:

(a) $\sum\limits_{i\in\bar{k}} \delta_i \leqslant \beta$. Putting in (15) $\Delta x_i = \delta_i$ we get (14);

(b) $\sum_{i\in\bar{k}} \delta_i > \beta$. Then there exists an integer $m$ such that

$$m\beta \leqslant \sum_{i\in\bar{k}} \delta_i < m\beta+\beta .$$

Putting in (15) $\Delta x_j = \delta_j\beta(\sum_{i\in\bar{k}} \delta_i)^{-1}$ we get

$$f(x_{\bar{k}}'', p/x_{\bar{k}}) \geqslant f\left(x_1''-\frac{\delta_1}{\sum \delta_i}, \ldots, x_k''-\frac{\delta_k}{\sum \delta_i}, p/x_{\bar{k}}\right). \tag{16}$$

Now let us put in (15)

$$x_j = x_j''-\frac{2\delta_j\beta}{\sum \delta_i}, \quad \Delta x_j = \frac{\delta_j\beta}{\sum \delta_i}, \quad j\in\overline{1,k}.$$

We obtain

$$f\left(x_1''-\frac{\alpha_1\beta}{\sum\alpha_i}, \ldots, x_k''-\frac{\alpha_k\beta}{\sum\alpha_i}, p/x_{\bar{k}}\right) \geqslant f\left(x_1''-\frac{2\alpha_1\beta}{\sum\alpha_i}, \ldots, x_k''-\frac{2\alpha_k\beta}{\sum\alpha_i}, p/x_{\bar{k}}\right). \tag{17}$$

Combining (16) with (17) we see that

$$f(x_{\bar{k}}'', p/x_{\bar{k}}) \geqslant f\left(x_1''-\frac{2\alpha_1\beta}{\sum\alpha_i}, \ldots, x_{\bar{k}}''-\frac{2\alpha_k\beta}{\sum\alpha_i}, p/x_{\bar{k}}\right).$$

Repeating this process $m$ times we get

$$f(x_{\bar{k}}'', p/x_{\bar{k}}) \geqslant f\left(x_1''-\frac{m\alpha_1\beta}{\sum\alpha_i}, \ldots, x_{\bar{k}}''-\frac{m\alpha_k\beta}{\sum\alpha_i}, p/x_{\bar{k}}\right). \tag{18}$$

Now, if $m\beta = \sum_{i\in\bar{k}} \delta_i$, then (18) is just (14). Suppose that $m\beta < \sum_{i\in\bar{k}} \delta_i$ and write $\sum_{i\in\bar{k}} \delta_i - m\beta = \beta^0$. Clearly $\beta^0 < \beta$. Put in (15)

$$x_j = x_j', \quad \Delta x_j = \frac{\delta_j\beta^0}{\sum\delta_i},$$

i.e.

$$x_j+\Delta x_j = x_j'+\frac{\delta_j\beta^0}{\sum\delta_i} = x_j''-\frac{m\delta_j\beta}{\sum\delta_i}.$$

It follows:

$$(19) \qquad f\left(x_1'' - \frac{m\delta_1\beta}{\sum \delta_i}, \ldots, x_k'' - \frac{m\delta_k\beta}{\sum \delta_i}, p/x_{\bar{k}}\right) \geqslant f(x_{\bar{k}}', p/x_{\bar{k}}).$$

Inequalities (18) and (19) give (14). The lemma is thus proved.

Let $\Pi_{k,q}^{(\lambda)}$ denote the class of functions $f$ satisfying inequality (1) for any $x_{\bar{k}}'' \geqslant x_{\bar{k}}'$, $y_{\bar{q}}'' \geqslant y_{\bar{q}}'$ such that

$$(20) \qquad 0 \leqslant \frac{1}{\lambda} \leqslant \frac{x_1'' - x_i'}{x_j'' - x_j'} \leqslant \lambda, \quad i, j \in \overline{1, n}.$$

Clearly $\Pi_{k,q}^{(\lambda_2)} \subset \Pi_{k,q}^{(\lambda_1)}$ for $\lambda_1 < \lambda_2$, i.e. the class $\Pi_{k,q}^{(\lambda)}$ increases when $\lambda$ decreases. For $\lambda = \infty$ the class $\Pi_{k,q}^{(\infty)}$ coincides with $\Pi_{k,q}$.

We shall write $\Pi_{k,q}^{(\lambda)}(Q)$ for functions defined in a domain $Q$. Assume $Q = [a_1, b_1; \ldots; a_n, b_n]$.

THEOREM 5. *Let $f \in \Pi_{k,q}^{(\lambda)}(Q)$ be a continuous function. Then the distance $E_f$ is expressed by the formula*

$$E_f = \tfrac{1}{4}[f(b_{\bar{k}}, b_{\bar{q}}) + f(a_{\bar{k}}, a_{\bar{q}}) - f(b_{\bar{k}}, a_{\bar{q}}) - f(a_{\bar{k}}, b_{\bar{q}})].$$

*If there exist at least two points $(x_{\bar{k}}^{00}, y_{\bar{q}}^{00})$ and $(x_{\bar{k}}^{0}, y_{\bar{q}}^{0})$ with $x_{\bar{k}}^{00} \geqslant x_{\bar{k}}^{0}$, $y_{\bar{q}}^{00} \geqslant y_{\bar{q}}^{0}$ and such that inequality* (1) *is strict at those points, then a best approximation is not unique.*

The proof of Theorem 5 is much the same as the proof of the analogous result concerning the class $\Pi_{k,q}(Q)$, due to Babaev [7]. Repeating that argument one just has to keep in mind that the coordinates of any pair of points $(x_{\bar{k}}'', y_{\bar{q}}'')$, $(x_{\bar{k}}', y_{\bar{q}}')$ at the parallelepiped $Q$ fulfil condition (20).

LEMMA 6. *If $f''_{x_{\bar{k}} y_{\bar{q}}, [\bar{a}\lambda]} \geqslant 0$, then $f \in \Pi_{k,q}^{(\lambda)}$.*

PROOF. By the definition of the mixed derivative

$$f''^{(\lambda)}_{x_{\bar{k}} y_{\bar{q}}} = (f'^{(\lambda)}_{x_{\bar{k}}})'^{(\lambda)}_{y_{\bar{q}}} \geqslant 0.$$

Thus, by Lemma 5, the function $f'^{(\lambda)}_{x_{\bar{k}}}$ is $\lambda$-increasing with respect to $y_{\bar{q}}$, i.e. for any $y_{\bar{q}}'' \geqslant y_{\bar{q}}'$ we have

$$0 \leqslant f'_{x_{\bar{k}}}(x_{\bar{k}}, y_{\bar{q}}'') - f'_{x_{\bar{k}}}(x_{\bar{k}}, y_{\bar{q}})$$
$$= \lim_{\substack{\Delta x_i \to 0 \\ i \in \bar{k}}} \frac{f((x+\Delta x)_k, y_{\bar{q}}'') + f(x_{\bar{k}}, y_{\bar{q}}') - f((x+\Delta x)_{\bar{k}}, y_{\bar{q}}') - f(x_{\bar{k}}, y_{\bar{q}}'')}{\sum_{i \in \bar{k}} \Delta x_i}.$$

The proof of Lemma 6 is then concluded analogously to that of Lemma 5.

REMARK. It can be seen from Lemma 6 that a sufficient condition for $f$ to belong to $\Pi_{k,q}^{(\lambda)}$ is the non-negativity of the mixed derivative with respect to vectors with positive coordinates only.

The following theorem is one of the results of the above quoted paper.

THEOREM 6. *If* $f \in C(Q)$, $f''_{x_{\bar{k}}, y_{\bar{q}}, [\bar{a}_\lambda]} \in C(Q)$, $Q = [a_1, b_1; \ldots; a_n, b_n]$, *then the following estimations hold and are exact*:

$$|L_f| \leqslant E_f \leqslant \frac{S}{2} \sum_{i=1}^{k} \sum_{j=k+1}^{n} (b_i - a_i)(b_j - a_j) + L_f, \tag{21}$$

*where*

$$L_f = \tfrac{1}{4}[f(b_{\bar{k}}, b_{\bar{q}}) + f(a_{\bar{k}}, a_{\bar{q}}) - f(b_{\bar{k}}, a_{\bar{q}}) - f(a_{\bar{k}}, b_{\bar{q}})],$$

$$S = 0 \quad \textit{if} \quad f''_{x_{\bar{k}} y_{\bar{q}}, [\bar{a}_\lambda]} \geqslant 0,$$

$$S = \sup_{[\bar{a}_1, \bar{a}_2] \subset [\bar{a}_\lambda]} \; \sup_{E\{p:\, f''_{x_{\bar{k}} y_{\bar{q}}, [\bar{a}_1, \bar{a}_2]}(p) < 0\}} |f''_{x_{\bar{k}} y_{\bar{q}}, [\bar{a}_1, \bar{a}_2]}(p)|$$

*if* $f''_{x_{\bar{k}} y_{\bar{q}}, [\bar{a}_\lambda]}$ *has also negative values.*

(*The supremum is taken with respect to all pairs of positive vectors* $\bar{a}_1 = [\nu_1, \ldots, \nu_k]$, $\bar{a}_2 = [\nu_{k+1}, \ldots, \nu_n]$ *satisfying condition* (10).)

PROOF. Put

$$g = S \sum_{r=1}^{k} \sum_{s=k+1}^{n} x_r x_s, \quad f_1 = f + g, \quad f_2 = -g.$$

Clearly $f = f_1 + f_2$.

We shall employ the following inequalities (see [7]):

$$|E_{f_1} - E_{f_2}| \leqslant E_f \leqslant E_{f_1} + E_{f_2}. \tag{22}$$

Let $\bar{a}_1 = [\nu_1, \ldots, \nu_k]$, $\bar{a}_2 = [\nu_{k+1}, \ldots, \nu_n]$ be any two vectors from the cone (10). Then, applying the formula for the mixed derivative with respect to systems of variables in the direction of several vectors (corollary to Theorem 4), we obtain

$$g''_{x_{\bar{k}}y_{\bar{q}},[\bar{a}_1,\bar{a}_2]} = \frac{\sum_{i=1}^{k}\sum_{j=k+1}^{n} \nu_i\nu_j\, g''_{x_i x_j}}{(\nu_1+\ldots+\nu_k)(\nu_{k+1}+\ldots+\nu_n)}$$

$$= S\frac{\sum_{i=1}^{k}\sum_{j=k+1}^{n} \nu_i\nu_j \left[\sum_{r=1}^{k}\sum_{s=k+1}^{n} x_r x_s\right]''_{x_i x_j}}{(\nu_1+\ldots+\nu_k)(\nu_{k+1}+\ldots+\nu_n)}$$

$$= S\frac{\sum_{i=1}^{k}\sum_{j=k+1}^{n} \nu_i\nu_j}{(\nu_i+\ldots+\nu_k)(\nu_{k+1}+\ldots+\nu_n)}$$

$$= S\frac{\sum_{i=1}^{k}\nu_i \sum_{j=k+1}^{n} \nu_j}{(\nu_1+\ldots+\nu_k)(\nu_{k+1}+\ldots+\nu_n)} = S.$$

Hence

$$(f_1)''_{x_{\bar{k}}y_{\bar{q}},[\bar{a}_1,\bar{a}_2]} = f''_{x_{\bar{k}}y_{\bar{q}},[\bar{a}_1,\bar{a}_2]} + g''_{x_{\bar{k}}y_{\bar{q}},[\bar{a}_1,\bar{a}_2]} = f''_{x_{\bar{k}}y_{\bar{q}},[\bar{a}_1,\bar{a}_2]} + S \geqslant 0,$$

$$(f_2)''_{x_{\bar{k}}y_{\bar{q}},[\bar{a}_1,\bar{a}_2]} = -S \leqslant 0.$$

Since $\bar{a}_1$ and $\bar{a}_2$ were arbitrary in the cone (10), thus

$$(f_1)''_{x_{\bar{k}}y_{\bar{q}},[\bar{a}_\lambda]} \geqslant 0, \qquad (f_2)''_{x_{\bar{k}}y_{\bar{q}},[\bar{a}_\lambda]} < 0.$$

Consequently, by Lemma 6, $f_1 \in \Pi^{(\lambda)}_{k,q}(Q)$. Hence, applying Theorem 5, we get $E_{f_1} = L_{f_1}$. It can be easily verified, that if $f''_{x_{\bar{k}}y_{\bar{q}},[\bar{a}_\lambda]} \leqslant 0$, then $E_f = -L_{k,q}(f)$. Applying this fact to the function $f_2$ we obtain

$$E_{f_2} = \frac{S}{4}\sum_{i=1}^{k}\sum_{j=k+1}^{n} (b_i-a_i)(b_j-a_j).$$

Substituting these values in (22) we get (21), concluding the proof.

REMARK. It is not difficult to see, that if, under conditions of Theorem 6, $f''_{x_{\bar{k}}y_{\bar{q}},[\bar{a}_\lambda]} \geqslant 0$, then by Theorem 5 inequalities (21) are actually equations. This proves the exactness of estimation (21).

COROLLARY 1. *Under conditions of Theorem* 6 *the following estimations hold and are exact*:

$$|L_f| \leqslant E_f \leqslant \frac{M}{2}\sum_{i=1}^{k}\sum_{j=k+1}^{n} (b_i-a_i)(b_j-a_j) - |L_f|,$$

*where*

$$M = \sup_{[\bar{a}_1, \bar{a}_2] \subset [\bar{a}_\lambda]} \sup_{p \in Q} |f''_{x_{\bar{k}} y_{\bar{q}}, [\bar{a}_\lambda]}(p)|.$$

COROLLARY 2. *If, under conditions of Theorem* 6, $f''_{x_{\bar{k}} y_{\bar{q}}, [\bar{a}_\lambda]}$ *has a constant sign, then*

$$E_f \leqslant \frac{M}{4} \sum_{i=1}^{k} \sum_{j=k+1}^{n} (b_i - a_i)(b_j - a_j),$$

*where M is the constant defined above.*

## REFERENCES


[1] M.-B. A. Babaev, *Dokl. Akad. Nauk Azerb.SSR* **23** (1967).

[2] M.-B. A. Babaev, *Dokl. Akad. Nauk Azerb.SSR* **23** (1967).

[3] M.-B. A. Babaev, *Izv. Akad. Nauk Azerb.SSR* **6** (1970).

[4] M.-B. A. Babaev, paper in the collection: *Special Problems of Differential Equations and Theory of Functions*, Baku 1970, 3–44.

[5] M.-B. A. Babaev, *Dokl. Akad. Nauk SSSR* **5** (1970) 193, 967.

[6] M.-B. A. Babaev, Proceedings of the Conference in Honour of the 50th Anniversary of the Establishment of Soviet Government in Azerbaijan and the 50th Anniversary of the Azerbaijan Communist Party (October 1970), Baku 1971, 31.

[7] M.-B. A. Babaev, *Mat. Zametki* (1972).

[8] S. P. Diliberto and E. G. Straus, *Pacific J. Math.* **1** (1951) 195.

[9] Leopold Flatto, *Amer. Math. Monthly* **73** (1966) 131.

[10] S. Ya. Havinson, *Izv. Akad. Nauk SSSR* **33** (1969) 650–666.

[11] I. I. Ibragimov and M.-B. A. Babaev, *Dokl. Akad. Nauk SSSR* **197** 166 (1971).

[12] I. I. Ibragimov and M.-B. A. Babaev, *Dokl. Akad. Nauk SSSR* **201** (1971) 1037.

[13] M. P. Šura-Bura, *Vyčisl. Matematika* **1** (1957) 3–19.

[14] V. M. Mordashev, *Dokl. Akad. Nauk SSSR* **183** (1968) 779.

[15] V. M. Mordashev, *Atomnaya Energiya* **22** (1967).

[16] V. M. Mordashev, *Atomnaya Energiya* **28** (1970).

[17] Yu. P. Ofman, *Izv. Akad. Nauk SSSR* **25** (1961) 239–252.

[18] T. J. Rivlin and R. J. Sibner, *Amer. Math. Monthly* **72** (1965) 1101.

[19] A. I. Vaindiner, *Dokl. Akad. Nauk SSSR* **192** (1970) 483.

# THE CONVERGENCE OF BERGMAN FUNCTIONS FOR A DECREASING SEQUENCE OF DOMAINS

T. IWIŃSKI and M. SKWARCZYŃSKI

*Warszawa*

In the present paper we shall be concerned with a bounded domain $D$ in the complex plane. In 1967 the following theorem was proved by I. Ramadanov [4]

THEOREM 1. *Let the bounded domain $D \subset \boldsymbol{C}$ be the union of an increasing sequence of domains $D_n$, $n = 1, 2, \ldots$ Then the sequence of Bergman functions $K_{D_n}(z, \bar{t})$ converges to $K_D(z, \bar{t})$ locally uniformly in $D \times D^*$, where $D^* = \{\bar{t}, t \in D\}$.*

The definition and properties of the Bergman function can be found in [1]. The above theorem suggests the study of the sequence $K_{D_n}$ for a decreasing sequence of domains $D_n$. In order to exclude slit domains from our considerations, e.g. a disc with one radius removed, we make an additional assumption

$$D = \operatorname{int}(\bar{D}).$$

We need two definitions.

DEFINITION 1. We say that the sequence $D_n$ *approximates a domain $D$ from outside*, if for each open $G$ such that $\bar{D} \subset G$ the inclusion

$$D \subset D_n \subset G$$

holds for sufficiently large $n$.

DEFINITION 2. A domain $D \subset \boldsymbol{C}$ is called *Ramadanov domain* if for every decreasing sequence $D_n$ approximating $D$ from outside the sequence $K_{D_n}(z, \bar{t})$ converges to $K_D(z, \bar{t})$, the Bergman function of $D$, locally uniformly in $D \times D^*$.

For a Lebesgue measurable set $E \subset \boldsymbol{C}$ we shall denote by $L^2H(E)$ the Hilbert space of all complex functions square integrable on $E$ and holomorphic in the interior of $E$. The set of functions defined in $E$, and possessing a holomorphic extension to an open neighborhood $G \supset E$ will be denoted by $H(E)$.

It is not difficult to prove the following lemma [6]

LEMMA 1. *For an arbitrary domain $D$ and $t \in D$ the function $K_D(\cdot, \bar{t})$ is uniquely characterized as an element $\varphi \in L^2H(D)$ which satisfies the conditions*

(a) $\varphi(t) \geqslant \|\varphi\|^2$,

(b) *If $f \in L^2H(D)$ and $f(t) \geqslant \|f\|^2$, then $\varphi(t) \geqslant f(t)$.*

Locally uniform convergence of the sequence $K_{D_n}$ is equivalent to the pointwise convergence on the "diagonal". Precisely

THEOREM 2. *Let $D_n$ be a decreasing sequence of domains and let $D$ be contained in every $D_n$. The sequence $K_{D_n}$ converges to $K_D$ locally uniformly in $D \times D^*$ if and only if for every $t \in D$*

$$\lim_{n\to\infty} K_{D_n}(t, \bar{t}) = K_D(t, \bar{t}).$$

PROOF. The necessity is obvious. We shall prove sufficiency of the condition. For an arbitrary compact $F \subset D$ there exists a constant $M$ such that $K_D(s, s) \leqslant M$ for $s \in F$. Therefore by Schwarz inequality for $(z, t) \in F \times F$

$$|K_{D_n}(z, \bar{t})| \leqslant K_{D_n}(z, \bar{z})^{1/2} K_{D_n}(t, \bar{t})^{1/2} \leqslant K_D(z, \bar{z})^{1/2} K_D(t, \bar{t})^{1/2} \leqslant M.$$

Hence $K_{D_n}$ is a Montel family on $D \times D^*$. It is now sufficient to show that every convergent subsequence of this family converges to $K_D$. With no loss of generality we consider the sequence $K_{D_n}$ itself, assuming that it is convergent to a function $k$, holomorphic in $D \times D^*$. For every compact $F \subset D$ and $t \in D$ we have

$$\int_F |k(\cdot, \bar{t})|^2 = \lim_{n\to\infty} \int_F |K_{D_n}(\cdot, \bar{t})|^2 \leqslant \lim_{n\to\infty} \int_{D_n} |K_{D_n}(\cdot, \bar{t})|^2 = \lim_{n\to\infty} K_{D_n}(t, \bar{t})$$
$$= k(t, \bar{t}).$$

Since $F$ is arbitrary $\|k(\cdot, \bar{t})\|^2 \leqslant k(t, \bar{t}) = K_D(t, \bar{t})$. By Lemma 1 $K_D(z, \bar{t}) = k(z, \bar{t})$, and the proof is completed.

Now we can prove that our problem is related to the following

CONDITION A. *$H(\bar{D})$ is dense in $L^2H(\bar{D})$.*

Next two lemmas explain this relation

LEMMA 2. *Suppose Condition* A *is satisfied. Then* $D$ *is Ramadanov domain.*

PROOF. Consider sequence $D_n$ approximating $D$ from outside. In view of Theorem 2 it is enough to show that $\lim K_{D_n}(t,\bar{t}) = K_D(t,\bar{t})$. Fix an arbitrary $t \in D$ and positive $\varepsilon$. Extend the function $K_D(\cdot, t)$ to an element in $L^2H(\bar{D})$ by setting $K_D(z,\bar{t}) = 0$ for $z \in \operatorname{bd} D$. Since Condition A holds we can find an element $h \in H(\bar{D})$ so close to $K_D(\cdot, t)$ that

$$\frac{|h(t)|^2}{\|h\|_{\bar{D}}^2} > K_D(t,\bar{t}) - \varepsilon/2.$$

Now for sufficiently large $n$

$$h \in L^2H(D_n) \quad \text{and} \quad \lim_{n\to\infty} \frac{|h(t)|^2}{\|h\|_{\bar{D}_n}^2} = \frac{|h(t)|^2}{\|h\|_{\bar{D}}^2}.$$

Hence

$$\frac{|h(t)|^2}{\|h\|_{\bar{D}_n}^2} > K_D(t,\bar{t}) - \varepsilon.$$

We combine this inequality with $|h(t)|^2 \leqslant K_{D_n}(t,\bar{t})\, \|h\|_{\bar{D}_n}^2$ to obtain

$$0 < K_D(t,\bar{t}) - K_{D_n}(t,\bar{t}) < \varepsilon$$

for large $n$, and the lemma is proved.

LEMMA 3. *Let* $D$ *be Ramadanov domain with boundary of the plane Lebesgue measure zero. Then Condition* A *is satisfied.*

PROOF. It is well known that the set $\{K_D(\cdot, t), t \in D\}$ is dense in $L^2H(D)$ and therefore in $L^2H(\bar{D})$. Choose the sequence $D_n$ approximating $D$ from outside such that $\bar{D} \subset D_n$, $n = 1, 2, \ldots$ The fact that $K_{D_n}(t,\bar{t})$ converges to $K_D(t,\bar{t})$ immediately implies that $K_{D_n}(\cdot,\bar{t})$ converges to $K_D(\cdot\,\bar{t})$ in $L^2$ norm. Since $K_{D_n}(\cdot, t) \in H(\bar{D})$ the proof is completed.

The class of domains for which Condition A is satisfied was investigated and recently characterized by S. Sinanian [5], W. Havin [2] and L. Hedberg [3]. We recall two following theorems

THEOREM 3 (Sinanian). *Consider a closed bounded set* $E \subset \mathbf{C}$. *If* $\mathbf{C} \setminus E$ *has finite number of components, then* $H(E)$ *is dense in* $L^2H(E)$.

THEOREM 4 (Havin). *Let $E$ be a bounded Borel set on the plane. The space $H(E)$ is dense in $L^2H(E)$ if and only if the set of all points at the boundary* bd $E$ *which do not belong to the fine closure of the complement $\mathbf{C} \setminus E$ has zero logarithmic capacity.*

These theorems yield our final

COROLLARY. *Consider a domain $D$ equal to the interior of its closure, with boundary of the zero Lebesgue measure. $D$ is Ramadanov domain if and only if the set of all points at the boundary* bd $D$ *which do not belong to the fine closure of the exterior of $D$ has zero logarithmic capacity.*

*In particular, if the complement of the closure of $D$ has finite number of components, then $D$ is Ramadanov domain.*

A well-known example of domain which does not satisfy Condition A is constructed from the unit disc by removing a closed segment and a sequence of pairwise disjoint closed discs which accumulate at every point of this segment, and such that the sequence of radii converges rapidly to zero. As our corollary shows this is also an example of non-Ramadanov domain.

## REFERENCES


[1] S. Bergman, 'Kernel Function and Conformal Mapping', *Amer. Math. Soc., Math. Surveys* **5** (1950).

[2] V. Havin, 'Approximation in Mean by Analytic Functions', *Dokl. Akad. Nauk* **178** (1968) 1025–1029.

[3] L. Hedberg, 'Approximation in the Mean by Analytic Functions', *Trans. Amer. Math. Soc.* **163** (1972) 157–171.

[4] I. Ramadanov, 'Sur une propriété de la fonction de Bergman', *C. R. Acad. Sci.* **20**, No 8 (1967).

[5] S. Sinanian, 'Approximation in Mean by Analytic Functions and Polynomials on the Complex Plane', *Mat. Sbor.* **69** (1966) 111.

[6] M. Skwarczyński, 'Bergman Function and Conformal Mapping in Several Complex Variables', Thesis, Warsaw University 1969.

# $n$-PARAMETRIGE HALBGRUPPEN VON OPERATOREN IN DER APPROXIMATIONSTHEORIE

WALTER KÖHNEN

*Neuß*

*Abstract.* Let $\{T(t); t \in \overline{E_n^+}\}$ be a uniformly bounded $n$-parameter semi-group of linear bounded transformations of class $C^0$ on the Banach space $X$. For $n = 1$ the degree of approximation of a given element by $T(t)f$ in the $X$-norm for small values of $t$ has been discussed by many authors, especially by P. L. Butzer [2]. Some of these results have been carried over to the case $n \geqslant 2$ by O. A. Ivanova [5], [6]. In the following note we show that parts of Ivanova's results can be much improved. We prove a theorem, which could be called a *saturation theorem*, as well as approximation theorems for Taylor and Riemann operators (Theorems 1 and 2). Furthermore we consider adjoint semi-groups, and analyse their approximation behaviour in the strong sense to the identity (Theorems 3 and 4).

**1. Einleitung.** Es sei $E_n$ der $n$-dimensionale euklidische Raum mit den Elementen $t, s, \ldots$ Wir schreiben $t = (t_1, t_2, \ldots, t_n)$ und bezeichnen mit $e_1, e_2, \ldots, e_n$ die Einheitsvektoren von $E_n$. Ferner setzen wir $\overline{E_n^+} = \{t \in E_n; t_k \geqslant 0, k = 1, 2, \ldots, n\}$. Mit $X$ bezeichnen wir einen komplexen Banachraum und mit $\mathfrak{E}(X)$ die Banachalgebra der Endomorphismen von $X$. Eine Funktion $T(t)$, definiert auf $\overline{E_n^+}$, mit Werten in $\mathfrak{E}(X)$ heißt gleichmäßig beschränkte $n$-parametrige Halbgruppe von Operatoren der Klasse $C^0$ (i.f. kurz $n$-parametrige Halbgruppe genannt), wenn gilt:

1) $T(t+s) = T(t)T(s) \quad (t, s \in \overline{E_n^+})$;
2) $T(0) = I$;
3) $s \cdot \lim\limits_{\substack{t \in \overline{E_n^+} \\ t \to 0}} T(t)f = f \quad (f \in X)$;
4) $\|T(t)\| < K \quad (t \in \overline{E_n^+})$.

Der Erzeuger $A(t)$ von $\{T(t); t \in \overline{E_n^+}\}$ im Punkte $t \in \overline{E_n^+}$ ist wie folgt erklärt: Der Definitionsbereich $D(A(t))$ von $A(t)$ besteht aus allen Elementen $f \in X$, für welche $s \cdot \lim\limits_{\eta \to 0^+} \big(T(\eta t)f - f\big)/\eta$ existiert. Für diese $f$

setzen wir

$$A(t)f = s \cdot \lim_{\eta \to 0^+} \big(T(\eta t)f - f\big)/\eta .$$

Zur Abkürzung schreiben wir $A_k$ anstelle von $A(e_k)$ für $k = 1, 2, \ldots, n$. Für den Fall $n = 1$ gibt es eine Vielzahl von Arbeiten, in denen das Verhalten von

$$\|T(t)f - f\|$$

für $t \to 0$ und damit verwandte Fragen in Abhängigkeit von den strukturellen Eigenschaften von $f$ untersucht werden. Sätze dieser Art sind vor allem von P. L. Butzer bewiesen worden. Man vergleiche dazu [2] und die dort zitierte Literatur.

Die ersten Approximationssätze für $n$-parametrige Halbgruppen ($n > 1$) stammen von O. A. Ivanova ([5] und [6]). In der ersten Arbeit [5] überträgt sie Resultate von P. L. Butzer und H. G. Tillmann [3] von $n = 1$ auf $n > 1$. In [6] beweist sie einen Approximationssatz für adjungierte $n$-parametrige Halbgruppen. Dieser Satz basiert auf einem Theorem von K. de Leeuw [7] für adjungierte einparametrige Halbgruppen.

Wir zeigen nun, daß die Ergebnisse von O. A. Ivanova wesentlich verallgemeinert werden können. An einigen Stellen wenden wir dazu eine Methode an, die H. Berens [1] bereits für $n = 1$ benutzt hat. Alle unsere Resultate formulieren wir nur für $n = 2$. Die Übertragung auf $n > 2$ ist problemlos.

**2. Approximationssätze.** Zuerst führen wir einige abkürzende Bezeichnungen ein. Für $t_j > 0$, $f \in D(A_j^{r-1})$, $j = 1, 2$ und $r = 1, 2, 3, \ldots$ setzen wir

$$B_{t_j}^r f = \frac{r!}{t_j^r}\left[T(t_j e_j) - \sum_{i=0}^{r-1} \frac{t_j^i}{i!} A_j^i\right] f.$$

Für $f \in D(A_1^{n-1}) \cap D(A_1^{n-1} A_2^{m-1})$, $n, m = 0, 1, 2, \ldots$ und $t_1, t_2 > 0$ definieren wir:

$$B_{t_1,t_2}^{n,m} f = \frac{n!\,m!}{t_1^n t_2^m}\left[T(t_1 e_1 + t_2 e_2) + \sum_{k=1}^{n-1}\sum_{i=0}^{m-1} \frac{t_1^k t_2^i}{k!\,i!} A_1^k A_2^i - \right.$$

$$\left. - \sum_{k=0}^{n-1} \frac{t_1^k}{k!} T(t_2 e_2) A_1^k - \sum_{i=0}^{m-1} \frac{t_2^i}{i!} T(t_1 e_1) A_2^i \right] f.$$

Für $t_j > 0, j = 1, 2$ und r = 0, 1, 2, ... setzen wir

$$C_{t_j}^r = t_j^{-r}[T(t_j e_j) - I]^r.$$

Mit $\tilde{D}(A_1^n A_2^m)$, $n, m = 0, 1, 2, \ldots$, bezeichnen wir die Menge der Elemente $f$ von $X$, für die es eine Folge $f_k$ gibt, so daß $s \cdot \lim_{k\to\infty} f_k = f$ und $\|A_1^n A_2^m f_k\| < C$ gilt.

Bevor wir unseren ersten Satz beweisen, zitieren wir noch ein Ergebnis aus [2], S.203, das wir oft benutzen.

Lemma 1. *Für ein* $f \in X$, $r = 1, 2, \ldots$ *und* $j = 1, 2$ *sind äquivalent*:

1) $\sup_{0<t_j<\infty} \|C_{t_j}^r f\| < L_1 < \infty$;

2) $f \in D(A_j^{r-1})$ *und* $\sup_{0<t_j<\infty} \|B_{t_j}^r f\| < L_2 < \infty$.

Hängt dabei $f$ von einem Parameter $\alpha$ ab und ist $L_1$ unabhängig von $\alpha$, dann ist auch $L_2$ unabhängig von $\alpha$ und umgekehrt.

Erwähnt sei schließlich noch, daß wir die Theorie der $n$-parametrigen Halbgruppenoperatoren, wie sie in [4], S. 327–336 dargestellt ist, als bekannt voraussetzen.

Satz 1. *Für ein* $f \in X$ *und alle* $n, m = 0, 1, 2, \ldots$ *sind äquivalent*:

1) $\sup_{0<t_1, t_2<\infty} \|C_{t_1}^n C_{t_2}^m f\| < \infty$;

2) $f \in \tilde{D}(A_1^n A_2^m)$;

3) $f \in D(A_1^n A_2^m)$, *falls* $X$ *reflexiv ist*.

Beweis. Es gelte 1). Wir bilden für $k = 1, 2, 3, \ldots$

$$f_k = k^{n+m} \int_0^{1/k} \ldots \int_0^{1/k} T\big((\xi_1 + \ldots + \xi_n)e_1 + (\eta_1 + \ldots + \eta_m)e_2\big) f d\xi_1 \ldots$$
$$\ldots d\xi_n d\eta_1 \ldots d\eta_m.$$

Dann gilt $s \cdot \lim_{k\to\infty} f_k = f$ sowie $f_k \in D(A_1^n A_2^m)$ für alle $k = 1, 2, \ldots$ und

$$A_1^n A_2^m f_k = C_{(1/k)_1}^n C_{(1/k)_2}^m f.$$

Damit erhalten wir 2).

Es gelte nun 2). Für $f_k \in D(A_1^n A_2^m)$ haben wir

$$C_{t_1}^n C_{t_2}^m f_k = t_1^{-n} t_2^{-m} \int_0^{t_1} \ldots \int_0^{t_1} \int_0^{t_2} \ldots \int_0^{t_2} T\big((\xi_1 + \ldots + \xi_n)e_1 +$$
$$+ (\eta_1 + \ldots + \eta_m)e_2\big) A_1^n A_2^m f_k d\xi_1 \ldots d\xi_n d\eta_1 \ldots d\eta_m.$$

Daraus folgt $\|C_{t_1}^n C_{t_2}^m f_k\| \leqslant K\|A_1^n A_2^m f_k\|$, woraus sich für $k \to \infty$ die Behauptung 1) ergibt.

Da die Operatoren $A_1^n A_2^m$ abgeschlossen sind und einen dichten Definitionsbereich in $X$ haben, ergibt sich die Äquivalenz von 2) und 3) unmittelbar aus einem Satz von H. Berens [1], S. 20.

BEMERKUNG. Da die Operatoren $C_{t_1}^n$ und $C_{t_2}^m$ kommutativ sind, erhalten wir eine Vielzahl weiterer Äquivalenzen. Insbesondere folgt, falls $X$ reflexiv ist, aus $f \in D(A_1^n A_2^m)$ ohne jede weitere Voraussetzung $f \in D(A_2^m A_1^n)$. Damit erhält man eine interessante Variante des Theorems 10.9.4 aus [4].

SATZ 2. *Für ein $f \in X$ und alle $n, m = 0, 1, 2, \ldots$ sind äquivalent:*

1) $\sup\limits_{0<t_1,t_2<\infty} \|C_{t_1}^n C_{t_2}^m f\| < \infty$;

2) $\sup\limits_{0<t_1,t_2<\infty} \|B_{t_1,t_2}^{n,m} f\| < \infty$;

BEWEIS. Es gelte 1). Wir zeigen zuerst, daß für alle $j$ und $k$, $0 \leqslant j \leqslant n$, $0 \leqslant k \leqslant m$,

$$\sup_{0<t_1,t_2<\infty} \|C_{t_1}^j C_{t_2}^k f\| < \infty \tag{1}$$

gilt. Aus Satz 1 folgt nämlich $f \in \tilde{D}(A_1^n A_2^m)$, woraus sich wegen der Abgeschlossenheit der Operatoren $A_1^j A_2^k$ die Beziehung $f \in \tilde{D}(A_1^j A_2^k)$ ergibt. Wendet man nun Satz 1 für $n = j$ und $m = k$ wieder an, so erhält man (1). Es gilt weiter $f \in D(A_1^{n-1}) \cap D(A_2^{m-1})$. Um dies zu zeigen, setzt man in (1) $j = 0$. Dann erhält man mit Hilfe von Satz 1 zunächst $f \in \tilde{D}(A_2^m)$ und dann $\sup\limits_{0<t_2<\infty} \|C_{t_2}^m f\| < \infty$. Aus der letzten Aussage erhalten wir aber wegen Lemma 1, daß $f \in D(A_2^{m-1})$ ist. Ganz entsprechend bekommen wir $f \in D(A_1^{n-1})$, wenn wir in (1) $k = 0$ setzen. Nun beweisen wir die gewünschte Aussage zunächst für $n = 2$ und $m = 2$. Aus Lemma 1 erhalten wir für alle $t_1$, $0 < t_1 < \infty$,

$$\sup_{0<t_2<\infty} \|B_{t_2}^2 C_{t_1}^2 f\| < M_1 < \infty,$$

wobei $M_1$ unabhängig von $t_1$ ist. Da $f \in D(A_2)$ gilt, können wir die Operatoren $B_{t_2}^2$ und $C_{t_1}^2$ vertauschen. Dann ergibt eine erneute Anwendung von Lemma 1 einmal

$$\sup_{0<t_1<\infty} \|B_{t_1}^2 B_{t_2}^2 f\| < M_2 < \infty$$

für alle $t_2$, $0 < t_2 < \infty$, und einem von $t_2$ unabhängigen $M_2$ und zum anderen $B_{t_2}^2 f \in D(A_1)$. Da $f \in D(A_1)$ gilt, ist somit auch $f \in D(A_1 A_2)$. Analog können wir auf $f \in D(A_2 A_1)$ schließen. Damit erhält man dann unmittelbar

$$\sup_{0<t_1, t_2<\infty} \|B_{t_1, t_2}^{2,2} f\| < \infty .$$

Es ist klar, wie man ganz entsprechend unsere Aussage für $n \leqslant 2$ und $m \leqslant 2$ erhält. Um 2) auch für die anderen $n$ und $m$ zu zeigen, gehen wir schrittweise vor. Ist $n = 3$ und $m = 2$, so können wir mit (1) und dem soeben bewiesenen Resultat zunächst auf $f \in D(A_2 A_1^2) \cap D(A_1^2 A_2)$ schließen. Genau dieselbe Schlußweise wie für den Fall $n = 2$, $m = 2$ liefert uns dann 2) für $n = 3$, $m = 2$. Analog findet man den Beweis für $n = 2$, $m = 3$. Die Anwendung der letzten beiden Ergebnisse liefert uns dann den Beweis für $n = 3$, $m = 3$. Damit ist aber nun klar, wie man schrittweise 2) für alle anderen $n$ und $m$ erhalten kann.

Es gelte nun 2). Wir können schreiben

$$B_{t_1, t_2}^{n,m} f = B_{t_1}^n B_{t_2}^m f .$$

Wenden wir Lemma 1 an, dann ergibt sich für alle $t_2$, $0 < t_2 < \infty$,

$$\sup_{0<t_1<\infty} \|C_{t_1}^n B_{t_2}^m f\| < N_1 < \infty ,$$

wobei $N_1$ unabhängig von $t_2$ ist. Die Operatoren $C_{t_1}^n$ und $B_{t_2}^m$ sind kommutativ. Nach ihrer Vertauschung ergibt eine erneute Anwendung von Lemma 1

$$\sup_{0<t_1, t_2<\infty} \|C_{t_1}^n C_{t_2}^m f\| < \infty .$$

Wir kommen nun zur Untersuchung der adjungierten Halbgruppe $\{T^*(t); t \in \overline{E_n^+}\}$. Die Hauptschwierigkeit dabei ist, daß nicht mehr $s\cdot \lim_{t \in \overline{E_n^+}, t \to 0} T^*(t) f^* = f^*$ für alle $f^*$ aus dem adjungierten Raum $X^*$, sondern lediglich $\lim_{t \in \overline{E^+},\, t \to 0} \langle T^*(t) f^*, f \rangle = \langle f^*, f \rangle$ für alle $f \in X$ und alle $f^* \in X^*$ gilt.

Das folgende Lemma von K. de Leeuw [7] wird in unseren Beweisen einige Male benutzt:

LEMMA 2. *Für ein* $f^* \in X^*$ *und* $j = 1, 2$ *sind äquivalent*:

1) $\sup_{0<t_j<\infty} \|(C_{t_j}^1)^* f^*\| < \infty$ (1);

2) $f^* \in D(A_j^*)$.

Nun können wir zeigen:

SATZ 3. *Aus* $f^* \in D(A_1^*) \cap D(A_2^*)$ *folgt*

$$\sup_{0<t_1,t_2<\infty} \|(C_{t_1}^1)^*(C_{t_2}^1)^* f^*\| < \infty \quad \textit{und} \quad \sup_{0<t_j<\infty} \|(C_{t_j}^1)^* f^*\| < \infty, \quad j = 1, 2.$$

BEWEIS. Aus Lemma 2 folgt unmittelbar der zweite Teil unseres Satzes. Aus der Voraussetzung ergibt sich weiterhin sofort $f \in D\big((A_2 A_1)^*\big) \cap \cap D\big((A_1 A_2)^*\big)$. Damit erhalten wir für alle $f \in X$

$$\begin{aligned}\langle (C_{t_1}^1)^* (C_{t_2}^1)^* f^*, f\rangle &= \langle f^*, C_{t_1}^1 C_{t_2}^1 f\rangle \\ &= \Big\langle f^*, A_2 A_1 \int_0^{t_1}\int_0^{t_2} T(\xi_1 e_1 + \xi_2 e_2) f d\xi_1 d\xi_2\Big\rangle \\ &= \Big\langle (A_2 A_1)^* f^*, \int_0^{t_1}\int_0^{t_2} T(\xi_1 e_1 + \xi_2 e_2) f d\xi_1 d\xi_2\Big\rangle.\end{aligned}$$

Daraus folgt

$$\|(C_{t_1}^1)^*(C_{t_2}^1)^* f^*\| \leqslant K \|(A_2 A_1)^* f^*\|,$$

womit Satz 3 bewiesen ist.

SATZ 4. *Gilt* $\sup_{0<t_1,t_2<\infty} \|(C_{t_1}^1)^*(C_{t_2}^1)^* f^*\| < \infty$ *sowie* $s \cdot \lim_{t_j \to 0_+} T^*(t_j e_j) f^* = f^*, j = 1, 2,$ *dann folgt* $f^* \in D(A_1^* A_2^*) \cap D(A_2^* A_1^*)$.

BEWEIS. Wir betrachten den folgenden Teilraum $X_0^*$ von $X^*$

$$X_0^* = \{f^*; f^* \in X^*, s \cdot \lim_{t \in \overline{E_n^+},\, t \to 0} T^*(t) f^* = f^*\}.$$

$X_0^*$ ist ein Banachunterraum von $X^*$. Die Restriktion $\{T_0^*(t)\}$ von $T^*(t)$ auf $X_0^*$ ist eine $n$-parametrige Halbgruppe von Operatoren der Klasse $C_0$ in $\mathfrak{E}(X_0^*)$. Mit $A_j^{0*}$ bezeichnen wir den Erzeuger von $\{T_0^*(t)\}$ im Punkte $e_j$ $(j = 1, 2)$. Aus den Voraussetzungen folgt nun leicht,

(1) Ist $P$ irgendein Operator in $X$, so bezeichnen wir mit $P^*$ seinen adjungierten Operator.

daß $f^* \in X_0^*$ ist. Damit liefert die Anwendung von Satz 1, daß $f^* \in \tilde{D}(A_1^{0*}A_2^{0*}) \cap \tilde{D}(A_2^{0*}A_1^{0*})$ ist. Daraus ergibt sich wie im Beweis von Satz 2, daß $f^* \in \tilde{D}(A_2^{0*}) \cap \tilde{D}(A_1^{0*})$ gilt. Da die Operatoren $A_j^*$ Erweiterungen der $A_j^{0*}$ sind, folgt $f^* \in \tilde{D}(A_1^*) \cap \tilde{D}(A_2^*)$. Aus einem Satz von H. Berens [1], S. 20, schließt man dann $f^* \in D(A_1^*) \cap D(A_2^*)$.

Andererseits folgt aus dem ersten Teil unserer Voraussetzungen mit Hilfe von Lemma 2, daß $(C_{t_1}^1)^* f^* \in D(A_2^*)$ und $(C_{t_2}^1)^* f^* \in D(A_1^*)$ für alle $t_1\ t_2$, $0 < t_1, t_2 < \infty$ gilt. Da ferner

$$A_j^* f^* = w^* - \lim_{t_j \to 0_+} \left(T^*(t_j e_j) f^* - f^*\right)/t_j, \qquad j = 1, 2$$

ist, erhalten wir

$$\sup_{0 < t_2 < \infty} \|A_1^* (C_{t_2}^1)^* f^*\| < R_1,$$

wobei $R_1$ unabhängig von $t_1$ ist, sowie

$$\sup_{0 < t_1 < \infty} \|A_2^* (C_{t_1}^1)^* f^*\| < R_2,$$

wobei $R_2$ unabhängig von $t_2$ ist. Weil nun $f^* \in D(A_1^*) \cap D(A_2^*)$ gilt, können wir die Operatoren $A_1^*$ und $(C_{t_2}^1)^*$ sowie $A_2^*$ und $(C_{t_1}^1)^*$ vertauschen. Wenden wir dann Lemma 2 wieder an, so erhalten wir $A_1^* f^* \in D(A_2^*)$ und $A_2^* f^* \in D(A_1^*)$. Also haben wir $f^* \in D(A_1^* A_2^*) \cap D(A_2^* A_1^*)$.

**3. Anwendungen.** Die Anwendungen von Satz 1 und Satz 2 laufen darauf hinaus, die Erzeuger $A_1$ und $A_2$ von bestimmten zweiparametrigen Halbgruppen zu berechnen. Um die Sätze 2 und 4 anwenden zu können, muß man zusätzlich die adjungierten Halbgruppen mit ihren Erzeugern bestimmen. In [5] und [6] sind solche Berechnungen durchgeführt worden. Dort werden die zweiparametrigen Halbgruppen von Abel–Poisson, Gauß–Weierstrass und der Translationen im periodischen Fall untersucht. Bei der Lektüre der Arbeiten [5] und [6] wird klar, welche Verbesserungen durch unsere Ergebnisse zu erzielen sind. Wir wollen uns hier deshalb darauf beschränken, die allgemeine Gruppe der Translationen zu betrachten.

Mit $L_p(E_2)$, $1 \leqslant p \leqslant \infty$, bezeichen wir den Raum der auf $E_2$ definierten, komplexwertigen, Lebesgue-meßbaren und zur $p$-ten Potenz integrierbaren bzw. wesentlich beschränkten Funktionen, versehen mit der üblichen Norm. Aus $L_p(E_2)$, $1 \leqslant p < \infty$, ist die zweiparametrige

Gruppe der Translationen $\{T(t); t \in E_2\}$ wie folgt erklärt:

$$[T(t)f](x) = f(x+t); \quad x = (x_1, x_2), \quad t = (t_1, t_2).$$

Wir erhalten $A_i = \frac{\partial f}{\partial x_i}$ mit $D(A_i) = \left\{f \in L_p(E_2); \frac{\partial f}{\partial x_i} \in L_p(E_2)\right\}$, $i = 1, 2$. Dabei sind die Ableitungen stets im Sobolevschen Sinne zu verstehen [2], S. 256. Die hierzu adjungierten Gruppen $\{T^*(t); t \in E_2\}$, die auf den Räumen $L_q(E_2)$, $1 < q \leqslant \infty$, definiert sind, haben die Gestalt

$$[T^*(t)f](x) = f(x-t).$$

Mit Hilfe einer von K. de Leeuw [7] schon im eindimensionalen Fall benutzten Methode erhält man $A_i^* = -\partial f/\partial x_i$ mit

$$D(A_i^*) = \left\{f \in L_q(E_2); \frac{\partial f}{\partial x_i} \in L_q(E_2)\right\}, \quad i = 1, 2.$$

Wie die Sätze 1 bis 4 aus Abschnitt 2 nun anzuwenden sind, ist evident und braucht im einzelnen nicht näher ausgeführt zu werden.

## LITERATUR

[1] H. Berens, 'Interpolationsmethoden zur Behandlung von Approximationsprozessen auf Banachräumen', *Lecture Notes in Mathematics* **64**, Berlin 1968.

[2] P. L. Butzer and H. Berens, 'Semi-Groups of Operators and Approximation, *Grundl. d. math. Wiss.* **145**, Berlin 1967.

[3] P. L. Butzer and H. G. Tillmann, 'Approximation Theorems for Semi-Groups of Bounded Linear Transformations', *Math. Ann.* **140** (1960) 256–262.

[4] E. Hille and R. S. Phillips, 'Functional Analysis and Semi-Groups', *Amer. Math. Soc. Colloq. Publ.* **31**, Providence 1957.

[5] O. A. Ivanova, '*n*-Parameter Semi-Groups of Bounded Linear Operators and Their Connection with the Approximation Theory', *Visnik, L'viv, Derz. Univ. Ser. Meh.-Mat.* **2** (1965) 44–52.

[6] O. A. Ivanova, 'Certain Theorems on *n*-Parameter Semi-Groups of Linear Bounded Operators and Their Applications to the Function Theory', *Teor. Funkcii Funkcional. Anal. i Prilozen.* **2** (1966) 35–41.

[7] K. de Leeuw, 'On the Adjoint Semi-Group and Some Problems in the Theory of Approximation', *Math. Z.* **73** (1960) 219–234.

# ON STRONG APPROXIMATION OF FOURIER SERIES

L. LEINDLER

*Szeged*

Let $f(x)$ be a continuous and $2\pi$-periodic function and let

$$f(x) \sim \frac{a_0}{2} + \sum_{n=1}^{\infty} (a_n \cos nx + b_n \sin nx) \tag{1}$$

be its Fourier series. Denote $s_n(x) = s_n(f; x)$ the $n$-th partial sum of (1), furthermore denote $\tilde{f}(x)$ and $\tilde{s}_n(x)$ the conjugate functions of $f(x)$ and $s_n(x)$.

In [2] (see Theorem IV) it is proved that the condition $f^{(r)}(x) \in \text{Lip}\,\alpha$ $(0 < \alpha \leqslant 1)$ ($f^{(r)}(x)$ denotes the $r$-th derivative of $f(x)$) does not imply that the series

$$\sum_{n=1}^{\infty} n^{(r+\alpha)p-1} |s_n(x) - f(x)|^p,$$

with some positive $p$, converges everywhere.

It is well known that if $\alpha < 1$, then the condition $f(x) \in \text{Lip}\,\alpha$ implies that $\tilde{f}(x) \in \text{Lip}\,\alpha$ but it is not the case if $\alpha = 1$.

Thus it is natural to ask whether the conditions $f^{(r)}(x) \in \text{Lip}\,1$ and $\tilde{f}^{(r)}(x) \in \text{Lip}\,1$ imply that the partial sums of the series

$$\sum_{n=1}^{\infty} n^{(r+1)p-1} |s_n(x) - f(x)|^p$$

for some positive $p$ are uniformly bounded.

In the present paper we give a negative answer to this problem.

We prove the following

THEOREM 1. *There exists a function* $f_0(x)$ *such that* $f_0^{(r)}(x)$ *and* $\tilde{f}_0^{(r)}(x)$ *belong to the class* Lip 1, *still for any positive* $p$

$$\sum_{n=1}^{m} n^{(r+1)p-1}|s_n(f_0, 0)-f_0(0)|^p \geqq K(r, p)\log m, \tag{2}$$

*where* $K(r, p)$ *is a positive constant depending on* $r$ *and* $p$.

Problems of converse type, that is, what follows from the condition

$$\sum_{n=1}^{\infty} n^{(r+\alpha)p-1}|s_n(x)-f(x)|^p \leqq K \text{ (}^1\text{)},$$

were treated in [3] and [4] in some special cases.

Next we strengthen and generalize the theorem given in [4].

THEOREM 2. *If* $r$ *is a non-negative integer and*

$$\sum_{n=1}^{\infty} n^r|s_n(x)-f(x)| \leqq K \tag{3}$$

*for all* $x$, *then we have*

$$f^{(r)}(x+h)-f^{(r)}(x) = O_x(h) \tag{4}$$

*for almost every* $x$, *furthermore*

$$|f^{(r)}(x+h)-f^{(r)}(x)| \leqq K_1 h\log\frac{1}{h} \tag{5}$$

*for all* $x$. *Estimate* (4) *holds with* $\tilde{f}^{(r)}(x)$ *instead of* $f^{(r)}(x)$, *too.*

*If* $r$ *is even, then* (3) *implies that*

$$\tilde{f}^{(r)}(x) \in \text{Lip}\,1, \tag{6}$$

*and if* $r$ *is an odd integer, then it implies that*

$$f^{(r)}(x) \in \text{Lip}\,1. \tag{7}$$

*Furthermore there exists a function* $F(x)$ *having the following properties:*

$$\sum_{n=1}^{\infty} n^r|s_n(F; x)-F(x)| \leqq K \tag{8}$$

---

[1] $K, K_1, \ldots$ denote positive constants.

*for all x, and if r is an even integer, then*

$$\left| F^{(r)}\left(\frac{\pi}{2^m}\right) - F^{(r)}(0) \right| > \frac{1}{4}\frac{\pi}{2^m}\log\frac{2^m}{\pi}, \tag{9}$$

$$\left| \tilde{F}^{(r)}\left(\frac{\pi}{2^m}\right) - \tilde{F}^{(r)}(0) \right| > \frac{1}{2}\frac{\pi}{2^m} \tag{10}$$

*for all* $m \geqslant 10$; *and if r is an odd integer, then* $F(x)$ *and* $\tilde{F}(x)$ *have to be interchanged in* (9) *and* (10).

PROOF OF THEOREM 1. Let

$$f_0(x) = \sum_{k=1}^{\infty} \frac{(-1)^k}{2^k} \sum_{l=2^{k-1}+1}^{2^k} \left( \frac{\cos(5\cdot 2^k - l)x}{(5\cdot 2^k - l)^r l} - \frac{\cos(5\cdot 2^k + l)x}{(5\cdot 2^k + l)^r l} \right).$$

First we show that $f_0^{(r)}(x)$ and $\tilde{f}_0^{(r)}(x)$ belong to the class Lip 1. If $r$ is an odd integer, then

$$f_0^{(r)}(x) = \pm \sum_{k=1}^{\infty} \frac{(-1)^k}{2^k} \sum_{l=2^{k-1}+1}^{2^k} \left( \frac{\sin(5\cdot 2^k - l)x}{l} - \frac{\sin(5\cdot 2^k + l)x}{l} \right)$$

$$\equiv \pm g(x)$$

and

$$\tilde{f}_0^{(r)}(x) = \pm \sum_{k=1}^{\infty} \frac{(-1)^k}{2^k} \sum_{l=2^{k-1}+1}^{2^k} \left( \frac{\cos(5\cdot 2^k - l)x}{l} - \frac{\cos(5\cdot 2^k + l)x}{l} \right)$$

$$\equiv \pm h(x);$$

and if $r$ is even, then

$$f_0^{(r)}(x) = \pm h(x) \quad \text{and} \quad \tilde{f}_0^{(r)}(x) = \pm g(x).$$

In [1] we have already proved that $g(x) \in \text{Lip}\,1$, thus we have only to prove that $h(x) \in \text{Lip}\,1$.

Setting

$$R_k(x) = \sum_{l=2^{k-1}+1}^{2^k} \left( \frac{\cos(5\cdot 2^k - l)x}{l} - \frac{\cos(5\cdot 2^k + l)x}{l} \right),$$

then

$$h(x)=\sum_{k=1}^{\infty}\frac{(-1)^k}{2^k}R_k(x).$$

Let $h$ be an arbitrary positive number and let $n$ be the smallest natural number such that $h \geqslant 2^{-n}$. By the aid of this $n$ we split $h(x)$ into two sums:

$$h(x)=\sum_{k=1}^{\infty}\frac{(-1)^k}{2^k}R_k(x)=\sum_{k=1}^{n}+\sum_{k=n+1}^{\infty}=h_{1,n}(x)+h_{2,n}(x).$$

Since $\max\limits_{-\pi\leqslant x\leqslant \pi}|R_k(x)|\leqslant K$, thus we have

$$|h_{2,n}(x+h)-h_{2,n}(x)|\leqslant\sum_{k=n+1}^{\infty}\frac{1}{2^k}2K\leqslant\frac{2K}{2^n}\leqslant 2Kh. \tag{11}$$

Next we show that $|h'_{1,n}(x)|\leqslant K_1$ for all $x$. A simple calculation gives that

$$h'_{1,n}(x)=10\sum_{k=1}^{n}(-1)^k\cos 5\cdot 2^k x\sum_{l=2^{k-1}+1}^{2^k}\frac{\sin lx}{l}+$$

$$+2\sum_{k=1}^{n}\frac{(-1)^k}{2^k}\sin 5\cdot 2^k x\sum_{l=2^{k-1}+1}^{2^k}\cos lx\equiv A_1^{(n)}(x)+A_2^{(n)}(x).$$

Since $h'_{1,n}(0)=0$ and $h'_{1,n}(x)$ is an odd function it is enough to prove that in the interval $0<x\leqslant\pi$ $h'_{1,n}(x)$ is uniformly bounded. Let $m$ be the smallest natural number such that

$$\frac{1}{x}\leqslant 5\cdot 2^m, \tag{12}$$

and let $\mu=\min(n,m)$. Furthermore we set

$$C_k(x)=\cos 5\cdot 2^k x\sum_{l=2^{k-1}+1}^{2^k}\frac{\sin lx}{l}$$

and

$$D_k(x) = \frac{1}{2^k} \sin 5 \cdot 2^k x \sum_{l=2^{k-1}+1}^{2^k} \cos lx.$$

By an elementary computation we get that if $k \leqslant \mu - 4$, then

$$(13) \qquad C_k(x) \leqslant C_{k+1}(x) \quad \text{and} \quad D_k(x) \leqslant D_{k+1}(x).$$

Consequently, by (12) and (13), we have

$$\tfrac{1}{10}|A_1^{(n)}(x)| \leqslant \left|\sum_{k=1}^{\mu-1} (-1)^k C_k(x)\right| + \left|\sum_{k=\mu}^{n} C_k(x)\right|$$

$$\leqslant K_2 + \sum_{k=\mu}^{n} \left|\sum_{l=2^{k-1}+1}^{2^k} \frac{\sin lx}{l}\right| \leqslant K_3$$

and

$$\tfrac{1}{2}|A_2^{(n)}(x)| \leqslant \left|\sum_{k=1}^{\mu-1} (-1)^k D_k(x)\right| + \left|\sum_{k=\mu}^{n} D_k(x)\right|$$

$$\leqslant K_4 + \sum_{k=\mu}^{n} \frac{1}{2^k} \left|\sum_{l=2^{k-1}+1}^{2^k} \cos lx\right| \leqslant K_5,$$

here the following well-known estimations are used

$$\left|\sum_{l=p}^{q} \frac{\sin lx}{l}\right| \leqslant \frac{4}{px} \quad \text{and} \quad \left|\sum_{l=p}^{q} \cos lx\right| \leqslant \frac{4}{x} \qquad (0 < x \leqslant \pi).$$

Collecting our results we obtain that

$$|h'_{1,n}(x)| \leqslant K_6$$

everywhere. Hence and from (11) the statement $h(x) \in \mathrm{Lip}\, 1$ follows obviously.

Now we prove inequality (2). Let $p$ be an arbitrary positive number and let $4\cdot 2^i \leqslant m < 4\cdot 2^{i+1}$. Then

$$(14)\qquad \sum_{n=1}^{m} n^{(r+1)p-1}|s_n(f_0,0)-f_0(0)|^p$$

$$\geqslant \sum_{\nu=1}^{i-1}\sum_{n=4\cdot 2^\nu+1}^{4\cdot 2^{\nu+1}} n^{(r+1)p-1}|s_n(f_0,0)-f_0(0)|^p$$

$$\geqslant \sum_{\nu=1}^{i-1}\sum_{n=5\cdot 2^\nu+2^{\nu-1}}^{6\cdot 2^\nu-1} n^{(r+1)p-1}|s_n(f_0,0)-f_0(0)|^p \equiv \sum_{\nu=1}^{i-1}\sigma_\nu .$$

Let

$$\gamma_k = \frac{1}{2^k}\sum_{l=2^{k-1}+1}^{2^k}\left(\frac{1}{(5\cdot 2^k-l)^r l}-\frac{1}{(5\cdot 2^k+l)^r l}\right).$$

It is easy to see that $\gamma_k \geqslant \gamma_{k+1}$, therefore if $5\cdot 2^\nu+2^{\nu-1}\leqslant n < 6\cdot 2^\nu$, then

$$|s_n(f_0,0)-f_0(0)| \geqslant |s_n(f_0,0)-s_{6\cdot 2^\nu}(f_0,0)|,$$

thus we obtain that

$$\sigma_\nu \geqslant \sum_{n=11\cdot 2^{\nu-1}}^{6\cdot 2^\nu-1} n^{(r+1)p-1}|s_n(f_0,0)-s_{6\cdot 2^\nu}(f_0,0)|^p$$

$$\geqslant \sum_{n=11\cdot 2^{\nu-1}}^{6\cdot 2^\nu-1} n^{(r+1)p-1}\left(\frac{1}{2^\nu}\sum_{l=n-10\cdot 2^{\nu-1}+1}^{2^\nu}\frac{1}{6^r 2^{\nu r} l}\right)^p$$

$$\geqslant K_1(r,p)2^{\nu(r+1)p}\cdot 2^{-\nu p}\cdot 2^{-\nu rp} = K_1(r,p) > 0.$$

Hence and from (14) inequality (2) follows immediately, which completes the proof of Theorem 1.

PROOF OF THEOREM 2. First we prove (4). Let

$$R_n(x) = f(x)-s_{n-1}(x)\qquad (n=1,2,\ldots)$$

and

$$t_n(r;x) = \sum_{k=1}^{n} k^{r+1}(a_k\cos kx+b_k\sin kx).$$

Then

$$(15)\qquad t_n(r;x) = \sum_{k=1}^{n} k^{r+1}\left(R_k(x)-R_{k+1}(x)\right)$$
$$= \sum_{k=1}^{n} R_k(x)\left(k^{r+1}-(k-1)^{r+1}\right)-n^{r+1}R_{n+1}(x).$$

Using (3) and (15) we obtain that

$$|t_n(r;x)| \leqslant K_1 + n^{r+1}|R_{n+1}(x)|,$$

whence

$$\left|\frac{1}{n}\sum_{k=1}^{n} t_k(r;x)\right| \leqslant K_1 + \frac{1}{n}\sum_{k=1}^{n} k^{\ +1}|R_{k+1}(x)| \leqslant K_2.$$

This inequality implies (see [6], p. 136) that

$$\sum_{k=1}^{\infty} k^{(r+1)2}(a_k^2+b_k^2) < \infty$$

and thus $f^{(r+1)}(x)$ and $\tilde{f}^{(r+1)}(x)$ are square integrable functions, therefore $f^{(r)}(x)$ and $\tilde{f}^{(r)}(x)$ are absolutely continuous, consequently (4) holds.

Next we prove (5). Setting

$$V_n(x) = \frac{1}{n}\sum_{k=n+1}^{2n} s_k(x) \qquad \text{and} \qquad U_n(x) = V_{2^n}(x) - V_{2^{n-1}}(x),$$

where $V_{2^{-1}}(x) \equiv 0$, we have

$$f(x) = \sum_{n=0}^{\infty} U_n(x).$$

To prove

$$(16)\qquad f^{(r)}(x) = \sum_{n=0}^{\infty} U_n^{(r)}(x)$$

first we verify the inequality

$$(17)\qquad |U_n^{(r)}(x)| \leqslant K2^{-n}.$$

Using

$$|U_n(x)| \leqslant |V_{2^n}(x)-f(x)| + |f(x)-V_{2^{n-1}}(x)|$$

and that by (3)

$$|f(x)-V_{2^n}(x)| \leqslant \frac{1}{2^n} \sum_{k=2^n+1}^{2^{n+1}} |f(x)-s_k(x)| \leqslant K2^{-n(r+1)},$$

we get (17) by the well-known Bernstein's inequality. By (17)

$$\sum_{n=0}^{\infty} U_n^{(r)}(x)$$

converges uniformly and thus (16) is verified.

By (16) and (17), if $2^{-m} < h \leqslant 2^{-m+1}$, we obtain

$$\begin{aligned} |f^{(r)}(x+h)-f^{(r)}(x)| &\leqslant \sum_{n=0}^{m-1} |U_n^{(r)}(x+h)-U_n^{(r)}(x)| + K_1 \sum_{n=m}^{\infty} 2^{-n} \\ &\leqslant h \sum_{n=0}^{m-1} \max_x |U_n^{(r+1)}(x)| + K_2 2^{-m} \\ &\leqslant K_3 h \log \frac{1}{h} \end{aligned}$$

which proves (5).

To prove (6) and (7) we use the following result of Zamansky [5]: $f^{(r)}(x) \in \operatorname{Lip} 1$ if and only if

$$\|f-R_n(r+1;f)\|_C = O\left(\frac{1}{n^{r+1}}\right) \qquad \text{for } r \text{ odd,}$$

$$\|\tilde{f}-R_n(r+1,\tilde{f})\|_C = O\left(\frac{1}{n^{r+1}}\right) \qquad \text{for } r \text{ even,}$$

where

$$R_n(r,f) = \sum_{k=0}^{n} \left[1-\left(\frac{k}{n+1}\right)^r\right](a_k \cos kx + b_k \sin kx).$$

We get by an easy computation, using an Abel-rearrangement, that

$$|R_n(r+1,f)-f(x)| \leqslant \frac{K_1}{n^{r+1}} \sum_{k=0}^{n} |s_k(x)-f(x)|k^r,$$

whence and from (3) by Zamansky's result (6) and (7) obviously follow.

The function having properties (8), (9) and (10) is a very simple one as follows:

(18) $$F(x) = \sum_{n=1}^{\infty} \frac{\sin nx}{n^{2+r}}.$$

To verify (8) we choose $N$ with $\frac{1}{N+1} \leqslant |x| < \frac{1}{N}$. Then

$$\left(\sum_{n=1}^{N} + \sum_{n=N+1}^{\infty}\right) n^r \left| \sum_{k=n+1}^{\infty} \frac{\sin kx}{k^{2+r}} \right| \leqslant \sum_{n=1}^{N} n^r \left| \sum_{k=n+1}^{N+1} \frac{\sin kx}{k^{2+r}} \right| +$$

$$+ \sum_{n=1}^{N} n^r \left| \sum_{k=N+2}^{\infty} \frac{\sin kx}{k^{2+r}} \right| + \sum_{n=N+1}^{\infty} n^r \left| \sum_{k=n+1}^{\infty} \frac{\sin kx}{k^{2+r}} \right|.$$

These sums are less than an absolute constant, namely

$$\sum_{n=1}^{N} n^r \left| \sum_{k=n+1}^{N+1} \frac{\sin kx}{k^{2+r}} \right| \leqslant \sum_{n=1}^{N} n^r x \sum_{k=n+1}^{N+1} \frac{1}{k^{r+1}} \leqslant K_1,$$

$$\sum_{n=1}^{N} n^r \left| \sum_{k=N+2}^{\infty} \frac{\sin kx}{k^{2+r}} \right| \leqslant \sum_{n=1}^{N} n^r \sum_{k=N+2}^{\infty} \frac{1}{k^{r+2}} \leqslant K_1,$$

furthermore using the well-known estimate

$$\left| \sum_{k=n+1}^{\infty} \frac{\sin kx}{k^{2+r}} \right| \leqslant \frac{1}{(n+1)^{2+r} \left| \sin \frac{x}{2} \right|} \leqslant \frac{4}{n^{2+r}|x|},$$

we obtain that

$$\sum_{n=N+1}^{\infty} n^r \left| \sum_{k=n+1}^{\infty} \frac{\sin kx}{k^{2+r}} \right| \leqslant \sum_{n=N+1}^{\infty} n^r \frac{4}{n^{2+r}|x|} \leqslant K_1.$$

Next we prove the following

LEMMA. *Let $\{\varrho_n\}$ be a non-increasing sequence of numbers such that $\sum_{n=1}^{\infty} n^{-1}\varrho_n < \infty$ and let*

$$\varrho(x) = \sum_{n=1}^{\infty} n^{-1} \varrho_n \sin nx.$$

*Then for any* $m > 2^9$

$$\varrho\left(\frac{\pi}{m}\right) > \frac{1}{2}\frac{1}{m}\sum_{n=1}^{m}\varrho_n.$$

PROOF. Denote $h_m = \pi/m$. We have

$$\varrho(h_m) = \sum_{n=1}^{\infty} n^{-1}\varrho_n \sin nh_m \tag{19}$$

$$= \left(\sum_{n=1}^{\frac{m}{4}} + \sum_{n=\frac{m}{4}+1}^{m} + \sum_{n=m+1}^{2m} + \sum_{k=2}^{\infty}\sum_{n=km+1}^{(k+1)m}\right)\varrho_n n^{-1}\sin nh_m \ (^2)$$

It is clear that for any $l \geqq 1$

$$\sum_{n=2lm+1}^{(2l+1)m} \varrho_n n^{-1}\sin nh_m > \left|\sum_{n=(2l+1)m+1}^{(2l+2)m} \varrho_n n^{-1}\sin nh_m\right|,$$

consequently the sum

$$\sum_{k=2}^{\infty}\sum_{n=km+1}^{(k+1)m} \varrho_n n^{-1}\sin nh_m$$

is positive. Next we show that

$$\sum_{n=\frac{m}{4}+1}^{m} \varrho_n n^{-1}\sin nh_m \geqq \left|\sum_{n=m+1}^{2m} \varrho_n n^{-1}\sin nh_m\right|. \tag{20}$$

Since

$$\sum_{n=\frac{m}{2}+1}^{m} \varrho_n n^{-1}\sin nh_m > \left|\sum_{n=m+1}^{\frac{3}{2}m-1} \varrho_n n^{-1}\sin nh_m\right|$$

---

(2) $\sum_{n=a}^{b}$, where $a$ or $b$ are not integers, means a sum over all integers between $a$ and $b$.

and if $m > 2^9$

$$\sum_{n=\frac{m}{4}+1}^{\frac{m}{2}} \varrho_n n^{-1} \sin nh_m > \frac{\sqrt{2}}{2} \varrho_m \frac{2}{m}\left(\frac{m}{4}-3\right)$$

$$\geqslant \varrho_m \left(\frac{m}{2}+2\right)\frac{2}{3m} \geqslant \varrho_m \sum_{n=\frac{3}{2}m}^{2m} n^{-1}$$

$$\geqslant \left|\sum_{n=\frac{3}{2}m}^{2m} \varrho_n n^{-1} \sin nh_m\right|,$$

thus (20) is proved.

Collecting our results, by (19), we have

$$\varrho(h_m) \geqslant \sum_{n=1}^{\frac{m}{4}} \varrho_n n^{-1} \sin nh_m \geqslant \frac{2}{\pi} h_m \sum_{n=1}^{\frac{m}{4}} \varrho_n \geqslant \frac{1}{2}\frac{1}{m}\sum_{n=1}^{m} \varrho_n$$

in accordance with the statement of lemma.

Using this lemma we can prove (9) and (10) easily. It is clear that if $r$ is even, then

$$\pm F^{(r)}(x) = \sum_{n=1}^{\infty} \frac{\sin nx}{n^2},$$

and

$$\pm \tilde{F}^{(r)}(x) = \sum_{n=1}^{\infty} \frac{\cos nx}{n^2}.$$

The statement (9) follows by lemma immediately, and since

$$\left|\tilde{F}^{(r)}\left(\frac{\pi}{2^m}\right) - \tilde{F}^{(r)}(0)\right| = \sum_{n=1}^{\infty} \frac{1}{n^2}\left(1-\cos n\frac{\pi}{2^m}\right)$$

$$= \sum_{n=1}^{\infty} \frac{1}{n^2} 2\sin^2 n \frac{\pi}{2^{m+1}} \geqslant \frac{1}{2}\frac{\pi}{2^m},$$

thus (10) is also proved.

In the case of odd $r$ (9) and (10) can be proved similarly.

## REFERENCES

[1] G. Alexits und L. Leindler, 'Über die Approximation im starken Sinne', *Acta Math. Acad. Sci. Hung.* **16** (1965) 27–32.
[2] L. Leindler, 'Über die Approximation im starken Sinne', *ibidem* **16** (1965) 255–262.
[3] L. Leindler, 'On Strong Summability of Fourier Series, II', *ibidem* **20** (1969) 347–355.
[4] L. Leindler and E. M. Nikisin, 'Note on Strong Approximation of Fourier Series', *ibidem* **23** (1972) 223-227
[5] M. Zamansky, 'Classes de saturation des procédés de sommation des séries de Fourier et applications aux séries trigonométriques', *Ann. Sci. Ecole Norm. Sup.* **67** (1950) 161–198.
[6] A. Zygmund, *Trigonometric Series*, Vol. I, Cambridge 1959.

# ON LINEAR JUXTAOPERATORS

I. MARUŞCIAC

*Cluj*

**1. Introduction.** Let $K$ be a compact pointset in the complex plane and $F, f_1, \ldots, f_n$ given complex functions defined and continuous on $K$. We denote by $\mathscr{P}(f)$ the set of generalized polynomials with respect to the system $f = (f_k)_1^n$:

$$p(z) = p(f; z) = a_1 f_1(z) + a_2 f_2(z) + \ldots + a_n f_n(z). \tag{1.1}$$

Sometimes we shall omit the word "generalized" and instead of saying "generalized polynomial" we shall simply say "polynomial".

Definition 1.1. A polynomial $\pi \in \mathscr{P}(f)$ is called a *juxtapolynomial* to $F$ on $K$ if there is no polynomial $p \in \mathscr{P}(f), p \neq \pi$, such that

$$F(z) - \pi(f; z) = 0, \quad z \in K \Rightarrow p(f; z) = \pi(f; z); \tag{1.2}$$

$$F(z) - \pi(f; z) \neq 0, \quad z \in K \Rightarrow |F(z) - p(f; z)| < |F(z) - \pi(f; z)|. \tag{1.3}$$

The set of generalized juxtapolynomials to $F$ on $K$ will be denoted by $\mathscr{J}(K; F; f)$.

If $f_k(z) = z^{n-k}$, then we obtain the set $\mathscr{J}_n(K; F)$ of algebraic juxtapolynomials of degree $n-1$ to the function $F$ on $K$, defined by T. S. Motzkin and J. L. Walsh [8].

The juxtapolynomials to a continuous function $F$ are a special case of the generalized infrapolynomials, introduced by the author [6].

If $F(z) = z^n$ and $f_k(z) = z^{n-k}$, $k = 1, 2, \ldots, n$, then

$$\mathscr{I}_n(K) = \{z^n - \pi(z) \mid \pi \in \mathscr{J}_n(K; z^n)\}$$

is the set of *infrapolynomials* of degree $n$ on $K$, introduced by M. Fekete and I. L. v. Neumann [2].

Remark. 1.1. If $A, B \subset K$ are two compact sets and $A \subset B$, then from $p \in \mathscr{J}(A; F; f)$ follows $p \in \mathscr{J}(B; F; f)$, because if there is no polynomial of $\mathscr{P}(f)$ satisfying (1.2)–(1.3) on $A$, then obviously there is no such polynomial satisfying the same conditions on $B$.

From Remark 1.1 it follows immediately that each generalized Lagrange interpolation polynomial $L(z_1, z_2, \ldots, z_n; f; F|z)$ to the function $F$ on the knots $\{z_j\}_1^n \subset K$ is a juxtapolynomial to $F$ on $K$, too.

Among the other well-known juxtapolynomials to a continuous function $F$ are those which realize the best approximation to $F$ in a given norm on the set $K$ [6].

The purpose of this note is to give a linear operator which assigns to each continuous function on $K$ one of its generalized juxtapolynomials on $K$.

In the second part of the note some applications are given to extremal solutions of the linear inconsistent systems.

**2. Linear juxtaoperators.** Throughout this section we assume that the compact set $K$ consists of at least $n+1$ points and $f = (f_k)_1^n$ is a Chebyshev system on $K$. In this case for every system $\{z_j\}_1^n \subset K$ there exists a generalized interpolation polynomial $L(z_1, z_2, \ldots, z_n; f; F|z) \in \mathscr{P}(f)$ satisfying the conditions

$$L(z_1, z_2, \ldots, z_n; f; F|z_j) = F(z_j), \qquad \forall j \in \{1, 2, \ldots, n\}$$

and it is unique.

This polynomial can be expressed in the form

$$(2.1) \quad L(z_1, z_2, \ldots, z_n; f; F|z) = \sum_{j=1}^{n} \frac{U(z_1, \ldots, z_{j-1}, z, z_{j+1}, \ldots, z_n; f)}{U(z_1, z_2, \ldots, z_n; f)} F(z_j),$$

where

$$U(z_1, \ldots, z_n; f) = \begin{vmatrix} f_1(z_1) \ldots f_n(z_1) \\ \cdots\cdots\cdots \\ f_1(z_n) \ldots f_n(z_n) \end{vmatrix}.$$

Using the properties of juxtapolynomials established in our previous papers (see for instance [6]–[7]) we are going to construct a linear operator defined over $C(K)$ (the space of functions continuous on $K$) with values on the set of generalized juxtapolynomials to the functions of $C(K)$. The main result is the following:

THEOREM 1.1. *Let $K$ be a compact set in the complex plane containing at least $n+1$ points and $f = (f_k)_1^n$ a Chebyshev system on $K$. The generalized polynomial $\pi \in \mathscr{P}(f)$, $\pi(f; z) \neq F(z)$, $z \in K$, is a juxtapolynomial to $F$ on $K$, i.e. $\pi \in \mathscr{J}(K; F; f)$, if and only if there exist a subset $Z_m = \{z_j\}_1^m \subset K (n+1 \leqslant m \leqslant 2n+1)$ and positive constants $d_j$ such that*

$$(2.2) \quad \pi(f; z) = \frac{\sum d_{j_1} \ldots d_{j_n} |U(z_{j_1}, \ldots, z_{j_n}; f)|^2 L(z_{j_1}, \ldots, z_{j_n}; f; F|(z)}{\sum d_{j_1} \ldots d_{j_n} |U(z_{j_1}, \ldots, z_{j_n}; f)|^2},$$

*where $\sum$ is taken for all $j_k$ from 1 to $m$.*

PROOF. *Necessity*. Assume that $\pi \in \mathscr{J}(K; F; f)$ and $\pi(f; z) \neq F(z)$, $z \in K$. Then ([6], Theorem 2) there exist a set $Z_m = \{z_j\}_1^m \subset K$ $(n+1 \leqslant m \leqslant 2n+1)$ and a system of positive constants $\delta_j$ such that

$$(2.3) \qquad \sum_{j=1}^{m} \delta_j \frac{f_k(z_j)}{F(z_j) - \pi(f; z_j)} = 0, \quad \forall k \in \{1, 2, \ldots, n\}.$$

But equalities (2.3) can be written in the form

$$(2.4) \quad \sum_{j=1}^{m} d_j \pi(f; z_j) \overline{f_k}(z_j) = \sum_{j=1}^{m} d_j F(z_j) \overline{f_k}(z_j), \qquad \forall k \in \{1, 2, \ldots, n\},$$

where

$$d_j = \frac{\delta_j}{|F(z_j) - \pi(f; z_j)|^2}.$$

If

$$\pi(f; z) = a_1 f_1(z) + a_2 f_2(z) + \ldots + a_n f_n(z),$$

then the system (2.4) is equivalent to

$$(2.5) \qquad a_1 L_{k1} + a_2 L_{k2} + \ldots + a_n L_{kn} = g_k, \quad \forall k \in \{1, 2, \ldots, n\},$$

where

$$L_{ki} = \sum_{j=1}^{m} d_j \overline{f_k}(z_j) f_i(z_j),$$

$$g_k = \sum_{j=1}^{m} d_j F(z_j) \overline{f_k}(z_j).$$

The system (2.5) to which we add the equality

$$a_1 f_1(z)+a_2 f_2(z)+ \dots +a_n f_n(z) = \pi(f; z)$$

permits us to eliminate the unknowns $a_1, a_2, \dots, a_n$, whence

$$\begin{vmatrix} \pi(f; z) & f_1(z) & \dots & f_n(z) \\ g_1 & L_{11} & \dots & L_{1n} \\ \dots & \dots & \dots & \dots \\ g_n & L_{n1} & \dots & L_{nn} \end{vmatrix} = 0.$$

Hence

$$\pi(f; z) = - \begin{vmatrix} 0 & f_1(z) & \dots & f_n(z) \\ g_1 & L_{11} & \dots & L_{1n} \\ \dots & \dots & \dots & \dots \\ g_n & L_{n1} & \dots & L_{nn} \end{vmatrix} : \begin{vmatrix} L_{11} & \dots & L_{1n} \\ \dots & \dots & \dots \\ L_{n1} & \dots & L_{nn} \end{vmatrix}. \tag{2.6}$$

But, if we write:

$$L_{kl} = \sum_{j_k=1}^{m} d_{j_k} \overline{f_k}(z_{j_k}) f_l(z_{j_k}),$$

then the denominator of (2.6) can be put in the form

$$\det(L_{ik}) = \frac{1}{n!} \sum d_{j_1} \dots d_{j_n} |U(z_{j_1}, \dots, z_{j_n}; f)|^2.$$

Similarly the determinant of the numerator of (2.6) can be written as

$$\frac{1}{n!} \sum d_{j_1} \dots d_{j_n} \overline{U}(z_{j_1}, \dots, z_{j_n}; f) V(z, z_{j_1}, \dots, z_{j_n}; f; F),$$

where

$$V(z, x_1, \dots, x_n; f; F) = \begin{vmatrix} 0 & f_1(z) & \dots & f_n(z) \\ F(x_1) & f_1(x_1) & \dots & f_n(x_1) \\ \dots & \dots & \dots & \dots \\ F(x_n) & f_1(x_n) & \dots & f_n(x_n) \end{vmatrix}$$

and $\overline{U}(x_1, \dots, x_n; f)$ is the imaginary conjugate of $U(x_1, \dots, x_n; f)$.

But in view of (2.1), we have

$$V(z, x_1, \ldots, x_n; f; F) = \sum_{k=1}^{n} (-1)^k F(x_k) U(z, x_1, \ldots, x_{k-1}, x_{k+1}, \ldots, x_n; f)$$

$$= -\sum_{k=1}^{n} U(x_1, \ldots, x_{k-1}, z, x_{k+1}, \ldots, x_n; f) F(x_k)$$

$$= -U(x_1, \ldots, x_n; f) L(x_1, \ldots, x_n; f; F|z).$$

Replacing the expression of the determinants in (2.6) we obtain (2.2).

*Sufficiency*. Now, assume that $\pi(f; z)$ is given by (2.2). First we remark that the numerator of (2.2) may be written in the form

$$\sum_{1}^{m} \frac{d_1 d_2 \ldots d_m}{d_{k_1} d_{k_2} \ldots d_{k_{m-n}}} |U(z_1, \ldots, z_{\hat{k}_1}, \ldots, z_{\hat{k}_2}, \ldots, z_{\hat{k}_{m-n}}, \ldots; f)|^2 \times$$

$$\times L(z_1, \ldots, z_{\hat{k}_1}, \ldots, z_{\hat{k}_2}, \ldots, z_{\hat{k}_{m-n}}, \ldots; f; F|z),$$

where $\hat{k}$ indicates that the index $k$ is missing.

Thus

$$\text{(2.7)}\quad F(z) - \pi(f; z) = \frac{1}{\Delta} \sum \frac{d_1 \ldots d_m}{d_{k_1} \ldots d_{k_{m-n}}} |U(z_1, \ldots, z_{\hat{k}_1}, \ldots, z_{\hat{k}_{m-n}}, \ldots; f)|^2 \times$$

$$\times [F(z) - L(z_1, \ldots, z_{\hat{k}_1}, \ldots, z_{\hat{k}_{m-n}}, \ldots; f; F|z)],$$

where $\Delta$ is the denominator of (2.2).

One verifies that from (2.7) follows

$$\Delta \sum_{j=1}^{m} d_j [F(z_j) - \pi(f;\, z_j)] \bar{f}_k = (z_j) = \sum \frac{d_1 \ldots d_m}{d_{k_2} \ldots d_{k_{m-n}}} D(z_1, \ldots, z_{\hat{k}_2}, \ldots$$

$$\ldots, z_m; f; F) \begin{vmatrix} \bar{f}_k(z_1) & \bar{f}_1(z_1) \ldots \bar{f}_n(z_1) \\ \cdots & \cdots \\ \bar{f}_k(z_{\hat{k}_2}) & \bar{f}_1(z_{\hat{k}_2}) \ldots \bar{f}_n(z_{\hat{k}_2}) \\ \cdots & \cdots \\ \bar{f}_k(z_{\hat{k}_{m-n}}) & \bar{f}_1(z_{\hat{k}_{m-n}}) \ldots \bar{f}_n(z_{\hat{k}_{m-n}}) \\ \cdots & \cdots \\ \bar{f}_k(z_m) & \bar{f}_1(z_m) \ldots \bar{f}_n(z_m) \end{vmatrix} = 0, \qquad \forall k \in \{1, 2, \ldots, n\},$$

where

$$D(z, x_1, \dots, x_n; f; F) = \begin{vmatrix} F(z) & f_1(z) & \dots & f_n(z) \\ F(x_1) & f_1(x_1) & \dots & f_n(x_1) \\ \dots & \dots & \dots & \dots \\ F(x_n) & f_1(x_n) & \dots & f_n(x_n) \end{vmatrix}.$$

Thus we are led to the further equations

$$\sum_{j=1}^{m} \delta_j \frac{f_k(z_j)}{w(z_j)} = 0, \qquad \forall k \in \{1, 2, \dots, n\},$$

where

$$\delta_j = d_j |w(z_j)|^2, \qquad w(z) = F(z) - \pi(f; z),$$

and so by Theorem 2 of [6] it follows that $\pi \in \mathscr{J}(K; F; f)$.

COROLLARY 2.1. *Under the conditions of Theorem* 2.1, *a polynomial* $\pi \in \mathscr{P}(f)$ *is a justapolynomial to f on K if and only if there exist a subset* $Z_m \subset K$ $(n+1 \leqslant m \leqslant 2n+1)$ *and positive constants* $\Lambda_{j_1, \dots, j_n}$ *with* $\sum \Lambda_{j_1, \dots, j_k} = 1$ *such that*

$$\pi(f; z) = \sum \Lambda_{j_1, \dots, j_n} L(z_{j_1}, \dots, z_{j_n}; f; F|z). \tag{2.8}$$

Indeed (2.8) follows from (2.2) if we write

$$\Lambda_{j_1, \dots, j_n} = \frac{d_{j_1} \dots d_{j_n} |U(z_{j_1}, \dots, z_{j_n}; f)|^2}{\sum d_{j_1} \dots d_{j_n} |U(z_{j_1}, \dots, z_{j_n}; f)|^2},$$

because obviously $\Lambda_{j_1, \dots, j_n} > 0$ and $\sum \Lambda_{j_1, \dots, j_n} = 1$.

The operator of the right-hand side of (2.8), which will be called a *juxtaoperator*, is an interesting linear operator which assigns to each continuous function on $K$ one of its generalized juxtapolynomials on $K$.

It is interesting to note that in virtue of (2.8) we can easily construct a juxtapolynomial to a given function $F$, continuous on a compact set $K$. It is sufficient to choose an arbitrary subset $Z_m \subset K$ $(n+1 \leqslant m \leqslant 2n+1)$ and by the arbitrary positive constants $\Lambda_{j_1, \dots, j_n}$ we construct the polynomial $\pi$ as in (2.8).

When $m = n+1$, this juxtaoperator has the following simpler form:

$$\pi(f; z) = \sum_{j=1}^{n} \Lambda_j L(z_0, \dots, \hat{z_j}, \dots, z_n; f; F|z). \tag{2.9}$$

This is the case [6] where $F$ and $f = (f_k)_1^n$ are real valued functions on $K$.

Using a characterization of the juxtapolynomials to $F$ on $K$, which are at the same time the best approximation polynomials to $F$ on $K$ (in uniform norm) given in [7], we have

COROLLARY 2.2. *Under the conditions of Theorem* 2.1, *the generalized polynomial* $\pi \in \mathscr{P}(f)$ *is a polynomial of best approximation to* $F$ *on* $K$ *if and only if there exist a subset* $Z_m \subset K$ $(n+1 \leqslant m \leqslant 2n+1)$ *and positive constants* $\Lambda_{j_1,\dots,j_n}$ *with* $\sum \Lambda_{j_1,\dots,j_n} = 1$ *such that* (2.8) *holds and, moreover,*

$$(2.10)\quad |F(z_j)-\pi(f;z_j)| = \varrho = \max_{z\in K}|F(z)-\pi(f;z)|, \quad \forall j \in \{1,2,\dots,m\}.$$

This result was established by the author in 1964 [5].

When $m = n+1$, from (2.8) and (2.10) the constants $\Lambda_j$ can be expressed explicitly.

If $K = \{z_j\}_1^q$ $(q \geqslant n+1)$ is a finite set, then in (2.2) (and therefore in (2.8)) we may take the sum $\sum$ for $j_k$ from 1 to $q$, putting $d_j = 0$ for $j > m$. So we have

COROLLARY 2.3. *If* $K = \{z_j\}_1^q$ $(q \geqslant n+1)$ *and* $f = (f_k)_1^n$ *is a Chebyshev system on* $K$, *then a polynomial* $\pi \in \mathscr{P}(f)$, $\pi(f;z) \neq F(z)$, $z \in K$, *is a juxtapolynomial to* $F$ *on* $K$ *if and only if*

$$(2.11)\qquad \pi(f;z) = \sum \Lambda_{j_1,\dots,j_n} L(z_{j_1},\dots,z_{j_n}; f; F|z),$$

*where*

$$\Lambda_{j_1,\dots,j_n} \geqslant 0, \qquad \sum \Lambda_{j_1,\dots,j_n} = 1.$$

Now let $\pi_0 \in \mathscr{P}(f)$ be a polynomial such that

$$\Big(\sum_{j=1}^{q} \mu_j |F(z_j)-\pi_0(f;z_j)|^2\Big)^{1/2} = \inf_{p\in\mathscr{P}(f)} \Big(\sum_{j=1}^{q} \mu_j |F(z_j)-p(f;z_j)|^2\Big)^{1/2},$$

where $\mu_j \geqslant 0$, $\sum \mu_j = 1$. Then $\pi_0$ satisfies the system

$$(2.12)\quad \sum_{j=1}^{q} \mu_j \pi_0(f;z_j)\overline{f_k}(z_j) = \sum_{j=1}^{q} \mu_j F(z_j)\overline{f_k}(z_j), \qquad \forall k \in \{1,2,\dots,n\}.$$

But (2.12) coincides with (2.4) for $d_j = \mu_j$, $j = 1, 2, \ldots, q$, so we have

COROLLARY 2.4. *$\pi_0 \in \mathscr{P}(f)$ is a polynomial of best weighted square approximation to $F$ on $K = \{z_j\}_1^q$ $(q \geqslant n+1)$ if and only if*

$$\pi_0(f; z) = \sum \Lambda^0_{j_1, \ldots, j_n} L(z_{j_1}, \ldots, z_{j_n}; f; F|z), \tag{2.13}$$

*where*

$$\Lambda^0_{j_1, \ldots, j_n} = \frac{\mu_{j_1} \ldots \mu_{j_n} |U(z_{j_1}, \ldots, z_{j_n}; f)|^2}{\sum \mu_{j_1} \ldots \mu_{j_n} |U(z_{j_1}, \ldots, z_{j_n}; f)|^2}. \tag{2.14}$$

First from (2.13)–(2.14) and Corollary 2.1 it follows that $\pi_0 \in \mathscr{J}(K; F; f)$. Then, using the form (2.8) of the juxtaoperator, we infer that a juxtapolynomial $\pi \in \mathscr{J}(K; F; f)$ becomes a weighted square best approximation polynomial to $F$ on $K$ if the constants $\Lambda_{j_1, \ldots, j_n}$ are given by (2.14), i.e. $d_j = \mu_j$, $j = 1, 2, \ldots, q$.

The juxtaoperator was constructed for the juxtapolynomials which do not coincide with the function $F$ on $K$. This assumption is not essential. It is possible [7] to give a similar juxtaoperator for the case.

**3. Extremal solutions of a linear system.** To find an approximate solution of an inconsistent linear system we frequently use the least squares method. Another method, called the *best approximation method*, was given by E. I. Remez [9]. Obviously, these two methods correspond to the two metrics considered. It is natural to study all such "extremal" solutions of an inconsistent linear system, introducing the notion of the juxtasolution of such a system.

Let us consider a system

$$y_\tau(z) = a_\tau z - b_\tau = \sum_{k=1}^{n} a_{\tau k} z_k - b_\tau = 0, \qquad \tau \in I, \tag{3.1}$$

where

$$a_\tau = (a_{\tau 1}, a_{\tau 2}, \ldots, a_{\tau n}), \qquad z' = (z_1, z_2, \ldots, z_n)$$

are complex vectors, $b_\tau \in \boldsymbol{K}$, $\tau \in I$, and $I$ is an index set such that

$$\operatorname{card} I \leqslant \aleph.$$

Sometimes this system will be denoted by

$$Az = b.$$

DEFINITION 3.1. An approximate solution $z \in K^n$ of the system (3.1) is called an *infrasolution* of the system (3.1) (abbreviated to $z \in \mathscr{J}(A; b)$) if there is no $u \in K^n$ satisfying the conditions

$$\exists \tau_0 \in I, \quad a_{\tau_0} u \neq a_{\tau_0} z; \tag{3.2}$$

$$\text{If } \tau \in I \text{ and } a_\tau z = b_\tau \Rightarrow a_\tau u = b_\tau; \tag{3.3}$$

$$\text{If } \tau \in I \text{ and } a_\tau z \neq b_\tau \Rightarrow |a_\tau u - b_\tau| < |a_\tau z - b_\tau|. \tag{3.4}$$

Obviously, each exact solution of the system (if such a solution exists) is an infrasolution, too.

Among the most important infrasolutions are those which minimize certain norms. For instance, the solution $z^* \in K^n$ obtained by the least squares method (where $I = \{1, 2, ..., m\}$) is an infrasolution of (3.1).

DEFINITION 3.2. An approximate solution $z^0 \in K^n$ of (3.1) is called the *best approximation* of the system (3.1) or a *Chebyshev point* of the system (3.1), if

$$\sup_{\tau \in I} |a_\tau z^0 - b_\tau| = \inf_{z \in K^n} \sup_{\tau \in I} |a_\tau z - b_\tau|. \tag{3.5}$$

We can immediately verify that every Chebyshev point $z^0$ of the system (3.1) is an infrasolution of (3.1), too.

Hence the set $\mathscr{J}(A; b)$ contains all the extremal solutions of the system (3.1) which appear in applications.

We are going to show that the infrasolutions of a linear system are a special case of the generalized juxtapolynomials to a given function on a certain pointset in the complex plane.

Thus let us write

$$A_j = \{a_{\tau j} \in K \mid \tau \in I\}, \qquad B = \{b_\tau \in K \mid \tau \in I\}$$

and let $M = \{\zeta_\tau \in K \mid \tau \in I\}$ be an arbitrary compact set in a complex plane. We consider the applications

$$f_j;\ M \to \overline{A}_j \quad (j = 1, 2, ..., n),$$

$$F\colon M \to \overline{B}$$

continuous on $M$ and satisfying the conditions

$$F(\zeta_\tau) = b_\tau, \quad f_j(\zeta_\tau) = a_{\tau j} \quad (j = 1, 2, ..., n), \tag{3.6}$$

where $\overline{A}_j$ and $\overline{B}$ are, respectively, the closures of $A_j$ and $B$.

Then it is clear that the approximation problem of the solutions of the system (3.1) is equivalent to the approximation problem of the function $F$ on the set $M$ by the generalized polynomials

$$p(z;\zeta) = z_1 f_1(\zeta) + \ldots + z_n f_n(\zeta). \tag{3.7}$$

From the definition of the juxtapolynomial to a function on a compact set, by virtue of Definition 3.1, it follows immediately that $z^* \in \mathcal{J}(A;b)$ if and only if $p(z^*;\zeta) \in \mathcal{J}(M;F;f)$. That is, all the properties established for juxtapolynomials can easily be transcribed for the infrasolutions of a linear system, too.

DEFINITION 3.3. The system (3.1) has *(T)-property* *(Chebyshev's property)* if the rank of every submatrix of $A$ of the type $(n, n)$ is equal to $n$.

For simplicity we assume that the system (3.1) is a real system. Then by Theorem 2.1 easily follows

THEOREM 3.1. *Let*

$$Ax = b \tag{3.8}$$

*be a real system possessing (T)-property. Then its approximate solution $x^0$ is an infrasolution, i.e. $x^0 \in \mathcal{J}(A;b)$, if and only if there exists a subsystem of* (3.8):

$$y_{\tau_j}(x) = 0, \quad j \in \{1, 2, \ldots, n+1\}$$

*and positive numbers $d_j$ such that*

$$x_k^0 = \frac{\sum_{j=1}^{n+1} d_1 \ldots \hat{d_j} \ldots d_{n+1} D_j(\tau_1, \ldots, \tau_{n+1}; A) D_j^{(k)}(\tau_1, \ldots, \tau_{n+1}; A; b)}{\sum_{j=1}^{n+1} d_1 \ldots \hat{d_j} \ldots d_{n+1} [D_j(\tau_1, \ldots, \tau_{n+1}; A)]^2}, \tag{3.9}$$

*where*

$$D_j(\tau_1, \ldots, \tau_{n+1}; A) = \begin{vmatrix} a_{\tau_1 1} & \ldots & a_{\tau_1 n} \\ \ldots & \ldots & \ldots \\ a_{\hat{\tau_j} 1} & \ldots & a_{\hat{\tau_j} n} \\ \ldots & \ldots & \ldots \\ a_{\tau_{n+1} 1} & \ldots & a_{\tau_{n+1} n} \end{vmatrix},$$

$$D_j^{(k)}(\tau_1, \ldots, \tau_{n+1}; A; b) = \begin{vmatrix} a_{\tau_1 1} & \ldots a_{\tau_1 k-1} & b_{\tau_1} a_{\tau_1 k+1} & \ldots a_{\tau_1 n} \\ \ldots & \ldots & \ldots & \ldots \\ a_{\tau_{\hat{j}} 1} & \ldots a_{\tau_{\hat{j}} k-1} & b_{\tau_{\hat{j}}} a_{\tau_{\hat{j}} k+1} & \ldots a_{\tau_{\hat{j}} n} \\ \ldots & \ldots & \ldots & \ldots \\ a_{\tau_{n+1} 1} & \ldots a_{\tau_{n+1} k-1} & b_{\tau_{n+1}} a_{\tau_{n+1} k+1} & \ldots a_{\tau_{n+1} n} \end{vmatrix}.$$

If we write

$$\Lambda_j = \frac{d_1 \ldots d_{\hat{j}} \ldots d_{n+1} [D_j(\tau_1, \ldots, \tau_{n+1}; A)]^2}{\sum_{j=1}^{n+1} d_1 \ldots d_{\hat{j}} \ldots d_{n+1} [D_j(\tau_1, \ldots, \tau_{n+1}; A)]^2},$$

then (3.9) can be written in the form

$$x_k^0 = \sum_{j=1}^{n+1} \Lambda_j \frac{D_j^{(k)}(\tau_1, \ldots, \tau_{n+1}; A; b)}{D_j(\tau_1, \ldots, \tau_{n+1}; A)}. \tag{3.10}$$

So we have the following interesting result:

COROLLARY 3.1. *In the conditions of Theorem* 3.1, *each infrasolution of the system* (3.8) *is a convex combination of the Cramerian solutions of a subsystem of* (3.8) *composed of* $n+1$ *equations.*

We have seen that every solution of best approximation of (3.1) in a uniform norm or square norm is an infrasolution of (3.1), too. By virtue of Corollaries 2.2 and 2.4, we conclude that for a certain choice of the constants $d_j$ in (3.9) we can obtain both of the above-mentioned approximate solutions of (3.8).

## REFERENCES

[1] L. Fejér, 'Über die Lage der Nullstellen von Polynomen die aus Minimumfolgerungen gewisser Art entspringen'. *Math. Ann.* **85** (1922) 42–48.

[2] M. Fekete und I. L. von Neumann, 'Über die Lage der Nullstellen gewisser Minimumpolynome', *Jahresberichte der Deutschen Mathematiker Vereinigung* **31** (1922 125–138.

[3] M. Fekete, 'On the Structure of Extremal Polynomials', *Proc. Nat. Acad. Sci. U.S.A.* **37** (1951) 95–103.

[4] I. Maruşciac, 'Sur certains infrapolynomes conditionnés', *Mahtematica* (*Cluj*) **4** (1962), 33–52, **7** (1965) 283–285.

[5] I. Maruşciac, 'Une forme explicite du polynome de meilleure approximation d'une fonction dans le domaine complexe', *Mathematica* (Cluj) **6** (1964) 257–263.

[6] I. Maruşciac, 'Generalized Infrapolynomials', *Mathematica (Cluj)* **7** (1965) 263–282.
[7] I. Maruşciac, 'On the Structure of Restricted Generalized Infrapolynomials', *Mathematica (Cluj)* **12** (1970) 111–125.
[8] T. S. Motzkin and J. L. Walsh, 'Underpolynomials and Infrapolynomials', *Illinois J. Math.* **1** (1957) 405–426.
[9] E. I. Remez, *Basis of the Numerical Methods in Tchebycheff's Approximation*, Kiev 1969.

# ON TRIGONOMETRIC INTERPOLATION WITH EQUIDISTANT KNOTS

G. P. NÉVAI

*Budapest*

Let $f(x)$ be a bounded function with period $2\pi$ defined on the real line; we denote by $\omega_R(f; t)$ ($R$ a non-negative integer) its modulus of smoothness of order $R$, and by $\Delta^R f(x)$ its $R$-th difference with step $h \equiv 2\pi/(2n+1)$ ($n = 1, 2, \ldots$) at the point $x$. Let, further, $S_n(x, f)$ denote the trigonometric polynomial of degree $n$ which coincides with the function $f(x)$ at the nodes

(1) $$x_{kn} = kh \qquad (|k| = 0, 1, 2, \ldots).$$

It has been shown by the author [1], [2], [5], that[1]

(2) $$|S_n(x, f) - f(x) - (-2)^{-R} S_n(x, \Delta^R f)| \leqslant C(R)\,\omega_R(f; h)$$

($R = 0, 1, 2, \ldots$; $n = 1, 2, \ldots$); if, further, $f(x)$ is $r$ times ($r = 0, 1, 2, \ldots$) differentiable ($f^{(r-1)}$ is continuous and $f^{(r)}$ is bounded), then

(3) $$|S_n(x, f) - f(x) - (-2)^{-R} S'_x(x, \Delta^R f)| \leqslant C(R, r) n^{-r} \omega_1(f^{(r)}; |\sin(n+\tfrac{1}{2})x| \cdot h)$$

($n > R = r+1, r+2, \ldots$), where $S'_n$ denotes thc corresponding interpolation sum with just one summand omitted, namely that one in which the node $x_{kn}$ lying closest to $x$ occurs.

Inequality (2) gives an immediate solution to the problem of finding the exact value of the constant $C(R)$ ($R = 1, 2, \ldots$) in the Lebesgue–Jackson inequality

(4) $$|f(x) - S_n(x, f)| \leqslant C(R) |\sin(n+\tfrac{1}{2})x|\, \omega_R(f; h) \log n + O_R[\omega_R(f; h)],$$

---

(1) Here $C(\ )$ denotes a finite positive constant depending only on the parameters occurring in the brackets.

if nothing is known about the periodic function $f(x)$ but boundedness (or, which is the same, continuity). Namely, this constant is equal $2^{-R+1}\pi^{-1}$ (see [2] and [5]). In the case $R = 1$ this problem was solved by Nikolskii in 1941 (see Doklady AN SSSR 31 (1941)).

Inequalities (2) and (3) give a way to improve the estimation (4) concerning its order if the function $f(x)$ satisfies certain additional conditions.

A function $f(x)$ will be said to *belong to the class* $M(N, R)$ ($N$, $R$ positive integers) whenever there exists a decomposition (depending on $f$) of the interval $[-\pi, \pi]$ into $N$ subintervals such that the $R$-th difference of $f(x)$ preserves sign, provided all arguments occurring in this difference are contained in one of the subintervals of the decomposition.

We now formulate the results concerning the approximation by $S_n(x, f)$ to functions of the class $M(N, R)$ (see [1] and [5]).

Let $r$ $(= 0, 1, 2, \ldots)$ be fixed, and let $f(x)$ be a function of the class $M(N, r+1)$, and let $f(x)$ be $r$ times differentiable. Then

$$|f(x) - S_n(x, f)| \leqslant C\big(r, \omega_1(f^{(r)})\big) n^{-r} \log(N+1) \omega_1\big(f^{(r)}; |\sin(n+\tfrac{1}{2})x| \cdot h\big),$$

provided the modulus of continuity of $f^{(r)}(x)$ fulfils the condition

$$\omega_1(f^{(r)}; nt) \leqslant C\big(\omega_1(f^{(r)})\big) n\, \omega_1(f^{(r)}; t) \qquad (0 \leqslant t \leqslant 2\pi;\ n = 1, 2, \ldots),$$

where $\theta \in (0, 1)$ is a fixed number.

If now a function $f(x)$ of the class $M(N, R)$ is $r$ $(< R-1)$ times differentiable, then

$$|f(x) - S_n(x, f)| \leqslant C(r, R) n^{-r} \log(N+1) \omega_1\big(f^{(r)}; |\sin(n+\tfrac{1}{2})x| \cdot h\big).$$

If $f(x)$ is any function of the class $M(N, R)$ with $R \geqslant 2$, then

$$|f(x) - S_n(x, f)| \leqslant C(R) \log(N+1) \omega_{R-1}(f; h).$$

The following application of formulae (2) and (3) concerns functions with bounded $(R, q)$-variation at the nodes (1) (see [3], [4], [5]).

If a function $f(x)$ has a bounded $(R_0, q)$-variation $(R_0 = 1, 2, \ldots;$ $1 \leqslant q < \infty)$ at the nodes (1), i.e. if

$$V = \sup_n \Big[\sum_{k=-n}^{n} |\Delta^{R_0} f(x_{kn})^q|\Big]^{1/q} < \infty,$$

then

$$|f(x) - S_n(x, f)| \leqslant C(R, q, V)|\log \omega_R(f, h)| \cdot \omega_R(f, h)$$

for all $R = R_0, R_0+1, \dots$

A local analogue of this theorem reads as follows. Let a function $f(x)$ be Riemann integrable on its period. If $f(x)$ is continuous at a point $y$ (or on $[a, b] \subset [-\pi, \pi]$) and if for some $\varepsilon > 0$

$$\sup_n \Bigg[ \sum_{\substack{|y - x_{kn}| \leqslant \varepsilon \\ (\text{resp. } a-\varepsilon \leqslant x_{kn} \leqslant b+\varepsilon)}} |\Delta^R f(x_{kn})|^q \Bigg]^{1/q} < \infty,$$

where $R$ $(= 1, 2, \dots)$ and $q \in [1, \infty)$ are fixed, then $S_n(y, f)$ converges (resp. converges uniformly on $[a, b]$) to $f(y)$ as $n \to \infty$.

Let $H(v, \omega)$ denote the class of bounded periodic functions with period $2\pi$ whose variation over the period does not exceed the given number $v$ $(0 < v \leqslant \infty)$ and whose modulus of continuity is majorized by a given modulus of continuity $\omega(t)$. Concerning such functions, the following asymptotic equality holds true:

$$\sup_{f \in H(v, \omega)} |f(x) - S_n(x, f)| = \frac{1}{\pi} |\sin(n+\tfrac{1}{2})x| \log\left[1 + \min\left\{n, \frac{v}{2\omega(h)}\right\}\right] \omega(h) +$$

$$+ O[\omega(|\sin(n+\tfrac{1}{2})x| \cdot h)].$$

All of the above results, except (3), can be extended to the theory of Fourier series. This has partially been done previously by several authors.

## REFERENCES


[1] G. P. Névai, 'On Certain Cases in Which Trigonometric Interpolations Gives an Approximation of the Best Order', *Studia Sci. Math. Hungar.* **7** (1972) 379–390.

[2] G. P. Névai ,'On the Deviation of the Trigonometric Interpolation Sums', *Acta Math. Acad. Sci. Hungar.* **23** (1972) 203–205.

[3] G. P.Névai, 'Notes on Trigonometric Interpolation and Fourier Sums', *Studia Sci. Math. Hungar.* **8** (1973) 113–122.

[4] G. P. Névai, 'An Asymptotic Formula for the Deviation of the Trigonometric Interpolation Sums', *Studia Sci. Math. Hungar.* **7** (1972) 391–394.

[5] G. P. Névai, 'On Trigonometric Interpolation with Equidistant Knots (in Hungarian) III', *Oszt. Közl.* **21** (1973) 449–484.

# KONSTRUKTIVE APPROXIMATIONSTHEORIE MIT HILFE VON JACOBI-POLYNOMEN UND KUGELFUNKTIONEN

SIEGFRIED PAWELKE

*Jülich*

*Abstract.* Let $P_n^{(\alpha,\beta)}(x)$ be the classical Jacobi polynomials and $w(x)$ the corresponding weight function on $(-1, 1)$. Then theorems of Jackson, Bernstein and Zamansky type and the converse of the latter are proved for the weighted approximation by polynomials in $L_w^p$. For that purpose a generalized translation due to R. Askey and St. Wainger and a generalized modulus of continuity will be used, which leads to Lipschitz classes in $L_w^p$ instead of the usual Lipschitz classes. There are similar results in function spaces on the $n$-sphere and the approximation by spherical harmonics.

**Einleitung.** Die Approximation periodischer Funktionen durch trigonometrische Polynome ist in der konstruktiven Approximationstheorie ausführlich untersucht worden. Klassische Ergebnisse sind die Sätze von D. Jackson (1912) und S. N. Bernstein (1913), weitere Ergebnisse stammen von A. Zygmund (1945), M. Zamansky (1949) und S. B. Stečkin (1951). In letzter Zeit bewies G. Sunouchi (1968) in diesem Zusammenhang ein wichtiges Ergebnis. Als Spezialfall aller dieser Ergebnisse erwähnen wir folgenden Äquivalenzsatz (siehe z.B. [6], [7], wo eine ausführlichere Darstellung zu finden ist).

SATZ A. *Es sei $f$ eine stetige, $2\pi$-periodische Funktion und $t_n$ das trigonometrische Polynom bester Approximation vom Grade $\leqslant n$ der Funktion $f$. Dann existieren Konstanten $M_1$, $M_2$ und $M_3$, die nicht von $n$ und $\theta$ abhängen, so daß folgende Aussagen äquivalent sind, falls* $0 < \gamma < 2$ *ist:*

(i) $|t_n(\theta)-f(\theta)| \leqslant M_1 n^{-\gamma}$;

(ii) $|f(\theta+h)+f(\theta-h)-2f(\theta)| \leqslant M_2 h^\gamma \quad (h \geqslant 0)$;

(iii) $|t_n''(\theta)| \leqslant M_3 n^{2-\gamma}$.

*Alle drei Aussagen gelten jeweils für* $0 \leqslant \theta \leqslant 2\pi$.

K. Scherer und P. L. Butzer [6], [7] haben diese Ergebnisse auf beliebige Banachräume übertragen. An die Stelle des Differentialoperators $d/d\theta$ tritt ein beliebiger abgeschlossener Operator, für den eine sogenannte Jacksonsche Ungleichung und eine Bernsteinsche Ungleichung erfüllt sein müssen. Das Verhalten des Stetigkeitsmoduls von $f$ oder die Lipschitzbedingung (wie Aussage (ii) in Satz A) wird durch das Verhalten des von J. Peetre [17] eingeführten $K$-Funktionals ersetzt.

In konkreten Funktionenräumen ist der Satz von Scherer und Butzer anwendbar, wenn man für einen gewissen Differentialoperator die im vorhergehenden Abschnitt erwähnten Ungleichungen herleiten und das äquivalente Verhalten eines verallgemeinerten Stetigkeitsmoduls und des $K$-Funktionals beweisen kann.

In dieser Arbeit wird die Approximation von Funktionen durch Polynome in gewissen gewichteten $L^p$-Räumen und auch für stetige Funktionen untersucht. Die Jacobi-Polynome sind die bzgl. dieser Gewichtsfunktionen orthogonalen Polynome. An die Stelle der gewöhnlichen Ableitung tritt dann der Differentialoperator zweiter Ordnung, dessen Eigenfunktionen die Jacobi-Polynome sind. Der Stetigkeitsmodul wird mit Hilfe einer verallgemeinerten Translation definiert. Man erhält unter anderem ein Analogon zu Satz A. Der Spezialfall der ultrasphärischen Polynome und, von diesem ausgehend, die Approximation durch Kugelfunktionen werden am Schluß der Arbeit kurz behandelt.

**1. Die Faltungsstruktur für Jacobi-Reihen.** Es sei $\alpha \geqslant \beta \geqslant -\frac{1}{2}$, $\alpha > -\frac{1}{2}$ und $P_n^{(\alpha,\beta)}(x)$ das Jacobi-Polynom vom Grade $n$ der Ordnung $(\alpha, \beta)$ (siehe [21]), normalisiert durch

$$P_n^{(\alpha,\beta)}(1) = 1\,. \tag{1.1}$$

Bezüglich der Gewichtsfunktion

$$w_{\alpha,\beta}(\theta) = \left(\sin\frac{\theta}{2}\right)^{2\alpha+1}\left(\cos\frac{\theta}{2}\right)^{2\beta+1}$$

bilden die Polynome $\{P_n^{(\alpha,\beta)}(\cos\theta)\}$ ein orthogonales System auf $[0, \pi]$. Es sei $L^p_{\alpha,\beta}$ $(1 \leqslant p < \infty)$ die Klasse der auf $(0, \pi)$ Lebesgue-messbaren Funktionen $f$, für die gilt

$$\|f\|_p := \left\{\int_0^\pi |f(\theta)|^p w_{\alpha,\beta}(\theta)\,d\theta\right\}^{1/p} < \infty\,,$$

und $C$ die Klasse der auf $[0, \pi]$ stetigen Funktionen versehen mit der Supremumnorm. Mit $X$ wird immer jeweils einer der Räume $C$ oder $L^p_{\alpha,\beta}$ bezeichnet und $\|\cdot\|$ ist die Norm des Raumes $X$.

Die Fourier–Jacobi-Koeffizienten $f^\wedge(n)$ einer Funktion $f \in X$ sind definiert durch

$$(1.2) \qquad f^\wedge(n) = \int_0^\pi f(\theta) P_n^{(\alpha,\beta)}(\cos\theta)\, w_{\alpha,\beta}(\theta)\, d\theta.$$

Damit hat $f$ die Fourier–Jacobi-Entwicklung

$$(1.3) \qquad f(\theta) \sim \sum_{n=0}^{n} h_n f^\wedge(n) P_n^{(\alpha,\beta)}(\cos\theta)$$

mit

$$h_n^{-1} = \int_0^\pi \{P_n^{(\alpha,\beta)}(\cos\theta)\}^2 w_{\alpha,\beta}(\theta)\, d\theta.$$

Wir werden im folgenden der Einfachheit halber stets die Indizes $(\alpha, \beta)$ bei der Gewichtsfunktion $w_{\alpha,\beta}$, den Polynomen $P_n^{(\alpha,\beta)}$ und den Räumen $L^p_{\alpha,\beta}$ weglassen. Zu bemerken ist jedoch, daß die jetzt definierten Translationen und Faltungen von $\alpha$ und $\beta$ abhängen.

Eine verallgemeinerte Translation $(T_u f)(\theta)$ $(0 \leqslant u \leqslant \pi)$ von $f \in X$ wird nun nach R. Askey und St. Wainger [1] definiert durch

$$(1.4) \qquad (T_u f)^\wedge(n) = P_n(\cos u) f^\wedge(n) \qquad (n = 0, 1, 2, \ldots).$$

Diese Translation hat die folgenden Eigenschaften.

Satz 1.1. *Ist $f \in X$, so ist $T_u f$ für fast alle $u \in [0, \pi]$ ebenfalls in $X$ und es gilt*

$$(1.5) \qquad \|T_u f\| \leqslant \|f\|,$$

$$(1.6) \qquad \lim_{u \to 0} \|T_u f - f\| = 0.$$

Die Ungleichung (1.5) wurde für $p = 1$ von R. Askey und St. Wainger [1], für $p > 1$ von C. Ganser [10] bewiesen. G. Gasper [11] leitete eine Integraldarstellung der Translation her und bewies die Positivität des Translationsoperators. $T_u$ hat also die Operatornorm $\|T_u\| = 1$. Die Beziehung (1.6) folgt aus (1.4), aus $P_n(1) = 1$, aus (1.5) und dem Satz von Banach–Steinhaus.

Für $f, g \in X$ kann nach R. Askey und St. Wainger [1] eine Faltung $f*g$ wie folgt definiert werden:

$$(1.7) \qquad (f*g)(\theta) = \int_0^\pi (T_u f)(\theta) g(u) w(u)\, du.$$

Die Funktion $f*g$ existiert in $X$ und es gilt

$$(1.8) \qquad \|f*g\| \leqslant \|f\|_1 \|g\|_1,$$

$$(1.9) \qquad (f*g)^\wedge(n) = f^\wedge(n) g^\wedge(n).$$

BEMERKUNG. Im hier ausgeschlossenen Fall $\alpha = \beta = -\frac{1}{2}$ ist $P_n(\cos\theta) = \cos n\theta$, die Translation hat die Form

$$(T_h f)(\theta) = \tfrac{1}{2}\{f(\theta+h)+f(\theta-h)\}$$

und die Faltung ist die gewöhnliche fouriersche Faltung speziell für gerade Funktionen.

**2. Verallgemeinerte Stetigkeitsmoduln und Funktionenklassen.** Für $f \in X$ kann man nun einen verallgemeinerten Stetigkeitsmodul wie in [10] definieren durch

$$(2.1) \qquad \omega(f; h) := \sup_{0 \leqslant u \leqslant h} \|T_u f - f\|.$$

Die Menge aller Funktionen $f \in X$, für die gilt

$$(2.2) \qquad \omega(f; h) \leqslant M_f h^\gamma \qquad (0 < \gamma \leqslant 2)$$

ist die verallgemeinerte Lipschitzklasse der Ordnung $\gamma$. Die Differenzierbarkeit von Funktionen wird durch folgenden Differentialoperator ausgedrückt. Die Jacobi-Polynome $P_n(\cos\theta)$ erfüllen die folgende Differentialgleichung

$$(2.3) \qquad \frac{d}{d\theta}\left\{w(\theta)\frac{d}{d\theta}P_n(\cos\theta)\right\} = -n(n+\alpha+\beta+1)\, w(\theta) P_n(\cos\theta),$$

d.h. sie sind Eigenfunktionen des formalen Differentialoperators

$$(2.4) \qquad \Delta := \{w(\theta)\}^{-1} \frac{d}{d\theta}\left\{w(\theta)\frac{d}{d\theta}\right\}.$$

Die Definitionsmenge von $\Delta$ wird so gewählt, daß $\Delta$ in $X$ ein abgeschlossener Operator ist.

DEFINITION 2.1. Man sagt, $f \in X$ ist in der Definitionsmenge $D(\Delta)$ des Operators $\Delta$ enthalten, falls eine Funktion $g \in X$ existiert, so daß gilt

$$(2.5) \qquad -n(n+\alpha+\beta+1)f^{\wedge}(n) = g^{\wedge}(n) \qquad (n = 0, 1, 2, \ldots).$$

Es ist dann $\Delta f = g$ in $X$.

Aus dieser Definition ergibt sich, daß $\Delta$ abgeschlossen ist. Der Definitionsbereich von $\Delta$ wird genauer durch folgendes Lemma charakterisiert. Es ist im Falle $\alpha = \beta$ in [14] zu finden.

LEMMA 2.1. *Für eine Funktion $f \in X$ sind die folgenden Aussagen äquivalent:*

(a) *es existiert eine Funktion $g \in X$, so daß gilt*

$$\lim_{h\to 0}\left\|\frac{4(\alpha+1)}{h^2}(T_h f - f) - g\right\| = 0;$$

(b) $f \in D(\Delta)$;

(c) *es existiert ein $g \in X$ mit $g^{\wedge}(0) = 0$, so daß für ein $a \in [0, \pi]$ gilt*

$$f(\theta) - f(a) = \int_a^\theta \frac{dv}{w(v)} \int_0^v g(u)\,w(u)\,du;$$

(d) *$f$ ist lokal absolut stetig auf $(0, \pi)$, die Funktion $w(\theta)f'(\theta)$ ist absolut stetig auf $[0, \pi]$, verschwindet für $\theta = 0$ und $\theta = \pi$, und es existiert eine Funktion $g \in X$, so daß in $X$ gilt*

$$\Delta f = g.$$

BEWEIS. Aus der Grenzwertbeziehung in (a) folgt für die Fourier–Jacobi-Koeffizienten

$$(2.6) \quad \lim_{h\to 0}\frac{4(\alpha+1)}{h^2}\{P_n(\cos h) - 1\}f^{\wedge}(n) = g^{\wedge}(n) \qquad (n = 0, 1, 2, \ldots)$$

und aus der Formel (4.21.2) in [21] erhält man

$$(2.7) \qquad \lim_{h\to 0}\frac{P_n(\cos h) - 1}{1 - \cos h} = -\frac{n(n+\alpha+\beta+1)}{2(\alpha+1)}$$

und damit

$$(2.8) \qquad -n(n+\alpha+\beta+1)f^{\wedge}(n) = g^{\wedge}(n),$$

also (b). Um von (b) auf (c) zu schließen, betrachten wir die Poissonschen Mittel von $f$, definiert durch

$$(V_r f)(\theta) = \sum_{n=0}^{\infty} r^n h_n f^{\wedge}(n) P_n(\cos\theta) \qquad (0 \leqslant r < 1).$$

Aus (b) folgt mit Hilfe der Differentialgleichung (2.3) die Beziehung

$$\Delta(V_r f)(\theta) = (V_r g)(\theta)$$

und durch Integration für $0 < \theta < \pi$

$$(V_r f)(\theta) - (V_r f)(a) = \int_a^{\theta} \frac{dv}{w(v)} \int_0^{v} (V_r g)(u) w(u) du. \tag{2.9}$$

Nach [1] gilt

$$\lim_{r \to 1+} (V_r f)(\theta) = f(\theta) \tag{2.10}$$

fast überall für $f \in L^p$ bzw. überall für $f \in C$ und

$$\lim_{r \to 1+} \|V_r f - f\| = 0 \qquad (f \in X). \tag{2.11}$$

Benutzt man (2.10) für die linke Seite und (2.11) für die rechte Seite von (2.9), dann folgt daraus für $r \to 1+$

$$f(\theta) - f(a) = \int_a^{\theta} \frac{dv}{w(v)} \int_0^{v} g(u) w(u) du \tag{2.12}$$

d.h. die Aussage (c).

Aus der Darstellung (2.12) folgt (d) durch Differentiation. Die Randbedingung an der Stelle $\theta = \pi$ ergibt sich aus

$$w(\pi) f'(\pi) = g^{\wedge}(0) = 0.$$

Aus (d) folgt nun

$$g^{\wedge}(n) = \int_0^{\pi} P_n(\cos\theta)(\Delta f)(\theta) w(\theta) d\theta.$$

Durch zweimalige partielle Integration oder mit Hilfe der Lagrangeschen Identität für selbstadjungierte Differentialoperatoren erhält man mit Rücksicht auf die Randbedingungen

$$g^{\wedge}(n) = \int_0^{\pi} \Delta\big(P_n(\cos\theta)\big) f(\theta) w(\theta) d\theta = -n(n+\alpha+\beta+1) f^{\wedge}(n).$$

Multipliziert man diese Gleichung mit $P_n(\cos u)$, so erhält man wie im zweiten Teil dieses Beweises die Identität

$$\{P_n(\cos h)-1\}f^\wedge(n) = \int_0^h \frac{dv}{w(v)} \int_0^v w(u)\, P_n(\cos u)\, g^\wedge(n)\, du$$

$$(n = 0, 1, 2, \ldots),$$

woraus mit dem Eindeutigkeitssatz die Beziehung

$$(2.13)\qquad T_h f - f = J_h g := \int_0^h \frac{dv}{w(v)} \int_0^v w(u)\, T_u g\, du$$

folgt. Mit

$$(2.14)\qquad C(h) := \int_0^h \frac{dv}{w(v)} \int_0^v w(u)\, du$$

ergibt sich nun

$$\frac{T_h f - f}{C(h)} - g = \frac{1}{C(h)} \int_0^h \frac{dv}{w(v)} \int_0^u w(u)\{T_u g - g\}\, du$$

und daraus

$$\left\| \frac{T_h f - f}{C(h)} - g \right\| \leqslant \sup_{0 \leqslant u \leqslant h} \|T_u g - g\|.$$

Wegen (1.6) geht die rechte Seite gegen Null für $h \to 0$. Da aber gilt

$$(2.15)\qquad \lim_{h \to 0} \frac{4(\alpha+1)}{h^2}\, C(h) = 1,$$

folgt die Aussage (a) aus (d), womit Lemma 2.1 vollständig bewiesen ist.

Wir betrachten nun den oben eingeführten Stetigkeitsmodul. Man kann ihn durch ein konkaves Funktional ersetzen, das von J. Peetre [17] eingeführt wurde. Dies ist wichtig in den späteren Ergebnissen. Für jedes $f \in X$ und $t \geqslant 0$ ist das $K$-Funktional definiert durch

$$(2.16)\qquad K(f; t) := \inf_{g \in D(\Delta)} (\|f-g\| + t\|\Delta g\|).$$

Für jedes feste $f \in X$ ist $K(f, t)$ eine monoton wachsende, konkave (daher stetige) Funktion in $t$ auf $(0, \infty)$ und für jedes feste $t$ eine Halbnorm auf $X$ (siehe [4]). Das Funktional ist keine Norm auf $X$, da $\|\Delta g\|$ nur eine Halbnorm auf $D(\Delta)$ bildet. Es gilt

SATZ 2.2. *Es existieren zwei positive Konstanten $C_1$ und $C_2$, die nur von $\alpha$ und $\beta$ abhängen, so daß*

$$C_1\omega(f;t) \leqslant K(f;t^2) \leqslant C_2\omega(f;t) \qquad (0 \leqslant t \leqslant \pi).$$

BEWEIS. Er verläuft ähnlich wie in [12] und [5]. Den rechten Teil der Ungleichung beweist man wie folgt. Für eine beliebige Funktion $f \in X$ erhält man wegen (1.5)

$$\omega(f;t) \leqslant 2\|f\|. \tag{2.17}$$

Ist $f \in D(\Delta)$, dann gilt wegen (2.13)

$$T_h f - f = J_h \Delta f$$

und daher mit (2.14) und (1.5)

$$\|T_h f - f\| \leqslant C(h)\|\Delta f\|.$$

Für $0 < h \leqslant \pi/2$ gilt

$$\frac{1}{\alpha+1}\sin^2\left(\frac{h}{2}\right) \leqslant C(h) \leqslant \frac{2^{\beta+1}}{\alpha+1}\sin^2\left(\frac{h}{2}\right),$$

wobei die rechte Seite der Ungleichung in [2] bewiesen wurde und der Beweis der linken Seite analog verläuft.

Für $\frac{\pi}{2} \leqslant h \leqslant \pi$ benutzt man eine Idee von Chernoff–Ragozin, die in [5] benutzt wurde. Es gilt nämlich, da $(\Delta f)^\wedge(0) = 0$ ist,

$$\int_0^\pi T_u \Delta f w(u)\,du = 0$$

und damit

$$\begin{aligned} T_h f - f &= \int_0^{\pi/2} \frac{dv}{w(v)} \int_0^v w(u) T_u \Delta f\,du - \int_{\pi/2}^h \frac{dv}{w(v)} \int_v^\pi T_u \Delta f w(u)\,du \\ &= J_{\pi/2}\Delta f + I(\Delta f). \end{aligned}$$

Nun ist wegen $h \geqslant \pi/2$

$$\|J_{\pi/2} f\| \leqslant C(\pi/2)\|\Delta f\| \leqslant \|\Delta f\| \frac{2^{\beta+1}}{\alpha+1}\sin^2(h/2)$$

und

$$\|I(\Delta f)\| \leqslant \|\Delta f\| \int_{\pi/2}^{h} \frac{dv}{w(v)} \int_{v}^{\pi} w(u)\,du \leqslant \|\Delta f\| \frac{2^{\alpha+1}}{\beta+1} \sin^2(h/2),$$

wobei die letzte Ungleichung ähnlich wie in [2] bewiesen wird. Also gilt auch für $\pi/2 \leqslant h \leqslant \pi$

$$\|T_h f - f\| \leqslant C_{\alpha,\beta} \|\Delta f\| \sin^2(h/2), \tag{2.18}$$

womit diese Ungleichung für $0 \leqslant h \leqslant \pi$ gilt.

Also hat man für $f \in D(\Delta)$ die Abschätzung

$$\omega(f; h) \leqslant C_{\alpha,\beta} \|\Delta f\| h^2 \tag{2.19}$$

und damit für beliebiges $g \in D(\Delta)$ mit (2.17) und (2.19)

$$\begin{aligned} \omega(f; t) &\leqslant \omega(f-g; t) + \omega(g; t) \\ &\leqslant C(\|f-g\| + t^2 \|\Delta g\|). \end{aligned}$$

Aus der Definition des $K$-Funktionals (2.16) ergibt sich

$$\omega(f; t) \leqslant C\, K(f; t^2),$$

womit der linke Teil der Ungleichung aus Satz 2.2 bewiesen ist.

Zum Beweis des rechten Teils der Ungleichung geht man wie in [12] vor und definiert für ein $t$ mit $0 < t \leqslant \pi/2$ ein $g \in X$ mit Hilfe von (2.13) und (2.14) durch

$$g = \frac{1}{C(t)} J_t(f).$$

Bildet man die Fourier–Jacobi-Koeffizienten von beiden Seiten und benutzt man die Differentialgleichung (2.3), dann folgt

$$-n(n+\alpha+\beta+1) g^{\wedge}(n) = \frac{1}{C(t)} \cdot \big(P_n(\cos t) - 1\big) \cdot f^{\wedge}(n),$$

also nach Lemma 2.1

$$\Delta g = \frac{1}{C(t)} (T_t f - f)$$

und daraus erhält man

$$t^2 \|\Delta g\| \leqslant M \omega(f; t) \qquad (0 < t \leqslant \pi/2).$$

Nun gilt

$$f-g = -\frac{1}{C(t)}\int_0^t \frac{dv}{w(v)}\int_0^u w(u)(T_u f-f)\,du$$

und damit

$$\|f-g\| \leqslant \omega(f;t).$$

Daraus folgt

$$K(f;t^2) \leqslant \|f-g\|+t^2\|\Delta g\|$$
$$\leqslant C_2\omega(f;t) \quad (0 \leqslant t \leqslant \pi/2).$$

Für $\pi/2 \leqslant t \leqslant \pi$ benutzen wir eine Methode aus [5]. Es gilt

$$K(f;t^2) = K\big(f;4(t/2)^2\big) \leqslant 4K\left(f;\frac{t^2}{4}\right)$$
$$\leqslant 4C_2\omega(f;t/2) \leqslant 4C_2\omega(f;t),$$

wobei die erste Ungleichung in [4] (ch. III) und [17] zu finden ist. Damit ist der Satz vollständig bewiesen.

Die Fejérschen Mittel $S_n^{(r)}f$ der Ordnung $r$ ($r \geqslant 0$) der Teilsumme $S_nf$ von (1.3) sind wie üblich definiert durch

$$(S_n^{(r)}f)(\theta) = (C_n^r)^{-1}\sum_{k=0}^{n} C_{n-k}^r h_k f^\wedge(k)\,P_k(\cos\theta)$$

mit

$$C_n^r = \binom{n+r}{n}$$

oder als Faltung

$$S_n^{(r)}f = K_n^{(r)} * f$$

mit dem Kern

$$K_n^{(r)}(\theta) = (C_n^r)^{-1}\sum_{k=0}^{n} C_{n-k}^r h_k P_k(\cos\theta).$$

Nun ist aber wegen [21] (§§ 3.1 und 9.4) die Folge

$$\{L_n^{(r)}\} = \{\|K_n^{(r)}\|_1\}.$$

Folge der Lebesgue-Konstanten der Ordnung $r$ von $(S_n^{(r)}f)$ (1), für die für $r > \alpha+1/2$ gilt [21] (§ 9.1, 9.41)

(2.20) $$L_n^{(r)} \leqslant M_r,$$

wobei $M_r$ unabhängig von $n$ ist. Daraus folgt ein in [1] erwähntes Ergebnis.

SATZ 2.3. *Für $r > \alpha+1/2$ gilt für jede Funktion $f \in X$*

$$\|S_n^{(r)}f\| \leqslant M_r\|f\|, \tag{2.21}$$

$$\lim_{n\to\infty}\|S_n^{(r)}f-f\| = 0. \tag{2.22}$$

BEWEIS. Mit (1.8) und (2.20) erhält man

$$\|S_n^{(r)}f\| = \|K_n^{(r)}*f\| \leqslant \|K_n^{(r)}\|_1\|f\| \leqslant M_r\|f\|.$$

Da (2.22) für alle Polynome gilt, folgt die Aussage (2.22) wegen (2.21) für alle $f \in X$ aus dem Satz von Banach–Steinhaus.

**3. Jacksonsche und Bernsteinsche Ungleichungen.** Zum Beweis der Jacksonschen Ungleichung benötigt man folgendes Ergebnis.

SATZ 3.1. *Zu jeder Funktion $f \in D(\Delta)$ gibt es eine Folge von algebraischen Polynomen $\{Q_n^{(\alpha,\beta)}f\}$, so daß gilt*

$$\|Q_n^{(\alpha,\beta)}f-f\| \leqslant M\|\Delta f\|n^{-2},$$

*wobei die Konstante $M$ nur von $\alpha$ und $\beta$ abhängt.*

BEWEIS. In Analogie zu D. J. Newman und H. S. Shapiro [13] definiert man eine Folge von positiven polynomialen Kernen $k_n$ vom Grade $\leqslant n$ und dann

$$Q_n^{(\alpha,\beta)}f = k_n * f;$$

$k_n$ ist definiert durch

$$k_{2n}(\theta) = k_{2n+1}(\theta) = c_n\{P_{n+1}(\cos\theta)/(\cos\theta-\cos\theta_{n+1})\}^2,$$

wobei $\cos\theta_{n+1}$ die größte Nullstelle von $P_{n+1}$ ist und $c_n$ so gewählt ist, daß $\|k_n\|_1 = 1$ ist. Die Polynome $k_{2n}$ haben die Eigenschaft (siehe [13], [19], [21])

$$\int_0^\pi \sin^2(\theta/2)k_{2n}(\theta)w(\theta)\,d\theta = \sin^2(\theta_{n+1}/2) \tag{3.1}$$

mit

$$\sin^2(\theta_{n+1}/2) = O(n^{-2}) \qquad (n \to \infty). \tag{3.2}$$

Das Integral (3.1) ist das zweite trigonometrische Moment des Kerns $k_{2n}$ (siehe [2]).

Für gerade $n = 2m$ hat man

$$Q_n^{(\alpha,\beta)}f-f = \int_0^\pi (T_u f-f)k_{2m}(u)w(u)\,du.$$

Die Ungleichung (2.18) führt zu der Abschätzung

$$\begin{aligned}\|Q_n^{(\alpha,\beta)}f-f\| &\leqslant C_{\alpha,\beta}\|\Delta f\| \int_0^\pi \sin^2(u/2)k_{2m}(u)w(u)\,du \\ &\leqslant C_{\alpha,\beta}\|\Delta f\| \sin^2(\theta_{m+1}/2) \\ &\leqslant M\|\Delta f\| n^{-2}\end{aligned}$$

wegen (3.1) und (3.2). Für ungerade $n$ verläuft der Beweis ähnlich wegen der Definition von $k_n$. Damit ist Satz 3.1 bewiesen.

BEMERKUNG. Für eine beliebige Funktion $f \in X$ kann man $\|Q_n^{(\alpha,\beta)}f-f\|$ durch den verallgemeinerten Stetigkeitsmodul von $f$ abschätzen. Dabei ist Satz 2.2 und die Konkavität des $K$-Funktionals wesentlich (siehe [15]).

Aus Satz 3.1 kann man Abschätzungen für die beste Approximation gewinnen. Es sei $\mathscr{T}_n$ der lineare Raum aller geraden trigonometrischen Polynome vom Grade $\leqslant n$ d.h. der trigonometrischen Polynome in $\theta$ der Form $p_n(\cos\theta)$, wobei $p_n$ ein algebraisches Polynom vom Höchstgrad $n$ ist. Die Minimalabweichung $E_n(X,f)$ von $f \in X$ bezüglich $\mathscr{T}_n$ und das Polynom bester Approximation $t_n^*$ von $f$ sind definiert durch

$$E_n(X,f) = \inf_{t_n \in \mathscr{T}_n} \|f-t_n\| = \|f-t_n^*\|.$$

Wegen $Q_n^{(\alpha,\beta)}f \in \mathscr{T}_n$ folgt nun $E_n(X,f) \leqslant \|Q_n^{(\alpha,\beta)}f-f\|$. Es ergibt sich also aus Satz 3.1 die nachstehende Folgerung.

FOLGERUNG 3.2. *Für jede Funktion $f \in D(\Delta)$ gilt*

$$E_n(X,f) \leqslant M\|\Delta f\| n^{-2}.$$

Wie üblich folgt hieraus

FOLGERUNG 3.3 (Jacksonsche Ungleichung). *Ist $f \in D(\Delta^m)$, wobei $m$ eine ganze positive Zahl ist, dann gilt*

$$E_n(X,f) \leqslant M^m n^{-2m}\|\Delta^m f\|.$$

Wir kommen nun zu einer Ungleichung vom Bernsteinschen Typ, einer Verallgemeinerung der klassischen Bernsteinschen Ungleichung für die Ableitung von trigonometrischen Polynomen.

SATZ 3.4 (Bernsteinsche Ungleichung). *Ist* $t_n(\theta) = p_n(\cos\theta)$ *ein gerades trigonometrisches Polynom in* $\theta$, *dann gilt*

$$\|\Delta t_n\| \leqslant B_{\alpha,\beta} n^2 \|t_n\|,$$

*wobei die Konstante* $B_{\alpha,\beta}$ *nur von* $(\alpha, \beta)$ *und der Norm des Raumes* $X$ *abhängt.*

BEWEIS. Für die Fejérschen Mittel $S_n^{(r)}f$ einer Funktion $f \in X$ gilt für $r > \alpha + 1/2$ die Ungleichung (2.21). Damit folgt Satz 3.4 aus einem Ergebnis von E. M. Stein [20].

Induktiv erhält man für die $m$-te Potenz des Operators $\Delta$ die

FOLGERUNG 3.5. *Unter den Voraussetzungen von Satz* 3.4 *gilt für jede ganze positive Zahl* $m$

$$\|\Delta^m t_n\| \leqslant (B_{\alpha,\beta})^m n^{2m} \|t_n\|.$$

**4. Approximationssätze.** Nach diesen Vorbereitungen und mit den Ergebnissen aus dem vorhergehenden Abschnitt kann man einen allgemeinen Approximationssatz aufstellen.

SATZ 4.1. *Für* $f \in X$ *und ganze Zahlen* $0 \leqslant k \leqslant l < m$ *existieren Konstanten* $M_1$, $M_2$, $M_3$ *und* $M_4$, *die nicht von* $n$ *abhängen, so daß die folgenden Aussagen äquivalent sind, falls* $0 < \gamma < 2$ *ist*:

(i) $E_n(X, f) \leqslant M_1 n^{-(2l+\gamma)}$;

(ii) $f \in D(\Delta^k)$ *und* $\|\Delta^k t_n^* - \Delta^k f\| \leqslant M_2 n^{-(2l+\gamma-2k)}$;

(iii) $\|\Delta^m t_n^*\| \leqslant M_3 n^{2m-2l-\gamma}$;

(iv) $f \in D(\Delta^l)$ *und* $\omega(\Delta^l f; h) \leqslant M_4 h^\gamma$ [1].

BEWEIS. Die Äquivalenz der Aussagen (i)–(iii) folgt aus einem allgemeinen Satz von Scherer und Butzer [6], [7], der die Gültigkeit von Ungleichungen vom Jacksonschen Typ und vom Bernsteinschen Typ

---

[1] Die Äquivalenz von (i) und (iv) wurde von H. Bavinck, 'Jacobi Series and Approximation', *Mathematical Centre Tracts* **39**, Mathematisch Centrum Amsterdam, 1972, bewiesen.

zur Voraussetzung hat. Diese Voraussetzungen sind hier wegen Folgerung 3.3 und Folgerung 3.5 erfüllt. Die Äquivalenz von (i) und (iv) beweist man analog zu [16], worauf hier verzichtet werden soll.

BEMERKUNG. Setzt man $\alpha = \beta = -1/2$ und außerdem $k = l = 0$, $m = 1$ und betrachtet den Raum $C$, dann stimmt Satz 4.1 mit Satz A in der Einleitung überein. Allerdings gilt Satz A für beliebige periodische Funktionen, Satz 4.1 jedoch nur für gerade Funktionen. Die Translation wurde für diesen Fall am Ende von § 1 angegeben und der Operator $\Delta$ ist die gewöhnliche zweite Ableitung.

Führt man die Substitution $\cos\theta = t$ $(0 \leqslant \theta \leqslant \pi)$ durch, dann ist Satz 4.1 ein Ergebnis über die Approximation von Funktionen, die auf $[-1, 1]$ definiert sind, durch algebraische Polynome.

Im Falle $\gamma = 2$ kann man noch die Äquivalenz von (iii) und (iv) für $l = m-1$ beweisen.

SATZ 4.2. *Es gilt* $\|\Delta^m t_n^*\| = O(1)$ $(n \to \infty)$ *genau dann, wenn* $f \in D(\Delta^{m-1})$ *ist und* $\omega(\Delta^{m-1}f; h) = O(h^2)$ $(h \to 0)$.

Dieses Ergebnis bedeutet, daß man aus der Beschränktheit der Folge der „Ableitungen der Ordnung $m$" $\{\Delta^m t_n^*\}$ nur darauf schließen kann, daß die Grenzfunktion $f$ der Folge $\{t_n^*\}$ „$(m-1)$-mal differenzierbar" ist d.h. $f \in D(\Delta^{m-1})$ und außerdem gilt $\omega(\Delta^{m-1}f; h) \leqslant M_4 h^2$. Der Beweis verläuft analog zu dem von Satz 4.4 in [16].

Wir kommen nun zu Spezialfällen und Kugelfunktionen.

A. Im Raum $X = C$ ist die Norm unabhängig von $\alpha$ und $\beta$. Da $\mathscr{T}_n$ unabhängig von $\alpha$ und $\beta$ ist, gilt dies auch für die Minimalabweichung $E_n(X, f)$. Der Operator $\Delta$, definiert in (2.4), die Translation, definiert durch (1.4) und damit der Stetigkeitsmodul hängen von $\alpha$ und $\beta$ ab. Die Aussagen (ii), (iii) und (iv) von Satz 4.1 sind daher im Raum $C$ auch äquivalent für verschiedene Wertepaare von $(\alpha, \beta)$.

B. Der Fall $\alpha = \beta$. In diesem Fall sind die Polynome $P_n$ die ultrasphärischen oder Gegenbauer-Polynome. Führt man die Transformation $\cos\theta = t$ $(0 \leqslant \theta \leqslant \pi)$ durch, dann hat die Gewichtsfunktion bis auf einen konstanten Faktor die Form

$$v(t) = (1-t^2)^\alpha$$

und man erhält Funktionen auf $[-1, 1]$. Die Translation $T_h f$ hat hier die Integraldarstellung

$$(4.1)\quad (T_h f)(t) = \begin{cases} c_\alpha \int_{-1}^{1} f\left(t \cdot \cos h + \sqrt{1-t^2} \cdot z \cdot \sin h\right)(1-z^2)^{\alpha-1/2} dz \\ \qquad\qquad\qquad\qquad\qquad\qquad (\alpha > -1/2), \\ \frac{1}{2}\left\{f\left(t \cdot \cos h + \sqrt{1-t^2} \cdot \sin h\right) + \right. \\ \qquad \left. + f(t \cdot \cos h - \sqrt{1-t^2} \cdot \sin h)\right\} \quad (\alpha = -\frac{1}{2}) \end{cases}$$

mit

$$c_\alpha^{-1} = \int_{-1}^{1} (1-z^2)^{\alpha-1/2} dz.$$

Diese Translation wurde von S. Bochner [3] eingeführt. Der Operator $\Delta$ kann jetzt geschrieben werden als

$$\Delta f(t) = (1-t^2)^{-\alpha} \frac{d}{dt}\left[(1-t^2)^{\alpha+1} \frac{d}{dt} f(t)\right].$$

In diesem Fall wurde Satz 4.1 vom Autor in [16] bewiesen. Spezialfälle sind bekannt (siehe z.B. [8] und die angegebene Literatur in [15]).

C. Kugelfunktionen. Die Ergebnisse dieser Arbeit sind auch für Funktionen gültig, die auf der Einheitssphäre $S^k$ im $k$-dimensionalen euklidischen Raum $R^k$ definiert sind. Der Operator $\Delta$ ist hier der Laplace-Beltrami-Operator auf der Sphäre. Seine Eigenfunktionen sind die Kugelfunktionen. Die verallgemeinerte Translation oder das „sphärische Mittel" einer Funktion $f \in L^1(S^k)$ ist definiert durch

$$(4.2)\qquad (T_h f)(x) = \frac{1}{O_{k-1}(\sin h)^{2\alpha+1}} \int_{(x,y)=\cos h} f(y)\, dt(y)$$
$$(2\alpha = k-3;\ 0 < h < \pi).$$

Hierbei sind $x, y \in S^k$, $(x, y)$ ist ihr euklidisches Skalarprodukt, $dt$ das $(k-2)$-dimensionale Oberflächenelement der Fläche $(x, y) = \cos h$ auf $S^k$ und $O_k$ die Oberfläche von $S^k$. Die Integration in (4.2) wird also auf einem Kreis auf $S^k$ um $x$ ausgeführt.

Betrachtet man zonale Funktionen auf $S^k$, dann geht für diese Funktionen die in (4.2) definierte Translation über in die Translation (4.1).

Untersucht man die Approximation von Funktionen aus $L^p(S^k)$ oder $C(S^k)$ durch Linearkombinationen von Kugelfunktionen, dann gelten zu den Sätzen 4.1 und 4.2 analoge Ergebnisse (siehe [5], [9], [16], [18], [19]).

## LITERATUR

[1] R. Askey and St. Wainger, 'A Convolution Structure for Jacobi Series', *Amer. J. Math.* **91** (1969) 463–485.

[2] H. Bavinck, 'On Positive Convolution Operators for Jacobi Series', *Tôhoku Math. J.* **24** (1972) 55–69.

[3] S. Bochner, 'Positive Zonal Functions on Spheres', *Proc. Nat. Acad. Sci. USA* **40** (1954) 1141–1147.

[4] P. L. Butzer and H. Berens, *Semi-Groups of Operators and Approximation*, Berlin–Heidelberg–New York 1967.

[5] P. L. Butzer and H. Johnen, 'Lipschitz Spaces on Compact Manifolds', *J. Funct. Anal.* **7** (1971), 242–266.

[6] P. L. Butzer und K. Scherer, 'Über die Fundamentalsätze der klassischen Approximationstheorie in abstrakten Räumen', in: *Abstract Spaces and Approximation, ISNM* **10**, Basel 1969, 113–125.

[7] P. L. Butzer and K. Scherer, 'On the Fundamental Approximation Theorems of D. Jackson, S. N. Bernstein and Theorems of M. Zamansky and S. B. Stečkin', *Aequationes Math.* **3** (1969) 170–185.

[8] A. S. Džafarov, 'Some Direct and Inverse Theorems in the Theory of Best Approximation of Functions by Algebraic Polynomials', *Soviet Math. Dokl.* **10** (1969) 916–919.

[9] A. S. Džafarov, 'On Spherical Analogs of the Classical Theorems of Jackson and Bernstein', *Soviet Math. Dokl.* **13** (1972) 373–377.

[10] C. Ganser, 'Modulus of Continuity Conditions for Jacobi series', *J. Math. Anal. Appl.* **27** (1969) 575–600.

[11] G. Gasper, 'Positivity and the Convolution Structure for Jacobi Series', *Ann. Math.* **93** (1971) 112–118.

[12] J. Löfström and J. Peetre, 'Approximation Theorems Connected with Generalized translations', *Math. Ann.* **181** (1969) 255–268.

[13] D. J. Newman and H. S. Shapiro, 'Jackson's Theorem in Higher Dimensions' in: *Über Approximationstheorie* (Proc. Conference on Approximation Theory Oberwolfach 1963), Basel 1964, 208–219.

[14] S. Pawelke, *Saturation und Approximation bei Reihen mehrdimensionaler Kugelfunktionen*, Dissertation, TH Aachen 1969.

[15] S. Pawelke, 'Ein Satz vom Jacksonschen Typ für algebraische Polynome', *Acta. Sci. Math* (*Szeged*) **33** (1972) 323–336.

[16] S. Pawelke, 'Über die Approximationsordnung bei Kugelfunktionen und algebraischen Polynomen', *Tôhoku Math. J.* **24** (1972) 473–486.

[17] J. Peetre, 'A Theory of Interpolation of Normed Spaces', *Notas de Matématica* **39**, Instituto de Matemática Pura e Aplicada, Rio de Janeiro 1968.

[18] D. L. Ragozin, 'Polynomial Approximation on Compact Manifolds and Homogeneous spaces', *Trans. Amer. Math. Soc.* **150** (1970) 41–53.

[19] D. L. Ragozin, 'Constructive Polynomial Approximation on Spheres and Projective Spaces', *Trans. Amer. Math. Soc.* **162** (1971) 157–170.

[20] E. M. Stein, 'Interpolation in Polynomial Classes and Markoff's Inequality', *Duke Math.* **24** (1957) 467–476.

[21] G. Szegö, 'Orthogonal Polynomials', *Amer. Math. Soc. Coll. Publ.* **23**, Providence, R. I., 1967.

# SOME DECOMPOSITIONS OF FUNCTIONAL SPACES BY MEANS OF QUASI-ANALYTIC FUNCTIONS OF SEVERAL VARIABLES

WIESŁAW PLEŚNIAK

*Kraków*

Let $E$ be a compact set in the space $\boldsymbol{C}^n$ of $n$ complex variables. Let $C(E)$ denote the Banach space of all complex continuous functions on $E$ with the norm

$$\|f\| = \sup\{|f(z)|: z \in E\}$$

and let $W(E)$ denote the Banach subspace of $C(E)$ formed by all functions uniformly approximable on $E$ by polynomials.

Consider the set $B(E)$ of all functions $f \in W(E)$ satisfying the following condition: There exist a sequence $\nu_k \to \infty$ of positive integers and a sequence of polynomials $p_k$, $\deg p_k \leqslant \nu_k$, $k = 1, 2, \ldots$, such that

$$\lim_{k\to\infty} \sqrt[\nu_k]{\|f-p_k\|} = \varrho < 1. \tag{1}$$

A function $f \in B(E)$ is called *quasi-analytic* on $E$ *in the sense of Bernstein.*

In the case when $E = [a, b] \subset \boldsymbol{R}^1$, quasi-analytic functions were introduced by Bernstein (cf. [1]). The term "quasi-analytic" is used due to the fact that, as Bernstein observed, if a function $f \in B(E)$ vanishes on a subinterval of $E$, then it vanishes on E. Szmuszkowiczówna [12] proved much more, viz.

(I) *The set of zeros of a function $f \in B(E)$, where $E = [a, b] \subset \boldsymbol{R}^1$, has the transfinite diameter equal to zero, unless $f = 0$ on $E$.*

One of the most interesting properties of quasi-analytic functions is the following property proved by Markuševič [5]:

(II) *Every continuous function on $E = [a, b] \subset \boldsymbol{R}^1$ is the sum of two quasi-analytic functions.*

Some years before this result Mazurkiewicz [6] showed that

(III) *The set $B(E)$ is residual in the Banach space $C(E)$, where $E = [a, b] \subset \boldsymbol{R}^1$.*

It appears that property (II) is the consequence of (III) since we have

LEMMA 1. *If $V$ is a residual subset of any Baire topological vector space $X$, then $X = V+V$ (the algebraic sum).*

This lemma follows from the trivial fact that for each point $a \in X$ the set $a-V$ is residual and, consequently, the set $(a-V)\cap V$ is non-empty.

In the case when $X$ is an $F$-space, Lemma 1 follows from (7.1) in [7].

Quasi-analytic functions of several complex or real variables were examined in [8], [9] and [10]. In particular, an extension of (I) to the $n$-dimensional case was given in [8]. That identity principle can be essentially strengthened for the set $B_0(E)$ of *quasi-entire functions* on $E$, i.e. such functions $f \in B(E)$ for which $\varrho = 0$ in (1). One can prove that a function $f \in B_0(E)$ cannot be too "flat" at any sufficiently regular point of $E$, unless $f = 0$ on $E$ (it will be published in [11]).

It appears that "quasi-entireness" of functions is of frequent occurrence in the Banach space $W(E)$. Indeed, we have

THEOREM 1. *The set $B_0(E)$ is residual in the Banach space $W(E)$ for any compact subset $E$ of $\boldsymbol{C}^n$.*

PROOF (similar to Mazurkiewicz's proof of (III)). Let $\{q_j\}$ be the set of all polynomials of $n$ variables with rational (complex) coefficients and let $\nu_j = \deg q_j$, $j = 1, 2, \ldots$ Then the set

$$V(E) = \bigcap_{i=1}^{\infty} \bigcup_{j=i}^{\infty} \{f \in W(E) \colon \|q_j - f\| \leqslant j^{-(j+\nu_j)}\}$$

is residual in $W(E)$. On the other hand, one can easily check that $V(E) \subset B_0(E)$. This completes the proof.

Comparing Theorem 1 and Lemma 1 we get a generalization as well as an extension of (II) to the $n$-dimensional case:

THEOREM 2. $W(E) = B_0(E)+B_0(E)$ *for any compact subset $E$ of $\boldsymbol{C}_n$.*

If $E$ is a compact subset of the space $\boldsymbol{R}^n$ of $n$ real variables, then, by the Stone–Weierstrass theorem, $C(E) = W(E)$. Hence by Theorem 2 we get

COROLLARY. *If $E \subset \boldsymbol{R}^n$, then $C(E) = B_0(E)+B_0(E)$.*

Now let $E$ be a compact interval in $\boldsymbol{R}^n$. Denote by $C^m(E)$, where $0 \leqslant m \leqslant \infty$, the set of all complex functions on $E$ having continuous derivatives on $E$ up to the order $m$. The set $C^m(E)$ can be considered as an $F$-space with the paranorm

$$p_m(f) = \sum_{k=0}^{m} 2^{-k} \frac{M_k(f)}{1+M_k(f)}, \tag{2}$$

where

$$M_k(f) = \sup\{|D^\alpha f(x)| : x \in E, |\alpha| = k\} \tag{3}$$

and

$$D^\alpha f = \frac{\partial^{|\alpha|} f}{\partial x_1^{\alpha_1} \ldots \partial x_n^{\alpha_n}}, \qquad |\alpha| = \alpha_1 + \ldots + \alpha_n.$$

The set $\{q_j\}$ of polynomials defined in the proof of Theorem 1 is dense in $C^m(E)$ (in the sense of paranorm (2)). This follows that the set

$$V^m(E) = \bigcap_{i=1}^{\infty} \bigcup_{j=i}^{\infty} \{f \in C^m(E) : p_m(q_j - f) \leqslant j^{-(j+\nu_j)}\} \tag{4}$$

is residual in the $F$-space $C^m(E)$, $0 \leqslant m \leqslant \infty$. Hence, since $V^m(E) \subset B_0(E)$ and since $D^\alpha f \in B_0(E)$ for $f \in V^m(E)$ and $|\alpha| \leqslant m$, by Lemma 1 we get

THEOREM 3. *For every function $f \in C^m(E)$, $0 \leqslant m \leqslant \infty$, there exist two functions $f_1, f_2 \in C^m(E)$ belonging to the set $B_0(E)$ with their derivatives up to the order $m$ and such that $f = f_1+f_2$ on $E$.*

REMARK 1. In the case when $E = [a, b] \subset \boldsymbol{R}^1$, Bernstein [1], p. 551, proved (by different methods) that $C^\infty(E) = B(E) \cap C^\infty(E) + B(E) \cap \cap C^\infty(E)$. Using Bernstein's method one can also show (cf. [13], p. 390) that if $f \in C^m(E)$, where $m < \infty$, and if the derivative $f^{(m+1)}$ exists and is bounded on $E$, then $f = f_1+f_2$, where $f_1, f_2 \in B(E) \cap C^m(E)$. It is seen that Theorem 3 (for $n = 1$) is stronger than these results.

Now consider the set $D(E)$ of all functions $f \in C^\infty(E)$, where $E$ is a compact interval in $\boldsymbol{R}^n$, for which

$$(5) \qquad J(f) = \int_0^\infty \frac{\ln\theta(r)}{1+r^2}\,dr = +\infty,$$

where

$$\theta(r) = \sup_\alpha \left(r^{|\alpha|}/\sup\{|D^\alpha f(x)|: x \in E\}\right) \quad \text{for } r > 0.$$

In the case of $n = 1$, a function $f \in D(E)$ is called *quasi-analytic on E in the sense of Denjoy–Carleman* (properties of such functions with the extensive bibliography may be found in [1], [4] or [13]).

If $E$ is a compact subset of $\boldsymbol{R}^n$ ($n \geqslant 1$), an identity principle for functions satisfying (5) was given by Lelong [2].

Let us return to the set $V^\infty(E)$ given by (4) when $m = \infty$. One can easily check that every function $f \in V^\infty(E)$ belongs to the set $D(E)$ with its all derivatives. Hence and by the proof of Theorem 3 we get the following two theorems:

THEOREM 4. *The set $B_0(E) \cap D(E)$ of functions which are quasi-entire on E in the sense of Bernstein and quasi-analytic on E in the sense of Denjoy–Carleman, simultaneously, is residual in the F-space $C^\infty(E)$.*

THEOREM 5. *Every function $f \in C^\infty(E)$ can be written in the form of the sum of two functions $f_1, f_2 \in C^\infty(E)$ belonging simultaneously to the sets $B_0(E)$ and $D(E)$ with their all derivatives.*

REMARK 2. Let $E = [a, b] \subset \boldsymbol{R}^1$. Mandelbrojt [3] proved (by rather elaborate way) that every function belonging to $C^\infty(E)$ is the sum of two functions *quasi-analytic on E in the sense of Denjoy*, i.e. such functions $f \in C^\infty(E)$ for which

$$(6) \qquad \sum_{k=1}^{\infty} [M_k(f)]^{-1/k} = \infty,$$

where $M_k(f)$ is defined by (3) for $n = 1$. Since, in the case $n = 1$, every function $f \in V^\infty(E)$ satisfies (6), Theorem 5 generalizes the quoted result of Mandelbrojt.

## REFERENCES

[1] S. N. Bernstein, *Collected Works*, v. **1**, 1952 (Russian).
[2] P. Lelong, 'Extension d'un théorème de Carleman', *Ann. Inst. Fourier, Grenoble* **12** (1962) 627–641.
[3] S. Mandelbrojt, 'Sur les fonctions indéfiniment dérivables', *Acta Math.* **72** (1940) 15–29.
[4] S. Mandelbrojt, *Séries adhérentes. Regularisation des suites. Applications*, Paris 1952.
[5] A. I. Markuševič, 'On the Best Approximation', *Doklady Akad. Nauk SSSR* **44** (1944) 290–292 (Russian).
[6] S. Mazurkiewicz, 'Les fonctions quasi-analytiques dans l'espace fonctionnel', *Mathematica* (*Cluj*) **13** (1937) 16–21.
[7] W. Orlicz and Z. Ciesielski, 'Some Remarks on the Convergence of Functionals on Bases', *Studia Math.* **16** (1958) 335–352.
[8] W. Pleśniak, 'Quasi-Analytic Functions of Several Complex Variables', *Zeszyty Nauk. Uniw. Jagiell.* **15** (1971) 135–145.
[9] W. Pleśniak,'On Superposition of Quasi-Analytic Functions', *Ann. Polon. Math.* **26** (1972) 75–86.
[10] W. Pleśniak, 'Characterization of Quasi-Analytic Functions of Several Variables by Means of Rational Approximation', *Ann. Polon. Math.* **27** (1973) 149–157.
[11] W. Pleśniak, *Quasi-Analytic Functions in the Sense of Bernstein*, Dissertationes Math. (in print).
[12] H. Szmuszkowiczówna, 'Un théorème sur les polynomes et son application à la théorie des fonctions quasi-analytiques', *C.R.Acad. Sci. Paris* **198** (1934) 1119–1120.
[13] A. F. Timan, *Theory of Approximation of Functions of a Real Variable*, Moscow 1960 (Russian).

# ZUR THEORIE DER SATURIERTEN APPROXIMATIONSVERFAHREN AUF LOKALKOMPAKTEN ABELSCHEN GRUPPEN

WALTER SCHEMPP

*Siegen*

*Abstract.* Let $G$ be a locally compact abelian group, $E$ a submodule of the Banach module $L^p(G), p > 1$, over the convolution algebra $\mathscr{M}^1(G)$ and $(\mu_t)_{t>0}$ an approximate unit on $G$ which generates an approximation process $(I_t)_{t>0}$ on $E$ according to the rule

$$I_t\colon E \ni f \rightsquigarrow \mu_t * f \quad (t > 0).$$

For any exponent $p > 2$, the Fourier transform of any function which is a member of the space $L^p(G)$ will be understood as a quasi-measure in the sense of Gaudry (1966). This generalized Fourier transformation makes it possible to prove for certain processes $(I_t)_{t>0}$ a $L^p$-saturation theorem which holds for all exponents $p > 1$ uniformly and which extends a result of Buchwalter (1960).

**1. Einleitung.** Im folgenden bezeichne stets $G$ eine abelsche lokalkompakte topologische Gruppe. Sei $dx$ ein festes Haar-Maß auf $G$ und $L^p(G) = L^p_{\mathbf{C}}(G; dx)$ der zum Exponenten $p \in [1, +\infty[$ gehörende Lebesgue-Raum aller Äquivalenzklassen auf $G$ in der $p$-ten Potenz $dx$-integrierbarer komplexwertiger Funktionen. Ferner sei $\mathscr{M}^1(G)$ die kommutative Banach-Algebra über $\mathbf{C}$ aller beschränkter komplexer Radon-Maße mit der Faltung $(\mu, \nu) \rightsquigarrow \mu * \nu$ als multiplikativer Verknüpfung. Unser Ziel besteht darin, auf den $\mathscr{M}^1(G)$-Moduln $L^p(G)$ bzw. auf gewissen Untermoduln von $L^p(G)$ Approximationsverfahren und deren Saturationsverhalten zu studieren. Der erste Ansatz in dieser Richtung stammt von H. Buchwalter [2]. Mit den Methoden der harmonischen Analyse, die schon J. Favard zur Untersuchung des Saturationsphänomens vorgeschlagen hat, führt er zum Begriff des Saturationsmaßes, der, geeignet modifiziert, in Abschnitt 2 eingeführt wird. Es ist jedoch eine bekannte Tatsache, daß sich, bei nicht kompakter Gruppe $G$, die „klassische" Fouriertransformation *nicht* auf die Räume

$L^p(G)$ mit $p > 2$ fortsetzen läßt. Aus diesem Grunde erläutern wir in Abschnitt 3 den Begriff der quasimaßwertigen Fouriertransformation und beweisen dann in Abschnitt 4 einen allgemeinen, für *alle* Exponenten $p > 1$ gültigen Saturationssatz. In Abschnitt 5 schließlich wenden wir dieses Ergebnis auf das Approximationsverfahren der sphärischen Mittel im $\boldsymbol{R}^n$ an.

Bezüglich der Saturationstheorie auf kompakten, nicht notwendig abelschen Gruppen sei auf Dreseler [3], [4] und die Arbeit [7] verwiesen.

**2. Saturationsstruktur und Saturationsmaß.** Eine Familie $(\mu_t)_{t>0}$ von Maßen aus $\mathscr{M}^1(G)$ heißt eine *approximative Einheit auf G*, falls

$$\lim_{t\to 0+} \|\mu_t * f - f\|_1 = 0$$

für jede Funktion $f \in L^1(G)$ zutrifft.

Sei $E$ ein Untermodul des $\mathscr{M}^1(G)$-Moduls $L^p(G)$ mit $p \in [1, +\infty[$. Aufgrund der Faktorisierung

$$L^1(G) * L^p(G) = L^p(G)$$

existiert zu jeder Funktion $f \in E$ ein Paar $(g, h) \in L^1(G) \times L^p(G)$ mit $f = g * h$. Es gilt demnach die Norm-Abschätzung

$$\begin{aligned} \|\mu_t * f - f\|_p &= \|\mu_t * (g * h) - g * h\|_p \\ &\leqslant \|\mu_t * g - g\|_1 \|h\|_p, \end{aligned}$$

und somit auch

$$\lim_{t\to 0+} \|\mu_t * f - f\|_p = 0$$

für jedes $f \in E$.

Unter dem von der approximativen Einheit $(\mu_t)_{t>0}$ *auf E erzeugten Approximationsverfahren* verstehen wir die Familie $(I_t)_{t>0}$ linearer Abbildungen

$$I_t : E \ni f \rightsquigarrow \mu_t * f \in E \qquad (t > 0)$$

von $E$ in sich. In diesem Falle sagen wir auch, das Verfahren $(I_t)_{t>0}$ operiere auf dem $\mathscr{M}^1(G)$-Modul $E$.

DEFINITION 1. *Wir sagen, das auf E operierende Approximationsverfahren* $(I_t)_{t>0}$ *weise die Saturationsstruktur* $(\varphi; E; V)$ *auf, falls die folgenden Bedingungen erfüllt sind:*

(I) *Es existiert eine stetige Abbildung* $\varphi$ *der offenen reellen Halbgeraden* $\boldsymbol{R}_+^* = \{t \in \boldsymbol{R} \mid t > 0\}$ *in sich mit der Eigenschaft* $\lim_{t \to 0+} \varphi(t) = 0$, *so daß die Bedingungen* $f \in E$ *und*

$$\|I_t(f) - f\|_p = o(\varphi(t)) \quad (t \to 0+)$$

*stets*

$$f(x) = \begin{cases} \textit{konstant}, & \textit{falls } G \textit{ kompakt ist;} \\ 0, & \textit{falls } G \textit{ nicht kompakt ist;} \end{cases}$$

$dx$ — *fast überall auf G implizieren.*

(II) *Es existiert ein Untervektorraum* $V \subset E$ *von* $L^p(G)$ *derart, daß die Bedingungen* $f \in E$ *und*

$$\|I_t(f) - f\|_p = O(\varphi(t)) \quad (t \to 0+) \tag{1}$$

*stets* $f \in V$ *implizieren.*

(III) *Die Bedingung* $f \in V$ *zieht stets die Gültigkeit von* (1) *nach sich.*

*In diesem Falle wird V der Favard-Raum der Saturationsstruktur* $(\varphi; E; V)$ *genannt.*

Ein wichtiges Hilfsmittel, um die Saturationsstruktur eines Approximationsverfahrens zu erkennen, ist der Begriff des Saturationsmaßes. Um ihn einführen zu können, bezeichnen wir mit $\hat{G}$ die Dualgruppe von $G$ unter der Pontrjagin-Topologie, d.h. unter der Topologie der gleichmäßigen Konvergenz auf den kompakten Teilen von $G$, und mit

$$(x, \hat{x}) \rightsquigarrow [\hat{x}, x] = \hat{x}(x)$$

die Dualitätsabbildung des Paares $(G, \hat{G})$. Die klassische Fouriertransformation wird dann durch die lineare Abbildung

$$\mathscr{F}_G : \mathscr{M}^1(G) \ni \mu \rightsquigarrow \left(\hat{G} \ni \hat{x} \rightsquigarrow \int_G \overline{[\hat{x}, x]}\, d\mu(x)\right)$$

gegeben.

Definition 2. *Eine approximative Einheit* $(\mu_t)_{t>0}$ *auf G wird ein Saturationsmaß vom Typ* $(\varphi; \psi)$ *auf G genannt, falls die folgenden Bedingungen erfüllt sind:*

(i) *Für ein geeignetes* $t_0 \in \boldsymbol{R}_+^*$ *ist die Abbildung*

$$]0, t_0] \times \hat{G} \ni (t, \hat{x}) \rightsquigarrow \mathscr{F}_G \mu_t(\hat{x}) \in \boldsymbol{C}$$

*stetig.*

(ii) *Es existiert eine stetige Abbildung* $\varphi\colon \boldsymbol{R}_+^* \to \boldsymbol{R}_+^*$, *welche der Bedingung* $\lim_{t\to 0+} \varphi(t) = 0$ *genügt und eine stetige Abbildung* $\psi\colon \hat{G} \to \boldsymbol{C}$, *welche höchstens im Neutralelement von* $\hat{G}$ *verschwindet, derart, daß*

$$\lim_{t\to 0+} \frac{\mathscr{F}_G \mu_t(\hat{x}) - 1}{\varphi(t)} = \psi(\hat{x})$$

*punktweise für alle Charaktere* $\hat{x} \in \hat{G}$ *gilt.*

(iii) *Es existiert eine Familie von Maßen* $(\nu_t)_{t>0}$ *in* $\mathscr{M}^1(G)$ *mit gemeinsamer endlicher Norm-Schranke*

$$M_0 = \sup_{t>0} \|\nu_t\|$$

*derart, daß die Identität*

$$\frac{\mathscr{F}_G \mu_t - 1}{\varphi(t)} = \psi \cdot \mathscr{F}_G \nu_t$$

*für alle Parameter* $t \in \boldsymbol{R}_+^*$ *gilt.*

In Abschnitt 5 geben wir ein Beispiel für ein Saturationsmaß im Sinne der Definition 2 auf der additiven Gruppe des $\boldsymbol{R}^n$.

**3. Quasimaße.** Es bezeichne $\hat{\mathfrak{R}}$ die Menge aller kompakten Teile von $\hat{G}$ und für jede Menge $\hat{K} \in \hat{\mathfrak{R}}$ sei $\mathscr{K}(\hat{G}; \hat{K})$ der $\boldsymbol{C}$-Vektorraum aller stetigen, komplexwertigen Funktionen auf $\hat{G}$ mit in $\hat{K}$ enthaltenem Träger unter der Topologie der gleichmäßigen Konvergenz. Der komplexe Banach-Raum

$$\mathscr{D}(\hat{G}; \hat{K}) = \varrho\big(\mathscr{K}(\hat{G}; \hat{K}) \mathbin{\hat{\otimes}_\pi} \mathscr{K}(\hat{G}; \hat{K})\big)$$

ist dann definiert als Bild des vervollständigten projektiven topologischen Tensorprodukts $\mathscr{K}(\hat{G}; \hat{K}) \mathbin{\hat{\otimes}_\pi} \mathscr{K}(\hat{G}; \hat{K})$ im Sinne von Grothendieck [10] unter der durch die Gleichung

$$\varrho(f \otimes g) = f * g$$

definierten linearen Abbildung

$$\varrho\colon \mathscr{K}(\hat{G}; \hat{K}) \mathbin{\hat{\otimes}_\pi} \mathscr{K}(\hat{G}; \hat{K}) \to \mathscr{K}(G; \hat{K}. \hat{K}).$$

Unter dem komplexen *Gaudryschen Grundraum* auf $\hat{G}$ verstehen wir den hausdorffschen lokalkonvexen topologischen Vektorraum

$$\mathscr{D}(\hat{G}) = \operatorname{ind.lim}_{\hat{K}\in\hat{\mathfrak{R}}} \mathscr{D}(\hat{G}; \hat{K}).$$

Die Elemente seines topologischen Dualraumes $\mathscr{D}'(\hat{G})$ werden komplexe *Quasimaße* auf $\hat{G}$ genannt; vgl. Gaudry [8], [9]. Man beachte, daß der Vektorraum $\mathscr{M}(\hat{G})$ aller komplexer Radon-Maße auf $\hat{G}$ mit einem Untervektorraum von $\mathscr{D}'(\hat{G})$ identifiziert werden kann.

Sei $p \in ]1, +\infty[$ und $p'$ der zu $p$ duale Exponent. Die Fouriertransformation

$$\mathscr{F}_{\hat{G}}: \mathscr{D}(\hat{G}) \to L^{p'}(G)$$

ist eine stetige lineare Abbildung. Die zugehörende transponierte Abbildung

$$\mathscr{F}_G = {}^t\mathscr{F}_{\hat{G}}: L^p(G) \to \mathscr{D}'(\hat{G})$$

heißt die *quasimaßwertige Fouriertransformation* auf $L^p(G)$.

**4. Ein Saturationssatz für** $p \in ]1, +\infty[$. Da die komplexen Radon-Maße auf $\hat{G}$ als Quasimaße aufgefaßt werden können, ist für jeden Exponenten $p \in ]1, +\infty[$ durch

$$C^p_{\mathbb{C}}(G) = \{f \in L^p_{\mathbb{C}}(G) \mid \mathscr{F}_G f \in \mathscr{M}(\hat{G})\}$$

ein Untervektorraum von $L^p(G)$ definiert. Offenbar nimmt $C^p(G)$ innerhalb der Gaudryschen Theorie der Quasimaße hinsichtlich der Fouriertransformation eine analoge Stelle ein, wie die Sobolev-Räume im Rahmen der Schwartzschen Distributionentheorie bezüglich der Differentialoperatoren. Man überzeugt sich, daß $C^p(G)$ hinsichtlich der Faltung ein $\mathscr{M}^1(G)$-Untermodul von $L^p(G)$ ist. Im Falle $q \in ]1, 2]$ gilt die Identität

$$C^q(G) = L^q(G),$$

im Falle $q \in ]2, +\infty[$ liegt $C^q(G)$ in $L^q(G)$ überall norm-dicht.

Wir wollen Approximationsverfahren betrachten, die auf $C^p(G)$ operieren. Für jede Funktion $\psi: \hat{G} \to \boldsymbol{C}$ führen wir den Untervektorraum

$$W^p_\psi(G) = \{f \in C^p(G) \mid \psi \cdot \mathscr{F}_G f \in \mathscr{F}_G L^p(G)\}, \quad p \in ]1, +\infty[$$

von $C^p(G)$ ein. Dann gilt der folgende Saturationssatz:

SATZ 1. *Das Approximationsverfahren* $(I_t)_{t>0}$ *operiere auf* $C^p(G)$, $p \in ]1, +\infty[$, *und werde vom Saturationsmaß* $(\mu_t)_{t>0}$ *vom Typ* $(\varphi; \psi)$ *auf* $G$ *erzeugt. Dann besitzt* $(I_t)_{t>0}$ *die Saturationsstruktur*

$$(\varphi; C^p(G); W^p_\psi(G)).$$

Es sei darauf hingewiesen, daß in Satz 1, im Gegensatz etwa zu Berens–Nessel [1], *keine* zusätzlichen Voraussetzungen über die Maße $\mu_t$ hinsichtlich Radialität oder Symmetrie erforderlich sind. Der Beweis von Satz 1 beruht hauptsächlich auf der folgenden Verallgemeinerung des Satzes von Bochner–Schoenberg–Eberlein:

SATZ 2. *Es sei* $p \in ]1, +\infty[$ *und* $\sigma \in \mathscr{D}'(\hat{G})$ *bezeichne ein komplexes Quasimaß auf der Dualgruppe* $\hat{G}$. *Die folgenden Aussagen sind äquivalent*:

(a) *Es existiert eine Funktion* $g \in L^p(G)$ *mit* $\|g\|_p \leqslant M$, *für die in* $\mathscr{D}'(\hat{G})$ *die Gleichung*

$$\mathscr{F}_G g = \sigma$$

*gilt.*

(*b*) *Für jede Funktion* $h \in \mathscr{D}(\hat{G})$ *gilt die Ungleichung*

$$\sigma(h) \leqslant M \, \|\mathscr{F}_{\hat{G}} h\|_{p'}.$$

Die Beweise finden sich in den Arbeiten [5], [6] und sollen hier nicht wiederholt werden.

**5. Ein Anwendungsbeispiel.** Wir wählen für $G$ die additive Gruppe des $\boldsymbol{R}^n$, $n \geqslant 1$. Bezeichnet $\sigma^{(n-1)}$ das normierte Oberflächenmaß der kompakten $(n-1)$-Sphäre $S^{n-1}$ und wird

$$\mu_t = H_t(\sigma^{(n-1)}) \qquad (t \in \boldsymbol{R}_+^*)$$

als Bildmaß von $\sigma^{(n-1)}$ unter der Streckung

$$H_t \colon \boldsymbol{R}^n \ni x \leadsto tx \in \boldsymbol{R}^n$$

definiert, so ist $(\mu_t)_{t>0}$ ein Saturationsmaß vom Typ

$$\left(t \leadsto t^2, \hat{x} \leadsto -\frac{1}{2n}|\hat{x}|^2\right)$$

auf dem $\boldsymbol{R}^n$. Hierbei bezeichnet $|\cdot|$ die euklidische Norm auf dem Raum $\hat{\boldsymbol{R}}^n$, der mit $\boldsymbol{R}^n$ identifiziert werden kann. Eine Anwendung von Satz 1 liefert die Saturationsstruktur des Approximationsverfahrens durch *sphärische Mittelbildung*. Die Details sind in [5] ausgeführt.

## LITERATUR

[1] H. Berens and R. J. Nessel, 'Contributions to the Theory of Saturation for Singular Integrals in Several Variables. V. Saturation in $L^p(E^n)$, $2 < p < \infty$', *Nederl. Akad. Wetensch. Proc. Ser. A*, **72** = *Indag. Math.* **31** (1969) 71–76.

[2] H. Buchwalter, 'Saturation sur un groupe abélien localement compact', *C.R. Acad. Sci. Paris* **250** (1960) 808–810.

[3] B. Dreseler, 'Saturationstheorie auf kompakten topologischen Gruppen', this volume, 55–61.

[4] B. Dreseler, *Saturation auf kompakten topologischen Gruppen und homogenen Räumen*, Dissertation, Universität Mannheim 1972.

[5] B. Dreseler and W. Schempp, 'Saturation on Locally Compact Abelian Groups', *Manuscripta Math.* **7** (1972) 141–174.

[6] B. Dreseler and W. Schempp, 'Saturation on Locally Compact Abelian Groups: an Extended Theorem', *Manuscripta Math.* **8** (1973) 271–286.

[7] B. Dreseler and W. Schempp, 'Central Approximation Processes', *Tôhoku Math. J.* **27** (1975).

[8] G. I. Gaudry, 'Quasimeasures and Operators Commuting with Convolution', *Pacific J. Math.* **18** (1966) 461–476.

[9] G. I. Gaudry, *Topics in Harmonic Analysis*, Lecture Notes, Department of Mathematics, Yale University, New Haven 1969.

[10] A. Grothendieck, 'Produits tensoriels topologiques et espaces nucléaires', *Memoirs of the Amer. Math. Soc.*, **16**, Providence, R. I., 1966.

# A COMPARISON APPROACH TO DIRECT THEOREMS FOR POLYNOMIAL SPLINE APPROXIMATION

KARL SCHERER

*Aachen*

**1. Introduction.** In several previous papers [15], [16], [17] the fundamental importance of Jackson- and Bernstein-type inequalities for approximation in Banach spaces by linear subspaces has been considered. First applications to best approximation by polynomial splines of lowest deficiency with fixed knots were given in [32], [33].

This paper is concerned exclusively with direct theorems or Jackson-type inequalities for polynomial spline approximation. At first the case of best approximation is considered, and the corresponding results in [32], [33] are completed to a form in which they are used in [36] in order to establish in a more general setting characterizations of Besov spaces.

The simplest and mostly treated *linear* spline operators seem to be spline interpolation operators. Their convergence behaviour contrasts favourably to that of algebraic or trigonometric interpolating polynomials. From a general point of view it is interesting here that the negative result of Lozinskii–Kharshiladse on bounded linear projections does not carry over to spline approximation in view of de Boor's famous counterexample [7]. Thus in spline interpolation there is no need for summation processes as is the situation in the theory of Fourier series, at least from the theoretical side. This advantage is partly lost by the fact that an interpolating spline depends on its boundary conditions, which is in turn reflected in the convergence behaviour, at least when the interpolating splines are smooth.

For these reasons comparison theorems concerning error bounds for interpolating splines with different boundary conditions are of definite interest. It is to be noted that the comparison aspect turned out to be of great advantage in the treatment of summation processes, see e.g. [13], [14], [22] and the literature cited there. In Section 3 three theorems of this type are established for odd-degree interpolating splines. The proofs are simple and of an elementary nature, and depend essentially upon Rolle's theorem and a Bernstein-type inequality for splines, obtained from Markov's inequality for algebraic polynomials. They reveal the importance of the first and second integral relations, originally developed in the Hilbert space framework, for the derivation of error bounds also in other $L_p$-norms, $p \geqslant 2$. Another advantage is that the interpolating spline used for comparison does not need to satisfy particular boundary conditions.

Just this fact enables one to study the case of *non*-equidistant partitions in Section 4 (Theorem 9). The results deal with interpolating splines satisfying Hermite–Birkhoff boundary conditions or satisfying the more general first or second integral relations. In the first case our results generalize those of Swartz [43], Swartz–Varga [44] for equidistant partitions and secondly results of Ahlberg–Nilson–Walsh [2], Schultz–Varga [39], Jerome–Varga [26] (obtained for $Lg$-splines) for $q = 2$, $p = 2, \infty$ to general $2 \leqslant q \leqslant p < \infty$.

In Section 5 the above results are carried over to even-degree splines. It is shown that in this case one obtains analogous error bounds if instead of point functionals in the knots one uses for interpolation integrals taken over each subsegment of the partition.

We reemphasize that in the present paper the comparison of error bounds of different interpolating splines is the important aspect. When actually considering the basic comparison spline we refer to known facts, see [1]. No attempt is made to improve such results (which depend on intricate estimates of certain matrix bounds), reflecting the fact that our final results in Sections 4 and 5 hold only in the case of asymptotically uniform meshes ($\sigma \ll 1$). But in view of the comparison theorems in Section 2 it would be desirable to have an interpolating spline with error bounds in the uniform norm under weaker restrictions upon the mesh $\Delta$. Though such error bounds are known in some cases without any restriction upon $\Delta$ (for cubic and quintic splines, see [40], [8]), one cannot expect this fact for our general theory which has to include the periodic case for which the well-known counterexample of Nord [30] gives the range of play.

With the exception of the best approximation case the error bounds here are formulated only in terms of derivatives. Many other and also more general types of error bounds formulated in terms of moduli of continuity can be derived from these. This is shown in [36] for general spline operators where such results are also used to obtain characterizations of Besov spaces.

The author would like to thank Professor Dr. P. L. Butzer and Dozent Dr. E. Görlich for valuable suggestions in connection with the critical reading of the manuscript.

**2. Jackson-type inequalities for best approximation.** Let $L_p(a, b)$ be the set of functions which are Lebesgue integrable to the $p$-th power on $(a, b)$ if $1 \leqslant p < \infty$, and essentially bounded on $(a, b)$ if $p = \infty$, endowed with the norms

$$\|f\|_p = \begin{cases} \left\{\int\limits_a^b |f(x)|^p dx\right\}^{1/p}, & 1 \leqslant p < \infty, \\ \operatorname{ess\,sup}\limits_{x \in [a,b]} |f(x)|, & p = \infty. \end{cases}$$

The set $C[a, b]$ of continuous functions on $[a, b]$ is the subspace of $L_\infty(a, b)$ for which

$$\|f\|_C = \sup_{x \in [a,b]} |f(x)| = \|f\|_\infty .$$

More generally, let $W_p^k(a, b)$, $k \in N$(1), be the Sobolev space of functions whose $(k-1)$-th derivative is absolutely continuous and for which $f^{(k)} \in L_p(a, b)$, and set $W_p^0(a, b) = L_p(a, b)$. The spaces $C^k[a, b]$ are defined similarly. We further introduce the norm $\|f\|_p^k = \|f\|_p + \|f^{(k)}\|_p$ on $W_p^k(a, b)$, $1 \leqslant p \leqslant \infty$.

To define the spline classes $\mathrm{Sp}(n, m, \Delta)$ let $\Delta$ be a partition of the finite interval $[a, b]$ of the form

(2.1) $$\Delta\colon a = x_0 < x_1 < \ldots < x_N = b$$

with nodes $x_i$, $1 \leqslant i \leqslant N-1$, where

(2.2) $$\overline{\Delta} = \max_{0 \leqslant i \leqslant N-1} (x_{i+1} - x_i), \quad \underline{\Delta} = \min_{1 \leqslant i \leqslant N-1} (x_{i+1} - x_i).$$

---

(1) $P$ and $N$ denote the sets of all integers which are non-negative and positive, respectively.

If $P_n$ denotes the set of algebraic polynomials of degree $n$, $n \in \boldsymbol{P}$, define

$$\text{(2.3)} \quad \mathrm{Sp}(n, m, \Delta) = \{s(x) : s(x) \in W_\infty^m(a, b),\; s(x) \in P_n,\, x \in (x_i, x_{i+1}),\, 0 \leqslant i \leqslant N-1\}$$

for all integers $n, m$ with $n \geqslant m \geqslant 0$. Then $\mathrm{Sp}(n, m, \Delta)$ is a finite-dimensional subspace of $W_\infty^m(a, b)$ since $\Delta$ is fixed.

THEOREM 1. *Defining the best approximation to $f \in L_p(a, b)$, $1 \leqslant p \leqslant \infty$, from the class* $\mathrm{Sp}(n, m, \Delta)$ *by*

$$\text{(2.4)} \qquad E_\Delta^{(n,m)}(f; p) = \inf_{s \in \mathrm{Sp}(n,m,\Delta)} \|f-s\|_p,$$

*then for $f \in W_q^{n+1}(a, b)$, $1 \leqslant q \leqslant \infty$, there holds the Jackson-type inequality*

$$\text{(2.5)} \qquad E_\Delta^{(n,m)}(f; p) \leqslant C\bar{\Delta}^{n+1+\min(0,1/p-1/q)} \|f^{n+1}\|_q$$

*$C$ (²) being a positive constant.*

The proof follows very rapidly by known results. Indeed, in case $n = 0$ define the step function

$$T(f; x) = f(x_i), \qquad x \in [x_i, x_{i+1}), \qquad 0 \leqslant i \leqslant N-1.$$

By Hölder's and Jensen's inequalities one has for $q \leqslant p$, $1/q+1/q' = 1$

$$\|T(f; x)-f(x)\|_p$$
$$= \Big\{\sum_{i=0}^{N-1} \int_{x_i}^{x_{i+1}} dx \Big| \int_{x_i}^{x} f'(u)du \Big|^p\Big\}^{1/p}$$
$$\leqslant \Big\{\sum_{i=0}^{N-1} \int_{x_i}^{x_{i+1}} dx \bar{\Delta}^{p/q'} \Big\{ \int_{x_i}^{x_{i+1}} |f'(u)|^q du\Big\}^{p/q}\Big\}^{1/p}$$
$$\leqslant \bar{\Delta}^{1/q'+1/p} \Big\{\Big[\sum_{i=0}^{N-1} \Big( \int_{x_i}^{x_{i+1}} |f'(u)|^q du\Big)^{p/q}\Big]^{q/p}\Big\}^{1/q}$$
$$\leqslant \bar{\Delta}^{1-1/q+1/p} \|f'\|_q,$$

(²) Here and in the following $C$ denotes any constant which is independent of the functions considered and independent of the maximum length $\bar{\Delta}$. However, this constant in general does depend on $n$, $m$, $a$, $b$ and the various norms and orders of derivatives considered, as well as upon the various constants connected with the mesh quantities such as $\bar{\Delta}/\underline{\Delta}$, if they are mentioned explicitely. $C$ may change its value from line to line.

establishing (2.5) for $n = 0$ and $q \leqslant p$. For $q \geqslant p$ we conclude by the case $q = p$ of the above result and Hölder's inequality

$$\|T(f; x)-f(x)\|_p \leqslant \bar{\Delta}\|f'\|_p \leqslant (b-a)^{1-p/q}\bar{\Delta}\|f'\|_q.$$

Thus we have shown that

$$E_\Delta^{(0,0)}(f; p) \leqslant C\bar{\Delta}^{\min(1,1-1/q+1/p)}\|f'\|_q. \tag{2.6}$$

As a second step one proves that for $f \in W_p(a, b)$ and $n = 0, 1, 2, \ldots$

$$E_\Delta^{(n+1,n+1)}(f, p) \leqslant CE_\Delta^{(n,n)}(f'; p). \tag{2.7}$$

This can be achieved by arguments in Popov–Sendov [31] and Butler–Richards [11] (see also [33]), where, in particular, the inequality

$$\left\|f-\int_a^x\left[s(t)+\sum_{j=0}^{N-n-1} A_j M_j(t)\right]dt\right\|_p \leqslant CE_\Delta^{(n,n)}(f; p) \tag{2.8}$$

is established for $f$ with $f(a) = 0$, $s_n(t)$ being a best approximating spline in $\mathrm{Sp}(n, n, \Delta)$ to $f$,

$$A_j = \int_{x_j}^{x_{j+1}} [f'(t)-s(t)]dt \qquad (j = 0, \ldots, N-1),$$

and $M_j(x)$ being the $B$-splines of degree $n$ with respect to $\Delta$.

Combining (2.6) and (2.7) yields inequality (2.5) for all $n$ with $n = m$. But then the general case $m < n$ also follows since $E_\Delta^{(n,m)}(f; p) \leqslant E_\Delta^{(n,n)}(f; p)$ for $m \leqslant n$.

We remark that the exponent of $\bar{\Delta}$ in (2.5) cannot be improved in general. This is shown by counterexamples in Schultz–Varga [39] and Subbotin [42], and also by saturation theorems for spline approximation (cf. [32], [21]).

The result of Theorem 1 can still be sharpened by arguments of the theory of intermediate spaces.

THEOREM 2. (a) *For each* $f \in L_p(a, b)$, $1 \leqslant p \leqslant \infty$, *one has*

$$E_\Delta^{(n,m)}(f; p) \leqslant C\omega_{n+1}(f; \bar{\Delta})_p. \tag{2.9}$$

(b) *For each* $f \in W_q^1(a, b)$, $1 \leqslant q \leqslant \infty$, *there holds for* $1 \leqslant p \leqslant \infty$.

$$E_\Delta^{(n,m)}(f; p) \leqslant C\bar{\Delta}^{\min(1,1-1/q+1/p)}\omega_n(f'; \bar{\Delta})_q. \tag{2.10}$$

*Here the n-th modulus of continuity* $\omega_n(t;f)_p$ *is defined by*

$$\omega_n(t;f)_p = \begin{cases} \sup\limits_{|h|\leqslant t}\left\{\int\limits_{x,x+nh\in(a,b)} |\Delta_h^n f(x)|^p dx\right\}^{1/p}, & 1 \leqslant p < \infty, \\ \sup\limits_{|h|\leqslant t}\ \operatorname*{ess\,sup}\limits_{x,x+nh\in(a,b)} |\Delta_h^n f(x)|, & p = \infty, \end{cases}$$

*where* $\Delta_h^n f(x) = \sum_{k=0}^{n} (-1)^k \binom{n}{k} f(x+kh)$ *is the n-th difference of f.*

The idea of the proof is standard (see e.g. [12]). To consider part (b), one first estimates in a simple manner, using Theorem 1 for $n = 0$ and general $n$,

$$E_\Delta^{(n,m)}(f;p) \leqslant C\bar{\Delta}^{1+\min(0,1/p-1/q)} K(\bar{\Delta}^n, f'; L_q(a,b), W_q^n(a,b)),$$

the $K$-functional being defined by

$$K(t,f;L_q(a,b),W_q^n(a,b)) = \inf_{g\in W_q^n(a,b)} (\|f-g\|_q + t\|g^{(n)}\|_q).$$

Then (2.10) follows in view of (cf. Johnen [27])

$$K(t^n, f; L_q(a,b), W_q^n(a,b)) \leqslant C\omega_n(f,t)_q. \tag{2.11}$$

We remark that a similar proof follows by arguments of Freud–Popov [19].

Among the numerous papers concerned with results analogous to Theorems 1 and 2 let us mention de Boor [7] who proved (2.5) for $p = q = \infty$, Subbotin [42] who proved (2.5) under the restriction $\bar{\Delta}/\underline{\Delta} \leqslant M$, and Freud–Popov [20], Popov–Sendov [31] who proved results similar to (2.9) for the case when the number of nodes (instead of $\Delta$ itself) is fixed. Of course in many particular cases more precise results have been obtained by estimates on spline interpolation.

From Theorem 2 there also follow estimates of $E_\Delta^{(n,m)}(f;p)$ in terms of norms of Besov-spaces, quoted in Hedström–Varga [24] for $p \geqslant 2$. But this direction of extension shall not be pursued any further here.

Finally we remark that the results of Theorems 1 and 2 can be carried over to the case of periodic functions as well as to functions defined on the whole real axis (under restriction $q \leqslant p$). This can be seen by modifying the above constructions, in particular those of Theorem 1, to these cases. Perhaps the greatest modification occurs in the proof of (2.8) for the periodic case. Here one approximates the periodic function $f^*$ on $[-\pi, \pi]$ with $f^*(-\pi) = 0$ by the periodic spline

$$S^*(x) = \int_{-\pi}^{x} \Big[g(t) - (1/2\pi) \int_{-\pi}^{\pi} g(u)\,du\Big]\,dt,$$

where $g(t) = s(t) + \sum_{j=0}^{N-n-1} A_j M_j(t)$ is defined as in (2.8). Then

$$f^*(x) - S^*(x) = \int_{-\pi}^{x} \Big[f'(t) - g(t)\Big]\,dt - (1/2\pi) \int_{-\pi}^{\pi} \Big[\int_{-\pi}^{x} [f'(u) - g(u)]\,du\Big]\,dt$$

can be estimated as in [31], [11].

**3. Comparison theorems for interpolating splines of odd degree.** Direct theorems for interpolating splines are obtained comparatively easily (via the Peano kernel theorem argument) if these are constructed by local interpolation, cf. [6], [44]. For example, a continuous interpolating spline of degree $n$ with respect to the mesh (2.1) is constructed by interpolating the given function locally via Lagrange interpolation at the nodes $x_{i,k} = x_i + (x_{i+1} - x_i)k/n$ of each subsegment $(x_i, x_{i+1})$, $0 \leqslant i \leqslant N-1$, of $\Delta$ (see e.g. [10]). If the degree is odd, $n = 2m+1$ say, the smoothest interpolating spline obtainable by local interpolation is constructed by two-point-Taylor interpolation with prescribed derivatives up to the order $m$ at $x_i$, $x_{i+1}$. This yields interpolating splines belonging to $\mathrm{Sp}(2m+1, m+1, \Delta)$. A precise formulation is given in Lemma 2 below. When even smoother piecewise polynomial interpolants have to be constructed ($n$, $\Delta$ being fixed), the number of interpolation conditions has to be reduced since the number of continuity requirements at the (interior) nodes of $\Delta$ increases. In this case the interpolation conditions at the boundary turn out to be of greater interest. In the following we shall study the influence of such conditions on the rate of convergence.

At first we establish a simple comparison theorem for splines $S_\Delta(f)$ in $\mathrm{Sp}(2n-1, 2n-1, \Delta)$ which interpolate a function $f \in W_2^{2n}[a, b]$ at the nodes of $\Delta$, i.e.

(3.1) $$S_\Delta(f; x_i) = f(x_i) \quad (1 \leqslant i \leqslant N-1)$$

and satisfy Hermite–Birkhoff boundary conditions, i.e.

$$S_\Delta^{(j_i)}(f; a) = \alpha_i \quad (1 \leqslant i \leqslant n), \qquad S_\Delta^{(k_i)}(f; b) = \beta_i \tag{3.2}$$

where $j_i$, $k_i$ are integers satisfying $0 \leqslant j_1 < \ldots < j_n \leqslant 2n-1$, $0 \leqslant k_1 < \ldots < k_n \leqslant 2n-1$, and $\alpha_i$, $\beta_i$ real numbers; $n \geqslant 2$.

For the proof of the comparison theorem we need the following Bernstein-type inequality.

THEOREM 3. *For each $s \in \mathrm{Sp}(n, 0, \Delta)$ there holds*

$$\|s^{(k)}\|_p \leqslant C^{+1} \underline{\Delta}^{k-k+\min(0,1/p-1/q)} \|s\|_q \qquad (k \in \boldsymbol{P},\ 1 \leqslant p, q \leqslant \infty) \tag{3.3}$$

*with constant C depending only upon p, q and n. The derivative $s^{(k)}$ is to be understood as equivalent to that function in $L_\infty(a, b)$ which is obtained by k-fold differentiation of s on each subsegment $(x_i, x_{i+1})$, $0 \leqslant i \leqslant N-1$.*

Concerning a proof we refer e.g. to [29]. We further need

LEMMA 1. *Let g be a function in $W_q^k(a, b)$, $k \in \boldsymbol{N}$, $1 \leqslant q \leqslant \infty$, satisfying*

$$g(x_i) = 0 \quad (1 \leqslant i \leqslant N-1).$$

*Then for $q \leqslant p \leqslant \infty$ one has*

$$\|g\|_p \leqslant C\bar{\Delta}^{k+1/p-1/q} \|g^{(k)}\|_q.$$

This is proved by a standard argument using Rolle's theorem (cf. [2], [39]) and the inequalities of Hölder and Jensen.

THEOREM 4. *Let $f \in W_q^k(a, b)$, $1 \leqslant q \leqslant \infty$ if $1 \leqslant k \leqslant 2n$ or $f \in C[a, b]$ if $k = 0$, and let for $N > n S_\Delta(f)$ be the interpolating spline in* Sp$(2n-1, 2n-1, \Delta)$ *satisfying* (3.1), (3.2), *where the $j_i$ and $k_i$ are such that for any function $g \in W_2^{2n}(a, b)$ satisfying the homogeneous system* (3.1), (3.2) *there holds*

$$\int_a^b [g^{(n)}(x)]^2 dx = (-1)^{(n)} \int_a^b g(x) g^{(2n)}(x)\, dx \ (^3). \tag{3.4}$$

(3) The proof of the corresponding theorem for $p = q = \infty$ in Scherer [34] is incomplete without this hypothesis.

*Furthermore let* $R_\Delta(f)$ *be any other interpolating spline in* $\mathrm{Sp}(2n-1, 2n-1, \Delta)$ *satisfying* (3.1). *Then under the assumption*

$$\min(x_1-a, b-x_{N-1}) \geqslant C\bar{\Delta} \tag{3.5}$$

*there holds the inequality*

$$\begin{aligned} \|S_\Delta(f)-f\|_p \leqslant C\{\|R_\Delta(f)-f\|_p+\bar{\Delta}^{k+1/p-1/q}\|f^{(k)}\|_q\}+ \\ +C\bar{\Delta}^{1/p} \sup_{0\leqslant j_i,k_i<\max(1,k)} \{\bar{\Delta}^{j_i}|f^{(j_i)}(a)-\alpha_i|+\bar{\Delta}^{k_i}|f^{(k_i)}(b)-\beta_i|\}+ \\ +C\bar{\Delta}^{1/p} \sup_{\max(1,k)\leqslant j_i,k_i\leqslant 2n-1} \{\Delta^{j_i}|\alpha_i|+\Delta^{k_i}|\beta_i|\} \end{aligned} \tag{3.6}$$

*for* $2 \leqslant p \leqslant \infty$.

PROOF. For simplicity set $u = R_\Delta(f)-S_\Delta(f)$. By Lemma 1

$$\|u\|_p \leqslant C\bar{\Delta}^{n+1/p-1/2}\|u^{(n)}\|_2 .$$

Squaring and integration by parts yield

$$\|u\|_p^2 \leqslant C\bar{\Delta}^{2n-1+2/p}P_n(u, u)|_a^b, \tag{3.7}$$

where the bilinear concomitant $P_n(u, v)$ is defined by

$$P_n(u, v) = \sum_{j=0}^{n-1} (-1)^j u^{(n-1-j)}(x)v^{(n+j)}(x).$$

In view of property (3.4), introduced by Swartz [43], Swartz–Varga [44], we have $P_n(g, g)|_a^b = 0$ for all $g \in W_2^{2n}(a, b)$ satisfying the homogeneous system (3.1), (3.2). This means that the numbers $j_i$, $k_i$ must be such that for each of the terms $g^{(n-1-j)}(a)g^{(n+j)}(a)$, $0 \leqslant j \leqslant n-1$, and $g^{(n-1-k)}(b)g^{(n+k)}(b)$, $0 \leqslant k \leqslant n-1$, the indices $j$ and $k$ are contained in the $j_i$'s and $k_i$'s, respectively. Otherwise one could construct a $g$ which satisfies the homogeneous system (3.1), (3.2) but $P_n(g, g)|_a^b \neq 0$. Thus we can write

$$P_n(g, g)|_a^b$$
$$= \sum_{i=1}^{n} [(-1)^{j_i}u^{(2n-1-j_i)}(a)u^{(j_i)}(a)]+[(-1)^{k_i}u^{(2n-1-k_i)}(b)u^{(k_i)}(b)]$$

and hence

$$\|u\|_p^2 \leqslant C\bar{\Delta}^{2n-1/2p}\sum_{i=1}^{n} \{|u^{(2n-1-j_i)}(a)u^{(j_i)}(a)|+|u^{(2n-1-k_i)}(b)u^{(k_i)}(b)|\}.$$

Applying (3.3) and observing (3.5) gives

$$\|u\|_p^2 \leqslant C\bar{\Delta}^{1/p}\|u\|_p \sum_{i=1}^{n} \{\bar{\Delta}^{j_i}|u^{(j_i)}(a)|+\bar{\Delta}^{k_i}|u^{(k_i)}(b)|\}. \tag{3.8}$$

Next we estimate by (3.2)

$$|u^{(j_i)}(a)| \leqslant \| R_\Delta^{(j_i)}(f)-f^{(j_i)} \|_\infty+|f^{(j_i)}(a)-\alpha_i| \qquad (0 \leqslant j_i < k),$$

$$|u^{(j_i)}(a)| \leqslant \| R_\Delta^{(j_i)}(f)\|_\infty+|\alpha_i| \qquad (k \leqslant j_i < 2n),$$

and similarly for $|u^{(k_i)}(b)|$. Then we use a result in [36] by which

$$C\bar{\Delta}_i^j \Delta_i^{-1/p}\|R_\Delta(f)-f\|_{p,i}+\Delta_i^{k-1/q}\|f^{(k)}\|_{q,i} \geqslant \begin{cases} \|R_\Delta^{(j)}(f)-f^{(j)}\|_{\infty,i}, & 0 \leqslant j < k, \\ \|R_\Delta^{(j)}(f)\|_{\infty,i}, & k \leqslant j, \end{cases} \tag{3.9}$$

where $\|\cdot\|_{q,i}$ denotes the $L_q$-norm with respect to the subsegment $(x_{i+1}-x_i)$ and $\Delta_i = x_{i+1}-x_i$, $0 \leqslant i \leqslant N-1$. Substituting this for $i = 0$, $N-1$ in (3.8) yields by (3.5)

$$\begin{aligned} \|u\|_p \leqslant\ & C\{\|R_\Delta(f)-f\|_p+\bar{\Delta}^{k+1/p-1/q}\|f^{(k)}\|_q\}+ \\ & +C\bar{\Delta}^{1/p} \sup_{0 \leqslant j_i,k_i < \max(1,k)} \{\bar{\Delta}^{j_i}|f^{(j_i)}(a)-\alpha_i|+\bar{\Delta}^{k_i}|f^{(k_i)}(b)-\beta_i|\}+ \\ & +C\bar{\Delta}^{1/p} \sup_{\max(1,k) \leqslant j_i,k_i \leqslant 2n-1} \{\bar{\Delta}^{j_i}|\alpha_i|+\bar{\Delta}^{k_i}|\beta_i|\}. \end{aligned}$$

Now inequality (3.6) immediately follows by noting that

$$\|S_\Delta(f)-f\|_p \leqslant \|u\|_p+\|R_\Delta(f)-f\|_p.$$

Next we establish comparison theorems in the case when the interpolating spline $S_\Delta(f)$, $n \geqslant 2$, satisfies so-called integral relations instead of (3.2). The first "integral relation" reads

$$\int_a^b [f^{(n)}(x)]^2 dx = \int_a^b [S_\Delta^{(n)}(f; x)]^2 dx+\int_a^b [f^{(n)}(x)-S_\Delta^{(n)}(f; x)]^2 dx \qquad (f \in W_2^n(a, b)), \tag{3.10}$$

whereas the second "integral relation" has the form

$$\int_a^b [f^{(n)}(x)-S_\Delta^{(n)}(f; x)]^2 dx = (-1)^n\int_a^b [f(x)-S_\Delta(f; x)]f^{(2n)}(x)\,dx \qquad (f \in W_2^{2n}(a, b)). \tag{3.11}$$

Since $S_\Delta(f)$ interpolates $f$ at the mesh points of $\Delta$ relations (3.10), (3.11) are equivalent by partial integration to (see [2], p. 155, 169)

$$P_n\big(S_\Delta(f)-f,\, S_\Delta(f)\big)\big|_a^b = 0 \quad (f \in W_2^n(a,b)) \tag{3.12}$$

or

$$P_n\big(S_\Delta(f)-f,\, S_\Delta(f)-f\big)\big|_a^b = 0 \quad (f \in W_2^{2n}(a,b)) \tag{3.13}$$

respectively. We note further that the validity of (3.10) is equivalent to a well-known minimum or orthogonality property of the interpolating spline which is, in the more abstract setting of spline theory, the starting point of all further development (see e.g. [9], [4], [3], [25], [28]).

In this context it is not necessary to have particular information on $S_\Delta(f)$ in form of boundary conditions of type (3.2), so that in this respect we consider here more general interpolating splines. On the other hand not all boundary conditions of type (3.2) must satisfy integral relations. But it can be shown (cf. [44]) that under condition (3.4) "exact" boundary conditions of the form

$$\begin{aligned} S_\Delta^{(j_i)}(f;a) &= f^{(j_i)}(a) \\ S_\Delta^{(k_i)}(f;b) &= f^{(k_i)}(b) \end{aligned} \qquad (1 \leqslant i \leqslant n) \tag{3.14}$$

satisfy the second integral relation. Examples for the first integral relation are given by the special case $j_i = k_i = i-1$ in (3.14) and the natural interpolating splines, i.e. the case $j_1 = k_1 = 0$, $\alpha_1 = f(a)$, $\beta_1 = f(b)$ and $j_i = k_i = n-2+i$, $\alpha_i = \beta_i = 0$ for $2 \leqslant i \leqslant n$ in (3.2).

THEOREM 5. *Let $S_\Delta(f)$ be a spline in* Sp$(2n-1, 2n-1, \Delta)$ *interpolating $f \in W_q^n(a,b)$, $1 \leqslant q \leqslant \infty$, and satisfying the first integral relation. Furthermore, let $R_\Delta(f)$ be given as in Theorem* 4. *Then one has for $2 \leqslant p \leqslant \infty$ under assumption* (3.5)

$$\|S_\Delta(f)-f\|_p \leqslant C\{\|R_\Delta(f)-f\|_p + \bar{\Delta}^{1/p-1/q+n}\|f^{(n)}\|_q\}. \tag{3.15}$$

PROOF. We proceed similarly as in the proof of Theorem 4. Let $v = S_\Delta(f)-\bar{S}_\Delta(f) = S-\bar{S}$, where $\bar{S}_\Delta(f)$ is the interpolating spline of $f$ in Sp$(2n-1, 2n-1, \Delta)$ satisfying (3.13) with $j_i = k_i = i-1$. Then we obtain, similarly to (3.7)

$$\|v\|_p^2 \leqslant C\bar{\Delta}^{2n-1+2/p} P_n(S-\bar{S},\, S-\bar{S})|_a^b. \tag{3.16}$$

By definition of $\bar{S}_\Delta(f)$ and since $S_\Delta(f)$ satisfies (3.12) it follows further that

$$P_n(S-\bar{S},\, S-\bar{S})|_a^b = P_n(S-f,\, S-\bar{S})|_a^b = -P_n(S-f,\, \bar{S})|_a^b$$
$$= -P_n(S-\bar{S},\, \bar{S})|_a^b.$$

An application of Theorem 3 yields similarly to (3.8)

$$\|v\|_p^2 \leqslant C\bar{\Delta}^{1/p}\|v\|_p \sum_{j=0}^{n-1} \bar{\Delta}^{n+j}[|\bar{S}_\Delta^{(n+j)}(f;\, a)|+|\bar{S}_\Delta^{(n+j)}(f;\, b)|].$$

Using (3.9) for $\bar{S}$ instead of $R$ gives in view of (3.5)

$$\|v\|_p \leqslant C\{\|\bar{S}_\Delta(f)-f\|_p+\bar{\Delta}^{1/p-1/q+n}\|f^{(n)}\|_q\}.$$

From this one obtains (3.15) by

$$\|S_\Delta(f)-f\|_p \leqslant \|\bar{S}_\Delta(f)-f\|_p+\|v\|_p$$
$$\leqslant C\{\|\bar{S}_\Delta(f)-f\|_p+\bar{\Delta}^{1/p-1/q+n}\|f^{(n)}\|_q\}$$

since applying Theorem 4 to $\bar{S}$ gives

$$\|\bar{S}_\Delta(f)-f\|_p \leqslant C\{\|R_\Delta(f)-f\|_p+\bar{\Delta}^{1/p-1/q+n}\|f^{(n)}\|_q\}.$$

For splines satisfying the second integral relation one has the following comparison theorem.

THEOREM 6. *Let $S_\Delta(f)$ be a spline in* $\mathrm{Sp}(2n-1, 2n-1, \Delta)$ *interpolating $f \in W_q^{2n}(a, b)$, $1 \leqslant q \leqslant \infty$, and satisfying the second integral relation. Furthermore let $R_\Delta(f)$ be given as in Theorem* 4. *Then under assumption* (3.5) *there holds for* $2 \leqslant p \leqslant \infty$

$$\|S_\Delta(f)-f\|_p \leqslant C\bar{\Delta}^{-1/q+1/p}\|R_\Delta(f)-f\|_q+\bar{\Delta}^{2n}\|f^{(2n)}\|_q. \tag{3.17}$$

PROOF. Again with $v = S-\bar{S}$ one arrives at (3.16). Since $S$ and $\bar{S}$ satisfy (3.13) it follows further

$$P_n(S-\bar{S},\, S-\bar{S})|_a^b = P_n(S-\bar{S},\, S-f)|_a^b+P_n(S-\bar{S}, f-\bar{S})|_a^b$$
$$= P_n(S-f,\, S-f)|_a^b+P_n(S-\bar{S}, f-\bar{S})|_a^b$$
$$= P_n(S-\bar{S}, f-\bar{S})|_a^b.$$

Applying Theorem 3 yields similarly to (3.8)

$$\|v\|_p^2 \leqslant C\bar{\Delta}^{1/p}\|v\|_p \sum_{j=0}^{n-1} \bar{\Delta}^{n+j}[|f^{(n+j)}(a)-\bar{S}^{(n+j)}(a)|+$$
$$+|f^{(n+j)}(b)-\bar{S}^{(n+j)}(b)|],$$

and using (3.9) for $\bar{S}$ gives

$$\|v\|_p \leqslant C\{\bar{\Delta}^{1/p-1/q+2n}\|f^{(2n)}\|_q + \|\bar{S}_\Delta(f)-f\|_p\}.$$

The rest of the proof follows in the same way as above.

We remark that Theorems 4–6 carry over with corresponding modifications to the more general *g*-, *L*- or *Lg*-splines (see e.g. [38], [39], [26]) provided an analogue of Theorem 3 is known.

As already mentioned above, the range of examples is not the same for the three comparison theorems but there is an overlapping. However, Theorem 6 can be used to extend Theorem 4 to more general boundary and interpolation conditions in the setting of a "stability" or "perturbation" theorem introduced by Swartz [43], Swartz–Varga [44]. To give a brief outline, consider the case that the spline $S_\Delta(f) \in \mathrm{Sp}(2n-1, 2n-1, \Delta)$ in question satisfies the "perturbed" interpolation data

$$S_\Delta(f, x_i) = r_i \quad (1 \leqslant i \leqslant N-1) \tag{3.18}$$

and the boundary conditions of the form

$$\lambda_j(S_\Delta(f)) = \gamma_j \quad (1 \leqslant j \leqslant 2n), \tag{3.19}$$

where the $\lambda_j$ are linear functionals and the $r_i$, $\gamma_j$ any reals.

The basic idea is to construct a smooth function $G(x) \in W_2^{2n}(a, b)$ satisfying

$$G(x_i) = r_i, \quad \lambda_j(G) = \gamma_j,$$

so that writing $S_\Delta(f) = \bar{S}_\Delta(G)$ this new interpolating spline $\bar{S}_\Delta$ satisfies the exact interpolation data and boundary conditions, i.e.

$$\bar{S}_\Delta(G; x_i) = G(x_i), \quad \lambda_j(\bar{S}_\Delta(G)) = \lambda_j(G).$$

Now, if the $\lambda_j$ are such that the second integral relation holds for $\bar{S}_\Delta(G)-G$, Theorem 6 applies, yielding,

$$\begin{aligned} \|S_\Delta(f)-f\|_p &\leqslant \|\bar{S}_\Delta(G)-G\|_p + \|G-f\|_p \\ &\leqslant C\{\|R_\Delta(G)-G\|_p + \bar{\Delta}^{2n}\|G^{(2n)}\|_p + \|G-f\|_p\}. \end{aligned} \tag{3.20}$$

Hence $G$ has to be constructed in such a way that the latter two terms can be bounded in a suitable manner, provided there holds the inequality

$$\|R_\Delta(G)-G\|_p \leqslant C\bar{\Delta}^{2n}\|G^{(2n)}\|_p \tag{3.21}$$

for the comparison spline $R_\Delta(G)$.

The first requirement can be shown to be satisfied for the boundary conditions (3.2) or even more general boundary condition of extended Hermite–Birkhoff type. This can be carried out by means of two-point-Taylor interpolation (for details see Swartz–Varga [44]). The final result would then correspond to that of Theorem 4.

However, this would not yield a true comparison theorem since (3.21) is needed. It will be the main task of the next section to justify this hypothesis to a range as far reaching as possible, i.e. for as general meshes $\Delta$ as possible.

**4. Jackson-type inequalities for odd-degree interpolating splines.** If $\Delta$ is an equidistant partition of $[a, b]$ there exist many interpolating splines of class $\mathrm{Sp}(2n-1, \Delta)$ which may serve as the comparison spline $R_\Delta(f)$ considered in the previous Sections 2 and 3. Among them one would prefer one for which property (3.21) is established in the simplest way. In this respect the periodic case seems to be best known (see e.g. Collatz–Quade [18], Ahlberg–Nilson–Walsh [1], Golomb [23]). In [1] also the case of non-equidistant partitions is treated. We state their result in a slightly different form.

THEOREM 7. *Let $f(x)$ be a function of class $C^{2n-2}[a, b]$ with period $b-a$, i.e. $f^{(j)}(a) = f^{(j)}(b)$ for $0 \leqslant j \leqslant 2n-2$, and let $S_\Delta(f)$ be the periodic spline interpolant to $f$ of class* $\mathrm{Sp}(2n-1, 2n-1, \Delta)$ *with respect to $\Delta$, i.e.*

$$(4.1)\qquad \begin{aligned} S_\Delta(f; x_i) &= f(x_i), & 0 \leqslant i \leqslant N,\\ S_\Delta^{(j)}(f; a) &= S_\Delta^{(j)}(f, b), & 1 \leqslant j \leqslant 2n-2. \end{aligned}$$

*As a measure of local mesh uniformity we introduce* ($\Delta_i = (x_i - x_{i-1})$)

$$(4.2)\qquad \lambda = \max_{1 \leqslant i \leqslant N-1} \left| \frac{\Delta_{i+1}}{\Delta_i + \Delta_{i+1}} - \frac{1}{2} \right|$$

*or equivalently*

$$(4.3)\qquad \sigma = \max_{1 \leqslant i \leqslant N-1} \left| \frac{\Delta_i}{\Delta_{i+1}} - 1 \right|.$$

*Then for $\Delta$ with $\bar{\Delta} \ll 1$ and $\sigma \ll 1$* (⁴) *there exists a constant $C$ independent of $f$ and $\Delta$ such that*

$$(4.4)\qquad \|S_\Delta^{(2n-2)}(f)\|_\infty \leqslant C \|f^{(2n-2)}\|_\infty .$$

(⁴) This means that $\bar{\Delta}$ and $\sigma$ have to be smaller than a certain constant depending only upon $a$, $b$ and $n$.

In particular, the precise bound of $\sigma$ depends on the value of the norm of an inverse matrix defined for equidistant meshes (cf. [1]).

Next we show how the result of Theorem 7 can be used to ensure the existence of an interpolating spline $R_\Delta(f)$ satisfying (3.21).

THEOREM 8. *Let $f$ be in $W_p^{2n}(a, b)$, $2 \leqslant p < \infty$. Then for $\Delta$ with $\bar{\Delta} \ll 1$ and $\sigma \ll 1$ there exists a spline $R_\Delta(f)$ in* $\mathrm{Sp}(2n-1, 2n-1, \Delta)$ *interpolating $f$ with respect to $\Delta$ and satisfying*

$$\|R_\Delta(f)-f\|_p \leqslant C\bar{\Delta}^{2n}\|f^{(2n)}\|_p. \tag{4.5}$$

PROOF. From (4.4) it follows that

$$\|S_\Delta^{(2n-2)}(g)-g^{(2n-2)}\|_\infty \leqslant C\|g^{(2n-2)}\|_\infty,$$

and hence by Lemma 1 for functions in $W_\infty^{2n-2}(a, b)$ with period $b-a$

$$\|S_\Delta(g)-g\|_\infty \leqslant C\bar{\Delta}^{2n-2}\|g^{(2n-2)}\|_\infty. \tag{4.6}$$

Now let $f \in W_p^{2n}(a, b)$, $2 \leqslant p \leqslant \infty$, not necessarily periodic. Then consider the derivative $f^{(2n-2)}$ and choose $p_1(f) \in P_1$ such that

$$\bar{f}^{(2n-2)} = f^{(2n-2)}+p_1(f) \tag{4.7}$$

has period $b-a$. Furthermore, introduce the linear operator

$$T(g; x) = \int_a^x \left[g(t)-\frac{1}{b-a}\int_a^b g(u)\,du\right] dt$$

for functions $g$ with period $b-a$. It is obvious that $T(g; x)$ has period $b-a$. We claim that

$$R_\Delta(f) = S_\Delta\left(T^{2n-2}(\bar{f}^{(2n-2)})\right)-T^{2n-2}(\bar{f}^{(2n-2)})+f \tag{4.8}$$

is the desired spline function $R_\Delta(f)$ satisfying (4.5). Indeed, since $T^{2n-2}(\bar{f}^{(2n-2)})$ can be represented by (4.7) in the form

$$T^{2n-2}(\bar{f}^{(2n-2)}) = f+q_{2n-1}(f),$$

where $q_{2n-1}(f)$ is a polynomial of degree $2n-1$, it follows that $R_\Delta(f)$ belongs to $\mathrm{Sp}(2n-1, 2n-1, \Delta)$. By (4.8) one concludes further that $R_\Delta(f)-f$ vanishes at the mesh points of $\Delta$, i.e. $R_\Delta(f)$ is interpolating. Furthermore one observes by (4.8) that

$$R_\Delta^{(j)}(f; a)-f^{(j)}(a) = R_\Delta^{(j)}(f; b)-f^{(j)}(b) \quad (1 \leqslant j \leqslant 2n-2),$$

which means that $R_\Delta(f)$ satisfies the second integral relation. By results of Schultz–Varga [39], Jerome–Varga [26], it follows that

$$\text{(4.9)} \qquad \|R_\Delta(f)-f\|_2 \leqslant C\bar{\Delta}^{2n}\|f^{(2n)}\|_2 \qquad (f\in W_2^{2n}(a,b)).$$

To obtain the corresponding error bound in the $L_\infty$-norm we use an argument of de Boor (see [43], p. 19). Let $t_\Delta$ be the polygonal line interpolating $\bar{f}^{(2n-2)}$ in the mesh points of $\Delta$, which implies the estimate

$$\|t_\Delta-\bar{f}^{(2n-2)}\|_\infty \leqslant C\bar{\Delta}^2\|\bar{f}^{(2n)}\|_\infty = C\bar{\Delta}^2\|f^{(2n)}\|_\infty .$$

Furthermore we have by (4.6) and (4.8), observing the projection property of $S_\Delta$,

$$\text{(4.10)} \quad \|R_\Delta(f)-f\|_\infty$$
$$= \|S_\Delta[T^{2n-2}(\bar{f}^{(2n-2)})-T^{2n-2}(t_\Delta)]-[T^{2n-2}(\bar{f}^{(2n-2)})-T^{2n-2}(t_\Delta)]\|_\infty$$
$$\leqslant C\bar{\Delta}^{2n-2}\|\bar{f}^{(2n-2)}(x)-t_\Delta(x)-\frac{1}{b-a}\int_a^b[\bar{f}^{(2n-2)}(u)-t_\Delta(u)]\,du\|_\infty$$
$$\leqslant C\bar{\Delta}^{2n}\|f^{(2n)}\|_\infty .$$

With the aid of interpolation theorems for bounded linear operators on Banach spaces (cf. Butzer–Berens [12]) one obtains the desired inequality (4.5) for $2 \leqslant p \leqslant \infty$ by (4.9), (4.10) (cf. Hedstrom–Varga [24]).

Before applying this theorem to the results of Section 3 we need the following auxiliary lemma on piecewise Hermite interpolation.

LEMMA 2. *Let $f$ be in $W_q^k(a,b)$, $1 \leqslant q \leqslant \infty$, if $1 \leqslant k \leqslant m+1$ with $m\in N$ and let $f\in C[a,b]$ if $k=0$. Then the element $H_{m,k,\Delta}(f)$ in* $\mathrm{Sp}(2m+1, m+1, \Delta)$, *uniquely determined by*

$$\text{(4.11)} \quad H_{m,k,\Delta}^{(j)}(f;x_i) = \begin{cases} f^{(j)}(x_i), & 0\leqslant j<\max(1,k),\\ 0, & \max(1,k)\leqslant j\leqslant m,\end{cases} \qquad 0\leqslant i\leqslant N,$$

*satisfies the estimates*

$$C\bar{\Delta}^{k-j-1/q+1/p}\|f^{(k)}\|_q \geqslant \begin{cases} \|H_{m,k,\Delta}^{(j)}(f)-f^{(j)}\|_p, & 0\leqslant j<\max(1,k),\\ \|H_{m,k,\Delta}^{(j)}(f)\|_p, & \max(1,k)\leqslant j\leqslant 2m+1\end{cases}$$

*for $1\leqslant q\leqslant p\leqslant\infty$ if $k\geqslant 1$ and $p=q=\infty$ if $k=0$ under the restriction*

$$\text{(4.12)} \qquad \bar{\Delta}/\underline{\Delta} < C.$$

A proof of this lemma can be found in Swartz–Varga [44] (cf. also Birkhoff–Schultz–Varga [6] in the case $k = m+1$).

THEOREM 9. *Under assumption* (4.12) *let* $\Delta$ *be a partition of* $[a, b]$ *with* $\bar{\Delta} \ll 1$ *and* $\sigma \ll 1$ *and let* $1 \leqslant q \leqslant \infty$, $\max(2, q) \leqslant p \leqslant \infty$. *Then the following assertions hold* ($N > n$):

(a) *If* $S_\Delta(f)$ *is the spline function in* $\mathrm{Sp}(2n-1, 2n-1, \Delta)$ *interpolating the "perturbed" data*

$$S_\Delta(f; x_i) = r_i \tag{4.13}$$

*instead of* (3.1), *then for* $f \in W_q^k(a, b)$ *if* $1 \leqslant k \leqslant 2n$ *or* $f \in C[a, b]$ *if* $k = 0$ *there holds*

$$\begin{aligned}\|S_\Delta(f)-f\|_p \leqslant{} & C\bar{\Delta}^{1/p-1/q}\Big\{\bar{\Delta}^k\|f^{(k)}\|_q+\Big[\sum_{i=1}^{N-1}|r_i-f(x_i)|^q\Big]^{1/q}\Big\}+\\ &+C\bar{\Delta}^{1/p}\sup_{0\leqslant j_i,k_i<\max(1,k)}\{\bar{\Delta}^{j_i}|f^{(j_i)}(a)-\alpha_i|+\bar{\Delta}^{k_i}|f^{(k_i)}(b)-\beta_i|\}+\\ &+C\bar{\Delta}^{1/p}\sup_{\max(1,k)\leqslant j_i,k_i^*\leqslant 2n-1}\{\bar{\Delta}^{j_i}|\alpha_i|+\bar{\Delta}^{k_i}|\beta_i|\}.\end{aligned} \tag{4.14}$$

(b) *If* $S_\Delta(f)$ *is an interpolating spline in* $\mathrm{Sp}(2n-1, 2n-1, \Delta)$ *satisfying* (3.1) *and the first integral relation, then for* $f \in W_q^n(a, b)$

$$\|S_\Delta(f)-f\|_p \leqslant C\bar{\Delta}^{1/p-1/q+n}\|f^{(n)}\|_q. \tag{4.15}$$

(c) *If* $S_\Delta(f)$ *is an interpolating spline in* $\mathrm{Sp}(2n-1, 2n-1, \Delta)$ *satisfying* (3.1) *and the second integral relation, then for* $f \in W_q^{2n}(a, b)$

$$\|S_\Delta(f)-f\|_p \leqslant C\bar{\Delta}^{1/p-1/q+2n}\|f^{(2n)}\|_q. \tag{4.16}$$

*In case* $p = q$ *assertion* (c) *and for* $k = 2n$, $r_i = f(x_i)$ *also assertion* (a) *are valid under the weaker condition* (3.5) *instead of* (4.12).

PROOF. At first the last statement is an immediate consequence of Theorems 4, 6 as well as Theorem 8. To prove (a) in its general form, let $\bar{H}_{2n,k,\Delta}(f)$ be the element in $\mathrm{Sp}(4n+1, 2n+1, \Delta)$ satisfying (4.11) with the exception that $f(x_i)$ is replaced by $r_i$. Hence, upon writing $S_\Delta(f) = \tilde{S}_\Delta(\bar{H}_{2n,k,\Delta}(f))$, this spline interpolates $\bar{H}_{2n,k,\Delta}(f)$ exactly so that by Theorems 4 and 8 and the very definition (4.11)

$$\begin{aligned}\|S_\Delta(f)-\bar{H}_{2n,k,\Delta}(f)\|_p \leqslant{} & C\bar{\Delta}^{2n}\|\bar{H}_{2n,k,\Delta}(f)\|_p+\\ &+C\bar{\Delta}^{1/p}\sup_{0\leqslant k_i,j_i<\max(1,k)}\{\bar{\Delta}^{j_i}|f^{(j_i)}(a)-\alpha_i|+\bar{\Delta}^{k_i}|f^{(k_i)}(b)-\beta_i|\}+\\ &+C\bar{\Delta}^{1/p}\sup_{\max(1,k)\leqslant j_i,\,k_i\leqslant 2n-1}\{\bar{\Delta}^{j_i}|\alpha_i|+\bar{\Delta}^{k_i}|\beta_i|\}.\end{aligned} \tag{4.17}$$

Furthermore we have by Lemma 2 and Theorem 3

$$(4.18)\quad \begin{aligned}&\bar{\Delta}^{2n}\|\bar{H}^{(2n)}_{2n,k,\Delta}(f)\|_p\\ &\leqslant \bar{\Delta}^{2n}\|\bar{H}^{(2n)}_{2n,k,\Delta}(f)-H^{(2n)}_{2n,k,\Delta}(f)\|_p+C\bar{\Delta}^{k-1/q+1/p}\|f^{(k)}\|_q\\ &\leqslant C\bar{\Delta}^{1/p-1/q}\{\|\bar{H}_{2n,k,\Delta}(f)-H_{2n,k,\Delta}(f)\|_q+\bar{\Delta}^k\|f^{(k)}\|_q\},\end{aligned}$$

and again by Lemma 2 and Theorem 3

$$(4.19)\quad \begin{aligned}&\|\bar{H}_{2n,k,\Delta}(f)-f\|_p\\ &\leqslant \|\bar{H}_{2n,k,\Delta}(f)-H_{2n,k,\Delta}(f)\|_p+C\bar{\Delta}^{k-1/q+1/p}\|f^{(k)}\|_q\\ &\leqslant C\bar{\Delta}^{1/p-1/q}\{\|\bar{H}_{2n,k,\Delta}(f)-H_{2n,k,\Delta}(f)\|_q+\bar{\Delta}^k\|f^{(k)}\|_q\}.\end{aligned}$$

Finally for $x \in [x_i, x_{i+1})$, $1 \leqslant i \leqslant N-2$, the representation

$$\begin{aligned}&\bar{H}_{2n,k,\Delta}(f;x)-H_{2n,k,\Delta}(f;x)\\ &=\Big\{[r_i-f(x_i)]\int_x^{x_{i+1}}Q_i(t)\,dt+[r_{i+1}-f(x_{i+1})]\int_{x_i}^{x}Q_i(t)\,dt\Big\}\Big\{\int_{x_i}^{x_{i+1}}Q_i(t)\,dt\Big\}^{-1}\end{aligned}$$

with $Q_i(t) = [(t-x_i)(x_{i+1}-t)]^{2n-1}$ is valid (with corresponding modifications for $[a, x_i)$ and $(x_{N-1}, b]$), so that

$$(4.20)\qquad \|\bar{H}_{2n,k,\Delta}(f)-H_{2n,k,\Delta}(f)\|_q \leqslant 2\Big\{\sum_{i=1}^{N-1}|r_i-f(x_i)|^q\bar{\Delta}\Big\}^{1/q}.$$

Combining inequalities (4.17)–(4.20) now yields assertion (a) in the general case. From this assertions (b) and (c) follow by Theorems 5 and 6, if one takes for the comparison spline $R_\Delta(f)$ not the spline of Theorem 8 but the interpolating spline $\bar{S}_\Delta(f)$ already mentioned in Theorems 5, 6 satisfying the boundary conditions (3.2) with $j_i = k_i = i-1$. By part (a) this spline satisfies

$$\|\bar{S}_\Delta(f)-f\|_p \leqslant C\bar{\Delta}^{1/p-1/q+k}\|f^{(k)}\|_q \qquad (f\in W^k_q(a,b),\ n\leqslant k\leqslant 2n).$$

Assertion (d) has already been proved for $\sigma = 0$, i.e. for uniform partitions, in Swartz [43], Swartz–Varga [44], see also [34]. Assertions (b) and (c) carry over the error bounds of Schultz–Varga [39], Jerome–Varga [26] for $q = 2$ and $p = 2, \infty$ to general $p, q$ with $2 \leqslant q \leqslant p \leqslant \infty$, however, for polynomial splines. An extension of the above results to the more general $Lg$-splines would require not only the already mentioned extension of Theorem 3 but also a corresponding one of Theorem 8.

Note that a slightly stronger but simpler form of the assumptions on the mesh $\Delta$ is given by $\bar{\Delta} \ll \sigma$ and $\sigma \leqslant M\underline{\Delta}$ for then the mesh ratio $\bar{\Delta}/\underline{\Delta}$ satisfies (see (4.3)), $M'$ and $M''$ being further constants $> 0$

$$\bar{\Delta}/\underline{\Delta} = \Delta_{i'}/\Delta_{i''} \leqslant (1+\sigma)^{|i'-i''|} \leqslant (1+M'/N)^N \leqslant M''.$$

Further remarks concern special instances of the above theorem. In case $q = 2$ we can prove the results without any hypothesis on $\sigma$ by observing that (4.9) is valid without any restriction upon $\Delta$ (cf. [26]). This is also true for cubic ($n = 2$) and quintic splines ($n = 3$) when one may take for $R_\Delta(f)$, e.g. the interpolating splines of Sharma–Meir [40] and de Boor [8], respectively. Furthermore, under the assumptions of both (b) and (c) it is possible to extend the estimates of (4.15), (4.16) to functions belonging to the intermediate spaces $W_q^k(a, b)$, $n \leqslant k \leqslant 2n$, by arguments of the interpolation theory of Banach spaces (cf. [24], [36]).

**5. Even-degree interpolating splines.** Even degree interpolating splines, the interpolation taking place halfway between the knots, seem hardly to be treated in the literature (see e.g. Ahlberg–Nilson–Walsh [2], where the periodic case is considered, and Subbotin [41], where error bounds for such splines defined on the whole real axis are obtained). Here we consider even-degree splines $T_\Delta(x)$ of class $\mathrm{Sp}(2n, 2n, \Delta)$, $n \geqslant 1$, interpolating data $Q_i$ in the sense that (see Anselone–Laurent [3], Varga [45])

$$\int_{x_i}^{x_{i+1}} T_\Delta(x)\,dx = Q_i \qquad (0 \leqslant i \leqslant N-1). \tag{5.1}$$

In general one chooses the $Q_i$ close or equal to the values

$$f_i = \int_{x_i}^{x_{i+1}} f(x)\,dx \qquad (0 \leqslant i \leqslant N-1), \tag{5.2}$$

for example one obtains $Q_i$ by some quadrature formula for $f_i$.

In order to ensure the uniqueness of $T_\Delta$ one has to impose further conditions on it. Here we consider boundary conditions of the form

$$\begin{aligned} T_\Delta^{(j_i)}(a) &= \alpha_i \\ T_\Delta^{(k_i)}(b) &= \beta_i \end{aligned} \qquad (1 \leqslant i \leqslant n), \tag{5.3}$$

where the $\alpha_i$, $\beta_i$ are given reals and where the $j_i$, $k_i$ are integers satisfying $0 \leqslant j_1 < \ldots < j_n \leqslant 2n-1$, $0 \leqslant k_1 < \ldots < k_n \leqslant 2n-1$.

Then the determination of $T_\Delta$ leads to a system of $N+2n$ linear equations with $N+2n$ unknowns. The existence and uniqueness questions for a solution of this system is settled by (cf. [3], [45])

LEMMA 3. *There exists a unique spline $T_\Delta$ in* Sp$(2n, 2n, \Delta)$, $n \geqslant 1$, *satisfying* (5.1) *and* (5.3) *provided $N \geqslant n$ and the $j_i$, $k_i$ are such that condition* (3.4) *is fulfilled for all $g \in W_2^{2n}(a, b)$ satisfying the homogeneous system* (5.3).

PROOF. One has to show that a solution $T_0$ of the homogeneous system corresponding to (5.1), (5.3) must be the trivial one. To see this integrate $\int_a^b [T_0^{(n)}(x)]^2 dx$ by parts to obtain by the above assumptions

$$\int_a^b [T_0^{(n)}(x)]^2\, dx = (-1)^n \int_a^b T_0(x)\, T^{(2n)}(x)\, dx.$$

Since $T_0^{(2n)}(x)$ is piecewise constant the homogeneous system (5.1) gives that $T_0$ is a polynomial of degree $n-1$ at most, and hence $\int_a^x T_0(u)\, du$ is a polynomial of degree $n$ at most vanishing at the $N+1$ nodes of $\Delta$. Thus $T_0$ must vanish identically.

The examples of boundary conditions (5.3) are just the same as in the case of splines of odd degree $2n-1$, for example one can take the natural boundary conditions in (5.3).

Error bounds for $T_\Delta$ are given in

THEOREM 10. *Let $\Delta$, $\bar{\Delta}$ and $\sigma$ be given as in Theorem* 9 *and let* $\max(2, q) \leqslant p \leqslant \infty$. *Let $T_\Delta(f)$ be the spline of class* Sp$(2n, 2n, \Delta)$, $N \geqslant n \geqslant 1$ *defined by* (5.1), (5.3) *such that condition* (3.4) *is satisfied for all $g \in W_2^{2n}(a, b)$ satisfying the homogeneous conditions* (5.3). *Then one has the inequality*

(5.4)
$$\begin{aligned}\|T_\Delta - f\|_p &\\ &\leqslant C\bar{\Delta}^{1/p}\Big\{\bar{\Delta}^{k-1/q}\|f^{(k)}\|_q + \Big[\sum_{i=0}^{N-1} (f_i - Q_i|^q (x_{i+1}-x_i)^{-q}\Big]^{1/q}\Big\} + \\ &\quad + C\bar{\Delta}^{1/p} \sup_{0\leqslant j_i, k_i < \max(1,k)} \{\bar{\Delta}^{j_i}|f^{(j_i)}(a) - \alpha_i| + \bar{\Delta}^{k_i}|f^{(k_i)}(b) - \beta_i|\} + \\ &\quad + C\bar{\Delta}^{1/p} \sup_{\max(1,k)\leqslant j_i, k_i \leqslant 2n-1} \{\bar{\Delta}^{j_i}|\alpha_i| + \bar{\Delta}^{k_i}|\beta_i|\}\end{aligned}$$

*for $f \in W_q^k(a, b)$ if $1 \leqslant k \leqslant 2n+1$ or for $f \in C[a, b]$ if $k = 0$.*

PROOF. We need two auxiliary spline functions. At first let $\tilde{S}_\Delta(g; x)$ be the spline of class $\mathrm{Sp}(2n+1, 2n+1, \Delta)$ interpolating $g \in W_q^{2n+2}(a, b)$ and satisfying the boundary conditions

$$\tilde{S}_\Delta(g; a) = g(a), \quad \tilde{S}_\Delta(g; b) = g(b),$$

$$\tilde{S}_\Delta^{(j)}(g; a) = g^{(j)}(a), \quad \tilde{S}_\Delta^{(j)}(g; b) = g^{(j)}(b) \quad (n+1 \leqslant j \leqslant 2n).$$

Then by Theorem 9 we know that

$$\|\tilde{S}_\Delta(g)-g\|_p \leqslant C\Delta^{1/p-1/q+2n+2}\|g^{(2n+2)}\|_q. \tag{5.5}$$

Next we construct with the help of Lemma 2 the piecewise polynomial function $h(f)$ of degree $4n$ by

$$h(f; x) = H'_{2n,k+1,\Delta}(F; x)+t(Q; x), \tag{5.6}$$

where $F(x) = \int_a^x f(u)\,du$ and the step-function $t(Q; x)$ is defined by

$$t(Q; x) = (x_{i+1}-x_i)^{-1}(Q_i-f_i) \quad (x \in [x_i, x_{i+1}), 0 \leqslant i \leqslant N-1). \tag{5.7}$$

Then one easily verifies that

$$\int_{x_i}^{x_{i+1}} h(f; x)\,dx = F(x_{i+1})-F(x_i)+Q_i-f_i = Q_i \quad (0 \leqslant i < N-1) \tag{5.8}$$

and that

$$f^{(j_i)}(a) = F^{(j_i+1)}(a) = H^{(j_i+1)}_{2n,k+1,\Delta}(F; a) = h^{(j_i)}(f; a) \qquad (1 \leqslant j_i \leqslant k-1), \tag{5.9}$$

as well as a corresponding relation for the point $b$.

After these preparations we estimate by setting $H(x) = \int_a^x h(f; u)\,du$

$$\|T_\Delta-f\|_p \leqslant \|T_\Delta-\tilde{S}'_\Delta(H)\|_p+\|h(f)-\tilde{S}'_\Delta(H)\|_p+\|h(f)-f\|_p. \tag{5.10}$$

The first term is treated similarly to Theorem 4. Since by (5.8)

$$\int_{x_i}^{x_{i+1}} \tilde{S}'_\Delta(H; x)\,dx = \tilde{S}_\Delta(H; x_{i+1})-\tilde{S}_\Delta(H; x_i) = H(x_{i+1})-H(x_i) \tag{5.11}$$

$$= Q_i = \int_{x_i}^{x_{i+1}} T_\Delta(x)\,dx,$$

the difference $\tilde{S}'_\Delta(H)-T_\Delta$ has at least one zero in each interval $(x_i, x_{i+1})$, and hence by repeated application of Rolle's theorem (cf. Lemma 1) with $u = T_\Delta-\tilde{S}'_\Delta(H)$

$$\begin{aligned}\|u\|_p &\leqslant C\bar{\Delta}^{n+1/p-1/2}\|u^{(n)}\|_2\\ &\leqslant C\bar{\Delta}^{n+1/p-1/2}\left\{P_n(u,u)|_a^b+(-1)^n\int_a^b u(x)u^{(2n)}(x)\,dx\right\}^{1/2}.\end{aligned}$$

The last term vanishes in view of (5.10) since $u^{(2n)}$ is piecewise constant. Arguing as in Theorem 4 one arrives at an analog of (3.8), namely

$$\|u\|_p^2 \leqslant C\bar{\Delta}^{1/p}\|u\|_p\sum_{i=1}^{n}\{\bar{\Delta}^{j_i}|u^{(j_i)}(a)|+\bar{\Delta}^{k_i}|u^{(k_i)}(b)|\}.$$

For $1 \leqslant j_i < k$ we estimate in view of (5.3), (5.9) and Theorem 3

$$\begin{aligned}|u^{(j_i)}(a)| &\leqslant |f^{(j_i)}(a)-\alpha_i|+\|h^{(j_i)}(f)-\tilde{S}^{(j_i+1)}(H)\|_\infty\\ &\leqslant |f^{(j_i)}(a)-\alpha_i|+\bar{\Delta}^{-j_i-1/p}\|h(f)-\tilde{S}'_\Delta(H)\|_p,\end{aligned}$$

and in case $j_1 = 0 < k$

$$|u(a)| \leqslant |f(a)-\alpha_i|+\|h(f)-f\|_\infty+\bar{\Delta}^{-1/p}\|h(f)-S'_\Delta(H)\|_p.$$

For $j_i \geqslant k$ we use a Bernstein-type inequality in [37] to obtain

$$\begin{aligned}|u^{(j_i)}(a)| &\leqslant |\alpha_i|+|\tilde{S}_\Delta^{(j_i+1)}(H;a)|\\ &\leqslant |\alpha_i|+C\bar{\Delta}^{-j_i-1/q}\omega_{j_i}(\tilde{S}'_\Delta(H);\Delta)_q\\ &\leqslant |\alpha_i|+C\bar{\Delta}^{-j_i-1/q}\{\|\tilde{S}'_\Delta(H)-h(f)\|_q+\|h(f)-f\|_q+\omega_{j_i}(f;\Delta)_q\}\\ &\leqslant |\alpha_i|+C\bar{\Delta}^{-j_i-1/q}\{\|\tilde{S}'_\Delta(H)-h(f)\|_q+\|h(f)-f\|_q+\bar{\Delta}^k\|f^{(k)}\|_q\}.\end{aligned}$$

We estimate $|u^{(k_i)}(b)|$ similarly, which gives upon substitution in (5.10)

$$\begin{aligned}(5.12)\quad &\|T_\Delta-f\|_p\\ &\leqslant C\bar{\Delta}^{1/p}\sup_{0\leqslant j_i,k_i<\max(1,k)}\{\bar{\Delta}^{j_i}|f^{(j_i)}(a)-\alpha_i|+\bar{\Delta}^{k_i}|f^{(k_i)}(b)-\beta_i|\}+\\ &+C\bar{\Delta}^{1/p}\sup_{\max(1,k)\leqslant j_i,k_i\leqslant 2n-1}\{\bar{\Delta}^{j_i}|\alpha_i|+\bar{\Delta}^{k_i}|\beta_i|\}+\\ &+C\{\|h(f)-\tilde{S}'_\Delta(H)\|_p+\|h(f)-f\|_p\}+\\ &+C\bar{\Delta}^{1/p}\{\|h(f)-\tilde{S}'_\Delta(H)\|_\infty+\|h(f)-f\|_\infty\}+\\ &+C\bar{\Delta}^{1/p-1/q}\{\|\tilde{S}'_\Delta(H)-h(f)\|_q+\|h(f)-f\|_q+\bar{\Delta}^k\|f^{(k)}\|_q\}.\end{aligned}$$

Next we observe that by Theorem 3 and (5.6)

$$\begin{aligned} D &= \bar{\Delta}^{1/p}\|h(f)-\tilde{S}'_\Delta(H)\|_\infty \leqslant C\|h(f)-\tilde{S}'_\Delta(H)\|_p \\ &\leqslant C\bar{\Delta}^{1/p-1/q}\|h(f)-\tilde{S}'_\Delta(H)\|_q \leqslant C\bar{\Delta}^{-1+1/p-1/q}\|H-\tilde{S}_\Delta(H)\|_q \\ &\leqslant C\bar{\Delta}^{-1+1/p-1/q}\|H_{2n,k+1,\Delta}(F)-\tilde{S}_\Delta(H_{2n,k+1,\Delta}(F))\|_q+ \\ &\quad + C\bar{\Delta}^{-1+1/p-1/q}\left\|\int_a^x t(Q;v)\,dv - S_\Delta\Big(\int_a^u t(Q;v)\,dv;\, x\Big)\right\|_q . \end{aligned}$$

Since $\tilde{S}_\Delta\big(\int_a^u t(Q;v)dv;x\big)$ is nothing but the natural interpolating spline of $\int_a^u t(Q;v)dv$, the above chain of inequalities can be continued by Theorem 9 and (5.5) by

$$D \leqslant C\bar{\Delta}^{1/p-1/q}\{\bar{\Delta}^{2n+1}\|H^{(2n+2)}_{2n,k+1,\Delta}(F)\|_q+\|t(Q)\|_q\},$$

and furthermore by a result in [37] and Lemma 2 in this paper

$$\begin{aligned} D &\leqslant C\bar{\Delta}^{1/p-1/q}\{\bar{\Delta}^{-1}\omega_{2n+2}(H_{2n,k+1,\Delta}(F);\Delta)_q+\|t(Q)\|_q\} \\ &\leqslant C\bar{\Delta}^{1/p-1/q}\{\bar{\Delta}^{-1}\|H_{2n,k+1,\Delta}(F)-F\|_q+\bar{\Delta}^{-1}\omega_{2n+2}(F;\Delta)_q+\|t(Q)\|_q\} \\ &\leqslant C\bar{\Delta}^{1/p-1/q+k}\|f^{(k)}\|_q+C\bar{\Delta}^{1/p}\left\{\sum_{i=0}^{N-1}\left|\frac{f_i-Q_i}{x_{i+1}-x_i}\right|^q\right\}^{1/q}. \end{aligned}$$

Finally it follows by Theorem 3 and Lemma 2 that

$$\begin{aligned} \|h(f)-f\|_p &\leqslant \|t(Q)\|_p+\|H'_{2n,k+1,\Delta}(F)-F'\|_p \\ &\leqslant C\bar{\Delta}^{1/p-1/q}\{\|t(Q)\|_q+\bar{\Delta}^k\|f^{(k)}\|_q\}. \end{aligned}$$

Similar estimates hold for $\|h(f)-f\|_\infty$ and $\|h(f)-f\|_q$, so that by substituting the last inequalities in (5.12) one obtains the assertion of the theorem.

Theorem 10 can be extended by assertions of type (b) and (c) in Theorem 9. Concerning further extensions and improvements the same remarks as for Theorem 9 apply.

A final remark concerns the sum on the right-hand side of (5.4). As already mentioned above, in concrete instances one may define $Q_i$ by some quadrature formula for

$$\int_{x_i}^{x_{i+1}} f(x)\,dx$$

such that it is of same order of magnitude as the term preceding it, i.e. comparable to $\bar{\Delta}^{k-1/q}\|f^{(k)}\|_q$.

An example is given by the composite Newton–Cotes formulae involving Lagrange interpolation at equidistant partitions of each segment $(x_i, x_{i+1})$, $0 \leqslant i \leqslant N-1$.

If e.g. $L_{k,i}(f)$ denotes the Lagrange interpolating polynomial of $f \in W_q^k(a, b)$, $k \geqslant 1$, at the nodes $y_j = x_i + (x_{i+1} - x_i)jk^{-1}$ $(0 \leqslant j \leqslant k)$, then with $Q_i = \int_{x_i}^{x_{i+1}} L_{k,i}(f; u)\,du$ one has (cf. [44], Corollary 4.2)

$$|f_i - Q_i| \leqslant \bar{\Delta}^{1-1/q}\left\{\int_{x_i}^{x_{i+1}} |f(u) - L_{k,i}(f; u)|\,du\right\}^{1/q}$$

$$\leqslant C\bar{\Delta}^{1-1/q+k}\left\{\int_{x_i}^{x_{i+1}} |f^{(k)}(u)|^q du\right\}^{1/q}.$$

Therefore it follows (observing (3.5))

$$\left\{\sum_{i=0}^{N-1} |f_i - Q_i|^q (x_{i+1} - x_i)^{-q}\right\}^{1/q} \leqslant C\left\{\sum_{i=1}^{N-1} \bar{\Delta}^{kq-1} \int_{x_i}^{x_{i+1}} |f^{(k)}(u)|^q du\right\}^{1/q} = C\bar{\Delta}^{k-1/q}\|f^{(k)}\|_q.$$

## REFERENCES


[1] J. H. Ahlberg, E. N. Nilson and J. L. Walsh, 'Best Approximation and Convergence Properties of Higher-Order Spline Approximation', *J. Math. Mech.* **14** (1965) 231–244.

[2] J. H. Ahlberg, E. N. Nilson and J. L. Walsh, *The Theory of Splines and Their Applications*, Academic Press, New York 1967.

[3] P. M. Anselone and P. J. Laurent, 'A General Method for the Construction of Interpolating or Smoothing Spline-Functions', *Numer. Math.* **12** (1968) 66–82.

[4] M. Atteia, *Etude de certains noyaux et théorie des fonctions "spline" en analyse numérique*, Dissertation, Univ. of Grenoble (1966).

[5] G. Birkhoff and C. de Boor, 'Error Bounds for Spline Interpolation', *J. Math. Mech.* **13** (1964) 827–836.

[6] G. Birkhoff, M. H. Schultz and R. S. Varga, 'Piecewise Hermite Interpolation in One and Two Variables with Applications to Partial Differential Fquations', *Numer. Math.* **11** (1968) 232–256.

[7] C. de Boor, 'On Uniform Approximation by Splines', *J. Approximation Theory* **1** (1968) 219–235.

[8] C. de Boor, 'On the Convergence of Odd-Degree Spline Interpolation', *J. Approximation Theory* **1** (1968), 452–463.

[9] C. de Boor and R. E. Lynch, 'On Splines and their Minimum Properties', *J. Math. Mech.* **15** (1966) 953–969.

[10] Yu. A. Brudnyi and I. E. Gopengauz, 'Approximation by Piecewise Polynomial Functions', *Dokl. Akad. Nauk SSSR* **141** (1961) 1283–1286.

[11] G. Butler and F. Richards, 'An $L$-Saturation Theorem for Splines', *Cand. J. Math.* **24** (1972) 957–966.

[12] P. L. Butzer and H. Berens, *Semi-Groups of Operators and Approximation*, Springer, New York 1967.

[13] P. L. Butzer and R. J. Nessel, *Fourier Analysis and Approximation I*, Academic Press, New York 1971.

[14] P. L. Butzer, R. J. Nessel and W. Trebels, 'On Summation Processes of Fourier Expansions in Banach Spaces, I, Comparison Theorems', *Tôhoku Math. J.* **24** (1972) 127–140; 'II, Saturation Theorems' **25** (1973) 551–569.

[15] P. L. Butzer and K. Scherer, 'On the Fundamental Approximation Theorems of D. Jackson, S. N. Bernstein and Theorems of M. Zamansky and S. B. Stečkin', *Aequationes Math.* **3** (1969) 170–185.

[16] P. L. Butzer, 'Approximation Theorems for Sequences of Commutative Operators in Banach Spaces', in: *Proc. of Conference on Constructive Function Theory*, Varna (1970) 137–146.

[17] P. L. Butzer, 'Jackson and Bernstein-Type Inequalities for Families of Commutative Operators in Banach Spaces', *J. Approximation Theory* **5** (1972) 308–342.

[18] L. Collatz und W. Quade, 'Zur Interpolationstheorie der reellen periodischen Funktionen', *Sitzber. Preuss. Akad. Wiss. Phys.-Math. Kl.* **30** (1938) 383–429.

[19] G. Freud and V. Popov, 'On Approximation by Spline Functions', in: *Proc. of Conference on Constructive Theory of Functions*, Budapest (1969) 163–172.

[20] G. Freud, 'Certain Questions Connected with Approximation by Spline Functions and Polynomials' (Russian), *Studia Sci. Math. Hungar.* **5** (1970) 161–171.

[21] D. Gaier, 'Saturation bei Spline-Approximation und Quadraturen', *Numer. Math.* **16** (1970) 129–140.

[22] E. Görlich, R. J. Nessel and W. Trebels, 'Bernstein-Type Inequalities for Families for Multiplier Operators in Banach Spaces with Cesàro Decompositions, I, General Theory; II, Applications', *Acta Sci. Math.* (Szeged) (to appear).

[23] M. Golomb, 'Approximation by Periodic Spline Interpolants on Uniform Meshes', *J. Approximation Theory* **1** (1968) 26–65.

[24] G. W. Hedstrom and R. S. Varga, 'Application of Besov Spaces to Spline Approximation', *J. Approximation Theory* **4** (1971) 295–327.

[25] J. W. Jerome and L. L. Schumaker, 'On $Lg$-Splines', *J. Approximation Theory* **2** (1969) 29–49.

[26] J. W. Jerome and R. S. Varga, 'Generalizations of Spline Functions and Applications to Non-Linear Boundary Value and Eigenvalue Problems', in: *Theory and applications of spline functions* (T.N.E. Greville, Ed.), Academic Press, New York (1969) 103–155.

[27] H. Johnen, 'Inequalities Connected with the Moduli of Smoothness', *Mat. Vestnik* **9** (1972) 289–303.

[28] T. R. Lucas, '*M*-splines', *J. Approximation Theory* **5** (1972) 1–14.
[29] J. Nitsche, 'Umkehrsätze für Spline Approximation', *Comp. Math.* **21** (1970) 400–416.
[30] S. Nord, 'Approximation Properties of the Spline Fit', *BIT* **7** (1967) 132–144.
[31] V. Popov and Bl. Kh. Sendov, 'Classes Characterized by Best Approximation by Spline Functions', *Mathematical Notes* **8** (1970) 550–557.
[32] K. Scherer, 'On the Best Approximation of Continuous Functions by Splines', *SIAM J. of Numer. Analysis* **7** (1970) 418–423.
[33] K. Scherer, 'Über die beste Approximation von $L_p$-Funktionen durch Splines', in: *Proc. of Conference on Constructive Function Theory*, Varna (1970) 277–286.
[34] K. Scherer, 'Über die Konvergenz von natürlichen interpolierenden Splines', in: *Linear Operators and Approximation* (edited by P. L. Butzer, J. P. Kahane and B. Sz. Nagy), Basel–Stuttgart, ISNM 20.
[35] K. Scherer, *Spline-Approximation und verallgemeinerte Lipschitzräume*, Habilitationsschrift, T. H. Aachen 1972/73. Part II.
[36] K. Scherer, *Spline-Approximation und verallgemeinerte Lipschitzräume*, Habilitationsschrift, T. H. Aachen 1972/73. Part III.
[37] K. Scherer, 'Über Ungleichungen vom Bernstein-Typ in Banachräumen' (to appear).
[38] I. J. Schoenberg, 'On the Ahlberg–Nilson Extension of Spline Interpolation, the *g*-Splines and Their Optimal Properties', *J. Math. Anal. Appl.* **21** (1968) 207–231.
[39] M. H. Schultz and R. S. Varga, '*L*-Splines', *Numer. Math.* **10** (1967) 345–369.
[40] A. Sharma and A. Meir, 'Degree of Approximation of Spline Interpolation', *J. Math. Mech.* **15** (1966) 759–767.
[41] Yu. N. Subbotin, 'On Piecewise-Polynomial Interpolation', *Mat. Zametki* **1** (1967) 63–70.
[42] Yu. N. Subbotin, 'A Certain Linear Method of Approximation of Differentiable Functions' (Russian), *Mat. Zametki* **7** (1970) 423–430.
[43] B. K. Swartz, '$O(h^{k-j}\omega(D^kf, h))$ Bounds on Some Spline Interpolation Errors', Los Alamos Scientific Laboratory Report LA-4477, 1970.
[44] B. K. Swartz and R. S. Varga, 'Error Bounds for Spline and *L*-Spline Interpolation', *J. Approximation Theory* **6** (1972) 6–49.
[45] R. S. Varga, 'Error Bounds for Spline Interpolation', in: Approximations with Special Emphasis on Spline Functions (edited by I. J. Schoenberg), Academic Press, New York 1969, 367–388.

# EINE SUBSTITUTIONSMETHODE ZUR BESTIMMUNG VON NIKOLSKIĬ-KONSTANTEN*

E. L. STARK

*Aachen*

*Abstract.* Concerning the approximation of function $f \in C_{2\pi}$ by means of positive singular integrals the measure of approximation with respect to Lipschitz classes is considered. In order to determine the leading Nikolskii constant in the corresponding asymptotic expansion some different methods are known for various classes of kernels. Here a far-reaching theorem is added which is of great advantage if a closed representation of the kernel is known.

Für die auf $(-\infty, \infty)$ stetigen, $2\pi$-periodischen Funktionen $f(x)$, d.h. $f \in C_{2\pi}$, wird als linearer positiver Operator das singuläre Faltungsintegral

(1) $$I_\varrho(\chi; f; x) = \frac{1}{\pi}\int_{-\pi}^{\pi} f(x-t)\chi_\varrho(t)dt \qquad (\varrho > 0; \varrho \to \varrho_0)$$

mit der geraden approximierenden Identität $\chi_\varrho(x) \geqslant 0$ als Kern (siehe [3], S. 30 f) betrachtet. Bezüglich der Klasse

$$\mathrm{Lip}_2^* \alpha = \{f \in C_{2\pi};\ |f(x+h)-2f(x)+f(x-h)| \leqslant 2|h|^\alpha\} \qquad (0 < \alpha \leqslant 2)$$

wird nach S. M. Nikolskiĭ als Approximationsmaß von (1) definiert (mit der üblichen sup-Norm):

(2) $$\Delta_\varrho^*(\chi; \alpha) = \sup_{f \in \mathrm{Lip}_2^* \alpha} \|I_\varrho(\chi; f; x) - f(x)\|;$$

zu bestimmen ist die asymptotische Entwicklung von (2) in der Form

(3) $$\Delta_\varrho^*(\chi; \alpha) = N^*(\chi; \alpha)\overline{\varphi}(\varrho) + o\big(\overline{\varphi}(\varrho)\big) \qquad (\varrho \to \varrho_0)$$

mit $\overline{\varphi}(\varrho) \to 0$ für $\varrho \to \varrho_0$ und $N^*(\chi; \alpha)$ als Nikolskiĭ-Konstanten (vgl. [3], S. 82f).

---

* Angekündigt in dem Vortrag, der unter dem Titel „Nikolskiĭ-Konstanten: ein Überblick, neueste Ergebnisse und offene Probleme" auf der 508 Sitzung der Polnischen Mathematischen Gesellschaft (24.8.1972) innerhalb der „Conference, Theory of Approximation", Poznań, 22.–26.8.1972, gehalten wurde.

Für einige Klassen derartiger Approximationsprozesse sind allgemeine Verfahren zur Berechnung von $N^*(\chi; \alpha)$ bekannt: 1° falls $\chi_\varrho(x)$ einer gewissen, über den Saturationsgrenzwert definierten Klasse $K^2$ angehört ($\alpha = 2$); 2° falls dem Kern eine bestimmte — im Sinne von P. P. Korovkin — erzeugende Funktion $\varphi$ zugeordnet werden kann ($0 < \alpha \leqslant 2$); 3° falls der Kern vom Fejérschen Typ ist ($0 < \alpha \leqslant 2$); 4° falls $\chi_\varrho \in K^2$ und vom gestörten Fejér-Typ ist ($0 < \alpha \leqslant 2$). Hierdurch werden fast alle klassischen Beispiele (zum Teil mehrmals) erfaßt; siehe [6], [8], [3] und die dort angeführte Literatur sowie [4].

Der hier bewiesene Satz reproduziert, allerdings in einheitlicher Weise, zahlreiche Ergebnisse, läßt sich jedoch auch auf einige bisher nicht erfaßte Fälle (u.a. Beispiel 2) anwenden.

SATZ. *Mit* $\varphi(\varrho) = \|\chi_\varrho(x)\|^{-1}$ *werde definiert*

$$\psi_\varrho(\chi; x) := \begin{cases} \varphi(\varrho)\chi_\varrho\big(x\varphi(\varrho)\big), & 0 \leqslant x \leqslant \dfrac{\pi}{\varphi(\varrho)}, \\ 0, & x > \dfrac{\pi}{\varphi(\varrho)}, \end{cases} \tag{4}$$

*und es existiere der Grenzwert*

$$\lim_{\varrho\to\varrho_0} \psi_\varrho(\chi; x) = \Psi(\chi; x) \quad f.ü.;$$

*weiterhin existiere eine (von* $\varrho$ *unabhängige) intergrierbare Majorante* $\psi^*(\chi; x)$, *so daß für alle* $\varrho > 0$ *gilt*

$$\psi_\varrho(\chi; x) \leqslant \psi^*(\chi; x) \quad f.ü.;$$

*für solche* $\alpha \in (0, 1]$, *für die das Moment (der Ordnung* $\alpha$)

$$m(\chi; \alpha) := \frac{2}{\pi}\int_0^\infty t^\alpha \Psi(\chi; t)\,dt \tag{5}$$

*existiert, gilt dann*

$$\lim_{\varrho\to\varrho_0} \frac{\Delta_\varrho^*(\chi; \alpha)}{[\varphi(\varrho)]^\alpha} = m(\chi; \alpha). \tag{6}$$

Die Nikolskiĭ-Konstante $N^*(\chi; \alpha)$ schließlich ist das Produkt aus $m(\chi; \alpha)$ und den noch in $[\varphi(\varrho)]^\alpha$ als Koeffizienten enthaltenen, von $\varrho$

unabhängigen Konstanten; sie ist abzulesen, wenn der Grenzwert (6) gemäß (3) entwickelt und somit die faktorfreie Ordnungsfunktion $\bar{\varphi}(\varrho)$ separiert wird.

BEWEIS. Für (2) gilt die hier fundamentale Beziehung (vgl. z.B. [3])

$$(7) \qquad \Delta_\varrho^*(\chi;\alpha) = \frac{2}{\pi}\int_0^\pi t^\alpha \chi_\varrho(t)\,dt \qquad (0 < \alpha \leqslant 1);$$

mit Hilfe der Substitution $t \to \varphi(\varrho)\cdot t$ erhält man über (4) sofort

$$\frac{\Delta_\varrho^*(\chi;\alpha)}{[\varphi(\varrho)]^\alpha} = \frac{2}{\pi}\int_0^\infty t^\alpha \psi_\varrho(\chi;t)\,dt.$$

Wegen (5) liefert der Grenzübergang, da die Voraussetzungen des Satzes von Lebesgue (dominated convergence) erfüllt sind, die Aussage (6).

BEMERKUNGEN. 1. Der Satz bleibt unter Voraussetzung (5) auch für $1 < \alpha \leqslant 2$ gültig; da allerdings (7) für diesen Bereich nicht mehr gilt, muß der Beweis geringfügig modifiziert werden; vgl. [6], [8].

2. Oft läßt sich $\|\chi_\varrho(x)\|$ unmittelbar durch $\chi_\varrho(0)$ ersetzen, da den meisten Kernen der Nullpunkt als „peaking point" ([3], p. 25) zuzuordnen ist (jedoch muß z.B. im Fall bestimmter Kerne von Jackson–Matsuoka (vgl. [5], p. 38) auf die ursprüngliche Formulierung zurückgegriffen werden).

3. In zwei Arbeiten benutzt V. A. Baskakov ähnliche Kriterien, um in [2], Transl. S. 346 ff, Nikolskiĭ-Konstanten im Spezialfall der Existenz einer erzeugenden $\varphi$-Funktion zu bestimmen bzw. um in [1] Sätze vom Voronovskaja-Typ für beschränkte Funktionen mit Sprungstellen aufzustellen.

BEISPIEL 1. Für das singuläre Integral von Abel–Poisson mit Kern

$$(8) \qquad p_r(x) = \frac{1}{2} + \sum_{k=1}^\infty r^k \cos kx = \frac{1}{2}\,\frac{1-r^2}{1-2r\cos x + r^2} \geqslant 0$$

$$(0 < r < 1;\ r \to 1-)$$

ist $\|p_r(x)\| = p_r(0) = \frac{1}{2}\frac{1+r}{1-r}$; für die Funktion gemäß (4) erhält man

$$\psi_r(P;x) = \begin{cases} \dfrac{1}{1+\dfrac{4rx^2}{(1+r)^2}\left(\dfrac{\sin\dfrac{x(1-r)}{1+r}}{\dfrac{x(1-r)}{1+r}}\right)^2}, & 0 \leqslant x \leqslant \dfrac{\pi(1+r)}{2(1-r)}, \\ 0, & x > \dfrac{\pi(1+r)}{2(1-r)}, \end{cases}$$

und somit $\Psi(P;x) = \frac{1}{1+x^2}$ sowie beispielsweise

$$\psi^*(P;x) = \frac{1}{1+R^2x^2}, \quad R^2 \equiv \frac{16r_0}{(1+r_0)^2\pi^2} \quad (0 < r_0 \leqslant r < 1).$$

Das Moment (5) existiert mit Ausnahme von $\alpha = 1$:

$$m(P;\alpha) = \frac{2}{\pi}\int_0^\infty \frac{t^\alpha}{1+t^2}\,dt = \frac{1}{\cos\dfrac{\alpha\pi}{2}} = N^*(P;\alpha) \quad (0 < \alpha < 1).$$

Für das Approximationsmaß ergibt sich die Entwicklung (bezüglich Literatur siehe [3])

$$\Delta_r^*(P;\alpha) = \frac{1}{\cos\dfrac{\alpha\pi}{2}}(1-r)^\alpha + o([1-r]^\alpha) \quad (0 < \alpha < 1;\ r \to 1-).$$

(Der Grenzfall $\alpha = 1$ bedarf einer gesonderten Behandlung; jedoch ist für $\Delta_r^*(P;1)$ bereits die vollständige asymptotische Entwicklung bekannt [9]).

Bemerkung. Eine Vereinfachung der Rechnungen wird vielfach ermöglicht, wenn die Funktion $\varphi(\varrho)$ aus der allgemeineren Relation $\lim_{\varrho\to\varrho_0}\|\chi_\varrho(x)\|\varphi(\varrho) = 1$ bestimmt wird. In diesem Beispiel wäre $\varphi(r) = 1-r$ zu setzen; vgl. auch (10).

BEISPIEL 2. Für den Kern von de La Vallée Poussin

$$(9)\qquad V_n(x) = \frac{1}{2} + \sum_{k=1}^{n} \frac{(n!)^2}{(n-k)!(n+k)!} \cos kx$$

$$= \frac{(n!)^2}{2(2n)!}\left(2\cos\frac{x}{2}\right)^{2n} \geqslant 0 \qquad (n = 1, 2, 3, \ldots, n \to \infty)$$

gilt

$$(10)\qquad \|V_n(x)\| = V_n(0) = \frac{2^{2n}(n!)^2}{2(2n)!}; \qquad \lim_{n\to\infty} \frac{2}{\sqrt{\pi n}} V_n(0) = 1$$

und somit

$$\Psi_n(V; x) = \begin{cases} \cos^{2n} \dfrac{(2n)!x}{2^{2n}(n!)^2}, & 0 \leqslant x \leqslant \pi V_n(0), \\ 0, & x > \pi V_n(0) \end{cases}$$

sowie

$$\Psi(V; x) = e^{-\frac{1}{\pi}x^2}; \qquad \psi^*(V; x) = e^{-Kx^2}$$

als Majorante mit einer geeigneten Konstanten $K > 0$ (vgl. [7], S. 218, no. 116; [1]). Schließlich erhält man aus

$$m(V: \alpha) = \pi^{(\alpha-1)/2}\Gamma\left(\frac{1+\alpha}{2}\right) \qquad (\alpha > 0)$$

über die Grenzwerte (6) und (10) nach Separation der Konstanten

$$\Delta_n^*(V; \alpha) = \frac{2}{\sqrt{\pi}}\Gamma\left(\frac{1+\alpha}{2}\right)n^{-\alpha/2} + o(n^{-\alpha/2}) \qquad (0 < \alpha \leqslant 1; n \to \infty),$$

also im Vergleich mit (3) die Nikolskiĭ-Konstante

$$N^*(V; \alpha) = 2^\alpha \pi^{-1/2}\Gamma\left(\frac{1+\alpha}{2}\right) \qquad (0 < \alpha \leqslant 1).$$

(Literaturhinweise: z.B. in [3]; bemerkenswert ist, daß $N^*(V; \alpha)$ mit der entsprechenden Konstanten der periodischen Version des singulären Integrals von Weierstrass mit $n = t^{-1}$ übereinstimmt, siehe z.B. [6], S. 20).

Wie die beiden Beispiele bereits erkennen lassen, ist die Anwendung des Satzes auf spezielle singuläre Integrale ohne große Komplikationen nur möglich, wenn der Kern in geschlossener Form (d.h. nicht nur als Fourierreihe, vgl. (8), bzw. als trigonometrisches Polynom, vgl. (9)) darstellbar ist. Diese Situation ist jedoch in vielen Fällen wie z.B. für die Kerne von Fejér, Fejér–Korovkin, Bohman–Zheng Wei-xing, Jackson und dessen zahlreiche Verallgemeinerungen (vgl. Bem. 2.), Anghelutza, Ghermanesco, etc. gegeben.

Herrn Karl Scherer sei für eine kritische Durchsicht des Manuskriptes gedankt.

## LITERATUR

[1] V. A. Baskakov, 'On the Approximation of Functions by Certain Singular Integrals' (Russ.), *Izv. Vysš. Učebn. Zaved. Matematika* **1** (68) (1968) 3–15.

[2] V. A. Baskakov, 'The Order of Approximation of Differentiable Functions by Certain Positive Linear Operators' (Russ.), *Mat. Sb. (N.S.)* **76** (118) (1968) 344–361 = Transl. V: *Math. USSR Sbornik* **5** (1968) 333–350.

[3] P. L. Butzer and R. J. Nessel, *Fourier Analysis and Approximation*, Vol. I, Basel-New York 1971, xvi+553.

[4] R. A. DeVore, 'The Approximation of Continuous Functions by Positive Linear Operators', *Lecture Notes in Math.* **293** (1972) viii+289.

[5] E. Görlich und E. L. Stark, 'Über beste Konstanten und asymptotische Entwicklungen positiver Faltungsintegrale und deren Zusammenhang mit dem Saturationsproblem', *Jber. Deutsch. Math.-Verein* **72** (1970) 18–61.

[6] R. J. Nessel, 'Über Nikolskii-Konstanten von positiven Approximationsverfahren bezüglich Lipschitz-Klassen', *Jber. Deutsch Math.-Verein* **73** (1971) 6–47.

[7] G. Pólya und G. Szegö, *Aufgaben und Lehrsätze aus der Analysis*, Band I, Berlin-Heidelberg-New York 1970, xiv+338.

[8] E. L. Stark, 'Nikolskiĭ Constants for Positive Singular Integrals of Perturbed Fejér-Type', in: *Linear Operators and Approximation* (Proc. Conf. Math. Res. Inst. Oberwolfach, Black Forest, 14.–22.8.1971; edited by P. L. Butzer, J. P. Kahane and B. Sz.-Nagy; ISNM 20) Basel-Stuttgart 1972, 506; 348–363.

[9] E. L. Stark, 'The Complete Asymptotic Expansion for the Measure of Approximation by Abel-Poisson Singular Integral of Functions From Lip 1' (Russ.), *Mat. Zametki* **13** (1973) 21–28 = Transl.: *Math. Notes* **13** (1973) 14–18.

# INTERPOLATING SPLINES

YU. N. SUBBOTIN

*Sverdlovsk*

This paper is concerned with a general study of the existence and convergence of interpolating and interpolating in mean spline functions at equidistant interpolation nodes.

DEFINITION 1. Let $f(x)$ be a function defined on the real line. A function $S_n(x, t, h; f)$ is called an *interpolating spline function* for $f(x)$ at the interpolation nodes $\{kh\}$, $h > 0, k = 0, \pm 1, \pm 2, \ldots, t$ fixed, $0 < t \leqslant 1$, if:

1. $S_n(x, t, h; f)$ is continuous in $x$ on $(-\infty, \infty)$ and is $n-1$ times continuously differentiable.

2. The $n$-th derivative exists and is piecewise constant in the intervals $(kh-h+th, kh+th]$, i.e.

$$S_n^{(n)}(x, t, h; f) = Z_k^{(n)} \quad \text{for} \quad kh-h+th < x \leqslant kh+th, \tag{1}$$

$k = 0, \pm 1, \pm 2, \ldots$, where $Z_k^{(n)}$ are constants.

3. $S_n(kh, t, h; f) = f(kh)$, $k = 0, \pm 1, \pm 2, \ldots$

DEFINITION 2. A function $S_n(x, \Delta, t; f)$ is called an *interpolating spline function* for $f(x)$ at the nodes $\Delta = \{x_k\}, x_k < x_{k+1}, k = 0, \pm 1, \pm 2, \ldots$, if Conditions 1 and 3 of Definition 1 are satisfied (with $kh$ replaced by $x_k$), and if

2′. The derivative $S_n^{(n)}(x, \Delta, t; f)$ is piecewise constant in the intervals $(x_{k-1}-tx_{k-1}+tx_k, x_k-tx_k+tx_{k+1})$.

These definitions can be altered so as to make sense also for a finite segment; in this case they have to be equipped, as usual, with appropriate boundary conditions.

Particular cases of splines of Definition 1, namely the cases $t = 1$, $n$ odd, and $t = \frac{1}{2}$, $n$ even, have been first considered by Schoenberg [14].

The solution of a certain extremal interpolation problem [18] improving a result of Riaben'kii ([10], see also [11]) has led the author to the consideration of splines introduced by Schoenberg.

There exists an extensive literature devoted to splines of odd degree with interpolation nodes at the knots of the spline, i.e. the points of possible discontinuity of the $n$-th derivative (see, e.g. [1], [5], [13]). However, just one paper [2] is known to the author, which investigates the existence of cubic interpolating splines ($n = 3$ in Definition 2) for arbitrarily distributed nodes.

The literature concerning splines of even degree is less extensive. Beyond the papers mentioned above we list here the papers by Subbotin [19], [20], Ahlberg, Nilson, Walsh [3], [1], Galkin [7] and Krinzessa [9]. They deal with splines corresponding to $t = 1$ and $t = \frac{1}{2}$ in Definition 1, in some cases imposing the condition of periodicity.

Clearly, there are infinitely many spline functions satisfying Definition 1, forming an $n$ parameter family. When studying the convergence of the interpolation process, one has to require additionally [17] the norm boundedness of the $n$-th derivative.

The following theorem holds true.

THEOREM 1. *Let $f(x)$ be a function continuous in $(-\infty, \infty)$ and such that the sequence $\Delta_h^n f(kh)$, $k = 0, \pm 1, \pm 2, \ldots$, is bounded for any fixed $h > 0$. Then there exists a unique interpolating spline function $S_n(x, t, h; f)$ whose n-th derivative is bounded on $(-\infty, \infty)$. An exception is made by the cases $t = 1$, $n$ even, and $t = \frac{1}{2}$, $n$ odd.*

PROOF. Write $S_n(x, t, h; f) = S_n(x)$ and compute the $n$-th differences with the jump $h$ at the point $kh$, requiring that $S_n(kh) = f(kh)$. We get

$$h^{-n}\Delta_h^n f(kh) = \sum_{l=0}^{n} Z_{k+l}^{(n)} \frac{1}{n} \sum_{s=0}^{l} (-1)^s C_{n+1}^s (l+t-s)^n, \tag{2}$$

$$k = 0, \pm 1, \pm 2, \ldots$$

If the characteristic polynomial

$$P_n(t, x) = \frac{1}{n} \sum_{l=0}^{n} x^l \sum_{s=0}^{l} (-1)^s C_{n+1}^s (l+t-s)^n \tag{3}$$

of the difference equation (2) is different from zero on the unit circle, then $1/P_n(t, e^{i\varphi})$ can, by the Wiener–Lévy theorem, be expanded into an absolutely convergent Fourier series

$$P_n^{-1}(t, e^{i\varphi}) = \sum_{k=-\infty}^{\infty} c_{k,n} e^{i\varphi k} \tag{4}$$

and, moreover, [8],

$$Z_k^{(n)} = \sum_{s=-\infty}^{\infty} c_{s,n} h^{-n} \Delta_h^n f(sh+mh) \tag{5}$$

and

$$\sup_k |Z_k^{(n)}| \leqslant h^{-n} \sup_k |\Delta_h^n f(kh)| \sum_{s=-\infty}^{\infty} |c_{s,n}|. \tag{6}$$

We shall show that polynomial (3) has, for a fixed $t$, only real negative zeros and that $P_n(t, -1) \neq 0$ provided $t$ is not an excluded value.

Put

$$g_n(t, x) = \sum_{l=0}^{\infty} (l+t)^n x^l, \tag{7}$$

then

$$P_n(t, x) = (1-x)^{n+1} g_n(t, x). \tag{8}$$

LEMMA 1. *Polynomials $P_n(t, x)$ fulfil the recurrence equation*

$$P_{n+1}(t, x) = x(1-x)P_n'(t, x) + [t + (n+1-t)x]P_n(t, x). \tag{9}$$

PROOF. We have

$$\begin{aligned} P_{n+1}(t, x) &= (1-x)^{n+2} x \sum_{l=0}^{\infty} l(l+t)^n x^{l-1} + (1-x)^{n+2} t g_n(t, x) \\ &= x(1-x)^{n+2} g_n'(t, x) + (1-x) t P_n(t, x), \end{aligned}$$

further, by (8) (the derivative is always taken with respect to $x$, unless otherwise explicitly stated),

$$P_n'(t, x) = -(n+1)(1-x)^n g_n(t, x) + (1-x)^{n+1} g_n'(t, x).$$

Computing $g_n'(t, x)$ from the last equality and substituting into the preceding one, we obtain (9).

LEMMA 2. *For any fixed $t$, $0 < t < 1$, the polynomial $P_n(t, x)$ has $n$ different negative zeros.*

PROOF. *Induction.* By definition $P_0(t, x) \equiv 1$. From the recurrence equation (9) we get

$$P_1(t, x) = t+(1-t)x$$

and the assertion holds with $n = 1$. Suppose that it holds for some $n = k$, i.e. the polynomial $P_k(t, x)$ has $k$ different zeros: $-\infty < x_k < \ldots < x_1 < 0$. Since all zeros of $P_k(t, x)$ are single and since, by (7) and (8), $P_k(t, 0) = t^k$ for any $k$, thus sign $P'_k(t, x_i) = (-1)^{i-1}$, $i = 1, 2, \ldots, k$. We have to show that $P_{k+1}(t, x)$ has $k+1$ different negative zeros. Since $P_k(t, x_i) = 0$ we have by (9) sign $P_{k+1}(t, x_i) = (-1)^i$, $i = 1, 2, \ldots \ldots, k$. Furthermore, as remarked above, $P_{k+1}(t, 0) = t^{k+1} > 0$. Let $a_k$ denote the coefficient with which $x$ occurs in $P_k(t, x)$ in the highest degree. Then, by (9),

$$a_{k+1} = -ka_k+(k+1-t)a_k = (1-t)a_k,$$

whence

$$a_{k+1} = (1-t)^{k+1},$$

as $a_0 = 1$. Consequently, there exists a point $x_{k+1}$ such that sign $P_{k+1}(t, x_{k+1}) = (-1)^{k+1}$. It follows that the polynomial $P_{k+1}(t, x)$ alters sign on $(-\infty, 0)$ $k+2$ times. The Lemma is thus proved.

REMARK 1. In view of (9), $P_{n+1}(1, x)$ has degree $n$, hence, similarly as above, this polynomial has $n$ different negative zeros.

According to Lemma 2 and Remark 1, all zeros of the polynomial $P_n(t, x)$ for $0 < t \leqslant 1$ are negative and different, and their number coincides with the degree of the polynomial in question. Therefore the solution of the difference equation (2) can be unbounded only in the case (see e.g. [12]), when $P_n(t, -1) = 0$.

LEMMA 3. *The polynomials* $P_n(1, x)$ *and* $P_n(\frac{1}{2}, x)$ *are symmetric ones, i.e.* $P_n(\frac{1}{2}, x) \equiv x^n P_n(\frac{1}{2}, 1/x)$ *and* $P_n(1, x) \equiv x^{n-1} P_n(1, 1/x)$.

We recall that $P_n(1, x)$ has degree $n-1$. The proof of Lemma 3 follows by induction on $n$ from the recurrence relation (9).

COROLLARY. *The polynomials* $P_{2r}(1, x)$ *and* $P_{2r-1}(\frac{1}{2}, x)$ *have value* 0 *at the point* $x = -1$.

LEMMA 4. *The zeros of polynomials* $P_n(t, x)$ *strictly decrease when* $t$ *increases,* $0 \leqslant t \leqslant 1$.

The proof of this lemma is contained in the proof of Theorem 2 of the author's paper [21]. The following lemma also is contained there.

LEMMA 5. *The polynomials* $P_{2r}(0, x)$ *and* $P_{2r}(1, x)$ *have equal negative zeros* $\{x_i\}$, $-\infty < x_{2r-1} < x_{2r-2} < \dots < x_1 < 0$, $P_{2r}(0, 0) = 0$. *If* $x_i(t)$ *denote the zeros of the polynomial* $P_{2r}(t, x)$, $i = 1, 2, \dots, 2r$, $0 < t < 1$, *then*

$$(10) \qquad x_{i-1} < x_i(t) \leqslant x_i(\tfrac{1}{2}) \leqslant x_i(1-t) < x_i \quad \text{for} \quad \tfrac{1}{2} \leqslant t < 1, \quad i = 1, 2, \dots, 2r,$$

*where* $x_0 = 0$ *and* $x_{2r} = -\infty$.

*The polynomials* $P_{2r-1}(0, x)$ *and* $P_{2r-1}(1, x)$ *have equal negative zeros* $\{y_i\}$, $i = 1, 2, \dots, 2r-2$. *If* $y_i(t)$ *denote the zeros of the polynomial* $P_{2r-1}(t, x)$, $i = 1, 2, \dots, 2r-1$, $0 < t < 1$, *then*

$$(11) \qquad y_{i-1} < y_i(t) < y_i(\tfrac{1}{2}) < y_i(1-t) < y_i \quad \text{for} \quad \tfrac{1}{2} < t < 1, \quad i = 1, 2, \dots, 2r-1,$$

*where again* $y_0 = 0$ *and* $y_{2r-1} = -\infty$.

Inequalities (10), (11) follow from Lemma 4.

LEMMA 6. *The polynomials* $P_{2r}(t, x)$ $(0 < t < 1)$ *and* $P_{2r-1}(t, x)$ $(0 < t \leqslant 1, t \neq \frac{1}{2})$ *have non-zero values at the point* $x = -1$.

This lemma follows from Lemma 5 and corollary to Lemma 3.

What was needed to conclude the proof of Theorem 1, was precisely to find out for which values of $t$, $0 < t \leqslant 1$, the polynomial $P_n(t, x)$ has value zero at the point $x = -1$, and for which ones it has not. Thus, by corollary and Lemma 6, the theorem is proved in its full extent, since in the excluded cases the difference equation (2) has an unbounded solution, in the remaining cases, however, equation (2) has a unique bounded solution $\{Z_m^{(n)}\}$, given by (5). So $S_n^{(n)}(x)$ is uniquely determined. Writing the Taylor formula for $S_n(x)$ with the integral form of the remainder, we can compute the off-integral summand from the interpolation conditions.

Since $P_n(t, x)$ has negative zeros only, we can prove, similarly as it is done in [18], that sign $c_{s,n} = (-1)^s$, whence

$$(12) \qquad \sum_{s=-\infty}^{\infty} |c_{s,n}| = \sum_{s=-\infty}^{\infty} (-1)^s c_{s,n} = P_n^{-1}(t, -1).$$

By the main result of [2]

$$|P_{2r-1}(1, -1)| \geqslant |P_{2r-1}(t, 1)|, \quad 0 \leqslant t \leqslant 1,$$

and

$$|P_{2r}(\tfrac{1}{2}, -1)| \geqslant |P_{2r}(t, -1)|, \quad 0 \leqslant t \leqslant 1.$$

The following more precise result actually holds:

LEMMA 7. *The function* $|P_{2r-1}(t, -1)|$ *strictly decreases in* $[0, \frac{1}{2}]$ *and strictly increases in* $[\frac{1}{2}, 1]$. *The function* $|P_{2r}(t, -1)|$ *strictly increases in* $[0, \frac{1}{2}]$ *and strictly decreases in* $[\frac{1}{2}, 1]$.

It follows from Lemma 7, that splines of an odd degree with interpolation nodes at the spline knots, and splines of an even degree with interpolation nodes at mean points between the spline knots, are a better tool for interpolation purpose than splines with nodes corresponding to other values of $t$.

We now formulate a theorem concerning the convergence of interpolating splines of Theorem 1.

THEOREM 2. *If the function* $f(x)$ *is continuous in* $(-\infty, \infty)$ *and if* $|\Delta^n f(kh)| < \infty$ *for any* $0 < h < \infty$, *then for any fixed* $t$ $(0 < t \leqslant 1)$ *different from the excluded values*

$$|f(x) - S_n(x, h, t; f)| \leqslant C(n, t)\omega_{n+1}(f, h). \tag{13}$$

*If, moreover,* $f(x)$ *is* $k$ *times continuously differentiable, then*

$$|f^{(i)}(x) - S_n^{(i)}(x, h, t; f)| \leqslant C(n, t, i)\omega_{n+1-i}(f^{(i)}, h), \quad i = 0, 1, \ldots, k. \tag{14}$$

Particular cases of this theorem are proved in papers [1], [3]–[5], [19], [20], [24]. Those papers, in general, deal with periodic functions, odd $n$, $t = 1$, and estimation in terms of the modulus of continuity of the $k$-th derivative of $f(x)$. In [22] the above theorem is proved for some specific values of $t$, see also [17]. In view of inequality (9) and equality (12) the proof of the theorem for the values of $t$ different from the excluded values runs precisely along the same scheme.

Let now a function $f(x)$ have a locally absolutely continuous $(l-1)$-th derivative $(l \geqslant 0)$. We shall say that $f(x) \in W^l H^p_\omega$ if

$$(15)\qquad \omega(f^{(l)}, h)_p = \sup_{|u| \leqslant h} \Big( \int_{-\infty}^{\infty} |f^{(l)}(x+u) - f^{(l)}(x)|^p dx \Big)^{1/p} \leqslant \omega(h)_p,$$

where $\omega(h)_p$ is a given modulus of continuity.

THEOREM 3. *Let $f(x) \in W^l H^p_\omega$ $(l \geqslant 1)$, let the sequence $|\Delta^n_h f(kh)|$ $(k = 0, \pm 1, \pm 2, \ldots)$ be bounded and let $S_n(x, t, h; f) = S_n(x)$ be interpolating spline functions of degree n for $f(x)$ at the interpolation nodes $\{kh\}$. Then, under assumptions of Theorem* 1 *on n and t, we have for $q \geqslant p \geqslant 1$,*

$$(16)\qquad \|f^{(i)}(x) - S^{(i)}_n(x)\|_{L_q(-\infty,\infty)} \leqslant C h^{l-i+1/q-1/p} \omega(f^{(l)}, h)_p, \qquad 0 \leqslant i \leqslant l-1,$$

*where C depends on l, t and l only.*

PROOF. Let $knh \leqslant x \leqslant (k+1)nh$. Then the following equality holds [23]:

$$(17)\qquad f(x) - S_n(x) = \int_{knh}^{(k+1)nh} \varphi_l(x, u) f^{(l)}(u)\, du - \int_{knh}^{(k+1)nh} \varphi_n(x, u) S^{(n)}_n(u)\, du,$$

where

$$(18)\qquad \varphi_r(x+knh, u+knh) = \psi_r(x, u)$$

$$= \frac{1}{(r-1)!} \Big\{ (x-u)^{r-1}_+ - \sum_{j=0}^{n} (-1)^{n-j} \frac{x(x-h)\ldots(x-nh)}{(x-jnh)\, j!\,(n-j)!} (jh-u)^{r-1}_+ \Big\},$$

$$r = l, n,\ 0 \leqslant x, u \leqslant nh.$$

In the sequel we base on the fact that, for $0 \leqslant u, x \leqslant nh$,

$$(19)\qquad |\psi_r(x, u)| \leqslant \frac{(nh)^{r-1}}{(r-1)!} \Big\{ 1 + \sum_{j=0}^{n} \frac{j^{r-1}}{j!(n-j)!} \max_{0 \leqslant x \leqslant n} \Big| \frac{x(x-h)\ldots(x-nh)}{x-j} \Big| \Big\}$$

and that, for $r = l, n$,

$$(20)\qquad \int_{knh}^{(k+1)nh} \varphi_r(x, u)\, du = 0.$$

According to (17) we have

$$(21)\qquad \|f(x)-S_n(x)\|_{L_q} \leqslant \left\{\sum_{s=-\infty}^{\infty} \int_0^{nh} \left| \int_0^{nh} \psi_l(x,u) f^{(l)}(snh+u)\,du \right|^p dx\right\}^{1/p} +$$

$$+\left\{\sum_{s=-\infty}^{\infty} \int_0^{nh} \left| \int_0^{nh} \psi_n(x,u)\, S_n^{(n)}(snh+u)\,du \right|^p dx\right\}^{1/p} = I_1 + I_2 .$$

To estimate $I_1$ we employ (17), (18) and the Minkowski and Hölder inequalities. We have

$$(22)\qquad I_1 = \left\{\sum_{s=-\infty}^{\infty} \int_0^{nh} \left| \int_0^{nh} \psi_l(x,u) \frac{1}{nh} \int_0^{nh} [f^{(k)}(snh+u) - \right.\right.$$

$$\left.\left. - f^{(k)}(snh+u+v) + f^{(k)}(snh+u+v) - f^{(k)}(snh+v)]\,dv \right|^q dx\right\}^{1/q}$$

$$\leqslant C(n,l) h^{l-2+1/q} \left\{\left[\sum_{s=-\infty}^{\infty} \left| \int_0^{nh} dv \int_0^{nh} |f^{(k)}(snh+u) - \right.\right.\right.$$

$$\left. \left. - f^{(k)}(snh+u+v)|\,du \right|^q \right]^{1/q} +$$

$$\left. + \left[\sum_{s=-\infty}^{\infty} \left| \int_0^{nh} du \int_0^{nh} |f^{(k)}(snh+u+v) - f^{(k)}(snh+v)|\,dv \right|^q \right]^{1/q} \right\}$$

$$\leqslant 2C(n,l) h^{l-1-1/p+1/q} \left\{\sum_{s=-\infty}^{\infty} \left| \int_0^{nh} dv \left[ \int_0^{nh} |f^{(l)}(snh+u+v) - \right.\right.\right.$$

$$\left.\left.\left. - f^{(l)}(snh+u)|^p du \right]^{1/p} \right|^q \right\}^{1/q}$$

$$\leqslant C(n,l) h^{l-2/p+1/q} \left( \int_0^{nh} dv \int_{-\infty}^{\infty} |f^{(l)}(u+v) - \right.$$

$$\left. - f^{(l)}(u)|^p du \right)^{1/p-1/q} \left[ \sum_{s=-\infty}^{\infty} \int_0^{nh} dv \int_0^{nh} |f^{(l)}(snh+u+v) - \right.$$

$$\left. - f^{(l)}(snh+u)|^p du \right]^{1/q}$$

$$\leqslant C(n,l) h^{l-1/p+1/q} \omega(f^{(l)}, h)_p, \qquad q \geqslant p,$$

where $C(n,l)$ are constants depending on the indicated quantities only.

The summand $I_2$ can be estimated similarly:

(23) $$I_2 \leqslant C(n)h^{n+1/q-1/p} \sup_{|v|\leqslant nh} \left( \int_{-\infty}^{\infty} |S_n^{(n)}(u+v)-S_n^{(n)}(u)|^p du \right)^{1/p}.$$

Applying formula (15) for $S_n^{(n)}(x)$ and the equality

$$\Delta_h^n f(kh) = \Delta_h^{n-l} \int_0^{lh} \varphi(kh+u) f^{(l)}(kh+u)\, du$$

we get

(24) $$I_2 \leqslant C(n)h^{l+1/q-1/p}\omega(f^{(l)}, h)_p, \qquad 1 \leqslant l \leqslant n.$$

The assertion is proved for $i = 0$.

Let $1 \leqslant i \leqslant l-1$. By the Rolle theorem, in each of the intervals

$$\alpha_s = 3s(i+1)h(n-i+1) < x < 3(s+1)\,(i+1)h(n-i+1) = \alpha_{s+1}$$

there are at least $3(n-i+1)$ points $\{\bar{x}_{s,j}\}$, $j = 1, 2, \ldots, 3(n-i+1)$, such that $f^{(i)}(\bar{x}_{s,j}) = S_n^{(i)}(\bar{x}_{s,j})$; in each interval $(kh, kh+ik)$ at least one such point. Thus we can choose $(n-i+1)$ points $x_{s,j}$, $j = 0, 1, 2, \ldots, n-i$ from the system $\{\bar{x}_{s,j}\}$, satisfying the inequalities:

(25) $$ih \leqslant |x_{s,j+1}-x_{s,j}| \leqslant 3ih, \qquad j = 0, 1, 2, \ldots, n-i-1.$$

As a consequence we can write for $\alpha_s \leqslant x \leqslant \alpha_{s+1}$ the analogue of equality (17):

(26) $$f^{(i)}(x)-S_n^{(i)}(x) = \int_{\alpha_s}^{\alpha_{s+1}} [\varphi_{l-i}(x, u) f^{(l)}(u) - \varphi_{n-i}(x, u) S_n^{(n)}(u)]\, du,$$

where

$$\varphi_r(x, u) = \frac{1}{(r-1)!} \left\{ (x-u)_+^{r-1} - \sum_{j=0}^{n-i} \frac{\omega(x)}{(x-x_{s,j})\omega'(x_{s,j})} (x_{s,j}-u)_+^{r-1} \right\},$$

$$r = l-i,\ n-i,\ \alpha_s \leqslant x, u \leqslant \alpha_{s+1}.$$

Again $\int_{\alpha_s}^{\alpha_{s+1}} \varphi_r(x, u)\, du = 0$ and so, by (25), inequality (19) holds with $\varphi_r(x, u)$. Therefore the proof can now be completed by repeating the argument of the case $i = 0$. The theorem is thus proved.

A formulation of this theorem is given in [23].

We shall now be concerned with approximation of locally integrable functions. Interpolating splines are of little use here. We shall therefore draw our attention to the approximation of such functions by splines interpolating in mean.

DEFINITION 3. Let $f(x)$ be a locally integrable function defined on the real line. A function $S_n(x, h, h_1; f)$ will be called an *interpolating in mean spline function* for $f(x)$, if conditions 1 and 2 of Definition 1 are satisfied with $t = \frac{1}{2}$ for $n$ even, $t = 1$ for $n$ odd, and if, moreover,

$$(27) \quad \frac{1}{h_1}\int_{-h_1/2}^{h_1/2} S_n(kh+u, h, h_1; f)\,du = \frac{1}{h_1}\int_{-h_1/2}^{h_1/2} f(kh+u)\,du = y_k$$

$$(0 < h_1 \leqslant h), \qquad k = 0, \pm 1, \pm 2, \ldots$$

This definition can be reformulated so as to make sense for arbitrary $t$, but we shall restrict ourselves to the cases indicated.

Let us remark that splines interpolating in mean and their generalizations occur inevitably in the solution of the following extremal problem.

Let a sequence of real numbers $Y = \{y_m\}$, $m = 0, \pm 1, \pm 2, \ldots$ be given. Write

$$\Delta^n y_m = \sum_{s=0}^{n} (-1)^{n-s} C_n^s y_{m+s}$$

and assume that $\|\Delta^n y_m\|_{l_p} < \infty$, where

$$\|y_m\|_{l_p} = \begin{cases} \left\{\sum\limits_{m=-\infty}^{\infty} |y_m|^p\right\}^{1/p}, & 1 \leqslant p < \infty, \\ \sup\limits_m |y_m|, & p = \infty. \end{cases}$$

Let $y_k$ denote the mean value of $f(x)$, i.e.

$$(28) \qquad y_m = \frac{1}{h}\int_{-h/2}^{h/2} f(m+x)\,dx \quad (m = 0, \pm 1, \pm 2, \ldots),$$

where $h$ is a fixed number, $0 \leqslant h \leqslant 1$. For $h = 0$ we put

$$y_m = \lim_{h\to 0} \frac{1}{h}\int_{-h/2}^{h/2} f(m+x)\,dx.$$

The problem consists in reproducing the function $f(x)$ from the given mean values $\{y_m\}$, requiring that the $L_p$-norm of its $n$-th derivative be minimal.

To give the problem a more precise formulation we introduce the following notation:

We shall write $\{y_m\} \in l_p^n$ if $\|\Delta^n y_m\|_{l_p} < \infty$; similarly, $f(x) \in L_p^n(-\infty, \infty)$ if the function $f(x)$ has a locally absolutely continuous $(n-1)$-th derivative and if

$$\|f^{(n)}(x)\|_{L_p} < \infty,$$

where

$$\|f^{(n)}(x)\|_{L_p} = \begin{cases} \left\{\int\limits_{-\infty}^{\infty} |f^{(n)}(x)|^p dx\right\}^{1/p}, & 1 \leqslant p < \infty, \\ \operatorname{ess\,sup}|f^{(n)}(x)|, & p = \infty. \end{cases}$$

Let $Y = \{y_m\}$ and let $F(Y)$ denote the class of functions $f(x)$ satisfying condition (28). In the introduced notation one problem reads as follows: find the quantity

$$(29) \qquad A_{n,p}(h) = \sup_{\|\Delta^n y_m\|_{l_p} \leqslant 1} \inf_{f \in F(Y)} \|f^{(n)}(x)\|_{L_p(-\infty,\infty)}.$$

A linear analogue of the problem (29) can also be considered, i.e. the problem of determining the operator with minimal norm in the class of all linear operators from $l_p^n$ to $L_p^n$ whose action satisfies condition (28). In the case $p = \infty$ this operator is constructed by means of polynomial spline functions, and for $1 < p < \infty$ by their generalizations. It has been shown that the norm of this operator equals $A_{n,p}(h)$, and that

$$(30) \qquad A_{n,p}(h) = (n-1)!\left\{\int\limits_0^1 |\psi_n(t,h)|^q dt\right\}^{-1/q} \quad \left(\frac{1}{p} + \frac{1}{q} = 1\right),$$

where

$$\psi_{2s}(t,h) = \begin{cases} \frac{1}{4s}\left[Q_{2s}^*\left(\frac{h}{2}+t\right) + Q_{2s}^*\left(\frac{h}{2}-t\right)\right], & 0 \leqslant t \leqslant \frac{h}{2}, \\ \frac{1}{4s}\left[Q_{2s}^*\left(\frac{h}{2}+t\right) - Q_{2s}^*\left(t-\frac{h}{2}\right)\right], & \frac{h}{2} \leqslant t \leqslant 1-\frac{h}{2}, \\ -\frac{1}{4s}\left[Q_{2s}^*\left(1+\frac{h}{2}-t\right) + Q_{2s}^*\left(-1+\frac{h}{2}+t\right)\right], & \\ & 1-\frac{h}{2} \leqslant t \leqslant 1, \end{cases}$$

$$\psi_{2s+1}(t,h) = \begin{cases} \dfrac{1}{4s+2}\left[Q^*_{2s+1}\left(\dfrac{h}{2}+t\right)-Q^*_{2s+1}\left(\dfrac{h}{2}-t\right)\right], & 0 \leqslant t \leqslant \dfrac{h}{2}, \\ \dfrac{1}{4s+2}\left[Q^*_{2s+1}\left(t+\dfrac{h}{2}\right)-Q^*_{2s+1}\left(t-\dfrac{h}{2}\right)\right], & \dfrac{h}{2} \leqslant t \leqslant 1-\dfrac{h}{2}, \\ \dfrac{1}{4s+2}\left[Q^*_{2s+1}\left(1+\dfrac{h}{2}-t\right)-Q^*_{2s+1}\left(-1+\dfrac{h}{2}+t\right)\right], & \\ & 1-\dfrac{h}{2} \leqslant t \leqslant t \leqslant 1, \end{cases}$$

and

$$Q^*_n(t) = \frac{(-1)^n 2^{n+1}}{n+1}\left[B_{n+1}\left(\frac{1-t}{2}\right)-(-1)^n B_{n+1}\left(\frac{t}{2}\right)\right],$$

where $B_n(t)$ are the Bernoulli polynomials. The corresponding problem for $h = 0$ is dealt with in [18], [21], see also [10]. Riaben'kii [10] proved that the numbers $A_{n,p}(h)$ are finite for $h = 0$ and $p = \infty$. Sobolev [16] showed that Riaben'kii's method can also be applied in the case $h = 0$, $1 \leqslant p < \infty$. In the case of a finite segment, $p = 2$, $h = 0$ and some additional restrictions on $f(x)$, a problem closely related to (29) has been intensely studied. We mean here the well known minimal norm property of splines of an odd degree, see e.g. [5].

The proof of relations (29), (30) follows the scheme presented in [18], [21], where the particular case $h = 0$, $1 \leqslant p \leqslant \infty$ is considered; this case was also dealt with by Schoenberg [15] for $p = 1, 2, \infty$.

The principal difficulty in this method lies in the investigation of properties of characteristic polynomials of corresponding difference equations. The following result concerns this topic.

THEOREM 4. *Suppose that an algebraic polynomial of an even degree can be written in one of the following two forms:*

$$P_{2s+2}(x) = \int_0^1 \varphi(t)\,dt\,\frac{1}{h}\int_{-h/2}^{h/2} \overline{P}_{2s+2}(t-u,x)\,du$$

*or*

$$P^*_{2s}(x) = \int_0^{1/2} \varphi(t)\,dt\,\frac{1}{h}\int_{-h/2}^{h/2} \overline{P}_{2s}(1+t-u,x)\,du+$$
$$+\int_{1/2}^{1} \varphi(t)\,dt\,\frac{1}{h}\int_{-h/2}^{h/2} \overline{P}_{2s}(t-u,x)\,du, \quad 0 < h \leqslant 1,$$

*where*

$$\overline{P}_{n+1}(t, x) = \sum_{l=0}^{n+2} \sum_{s=0}^{n+1} (-1)^s C_{n+1}^s (l-1-s+t)_+^n x^l \tag{31}$$

*and* $\varphi(t)$ *is a non-negative function, non-zero on a set of a positive measure, non-decreasing on* $[0, \frac{1}{2}]$, *and* $\varphi(t) = \varphi(1-t)$. *Then all zeros of this polynomial are negative, distinct, and the polynomial is symmetric.*

The proof of this theorem involves the connection between polynomials (31) and (3), as well as Lemmas 1–5.

Spline functions of Definition 3 can be applied to the approximation of locally integrable functions. The following theorems hold true.

THEOREM 5. *If a function* $f(x)$ *is locally integrable and if the sequence*

$$\Delta^n y_k = h_1^{-1} \int_{-h_1/2}^{h_1/2} \Delta_h^n f(kh+t)\,dt$$

*is bounded,* $k = 0, \pm 1, \pm 2, \ldots$, $0 < h_1 \leqslant h < \infty$ *and* $h_1/h \geqslant C$, *then*

$$\|f(x) - S_n(x, h, h_1; f)\|_{L_p(-\infty,\infty)} \leqslant C_1 \omega(f, h)_p, \tag{32}$$

*where* $C$ *and* $C_1$ *are absolute constants.*

THEOREM 6. *If a function* $f(x)$ *has a locally absolutely continuous l-th derivative,* $0 \leqslant l \leqslant n-1$, *and if the sequence* $\Delta^n y_k$ *is bounded, then*

$$\|f^{(i)}(x) - S_n^{(i)}(x, h, h_1; f)\|_{L_q} \leqslant C_2 h^{l-i+1/q-1/p} \omega(f^{(l)}, h)_p \tag{33}$$

*holds for* $q \geqslant p$, $i = 0, 1, 2, \ldots, l$. *If* $l = n$, *then* (33) *holds also for* $q = p$ *and* $h_1/h \geqslant C > 0$.

As follows from the results of paper [6], inequalities (32) and (33) cannot hold for $1 \leqslant q < p$.

Some questions concerning the functional interpolation on a finite segment are investigated by Varga [25].

## REFERENCES

[1] I. H. Ahlberg, E. N. Nilson and J. L. Walsh, *The Theory of Splines and Their Applications*, New York and London 1967.

[2] I. H. Ahlberg, E. N. Nilson and J. L. Walsh, 'Cubic Splines on the Real Line', *J. Approx. Theory* **1** (1968) 5–10.

[3] I. H. Ahlberg, E. N. Nilson and J. L. Walsh, 'Best Approximation and Convergence Properties of Higher-Order Spline Approximations', *J. Math. Mech.* **14** (1965) 231–243.

[4] I. H. Ahlberg and E. N. Nilson, 'Polynomial Spline on the Real Line', *J. Approx. Theory* **3** (1970) 398–409.

[5] I. H. Ahlberg, E. N. Nilson and J. L. Walsh, *The Theory of Splines and Their Approximations*, Moskva 1972.

[6] V. V. Arestov and V. N. Gabushin, 'Approximation of Classes of Differentiable Functions', *Mat. Zametki* **9** (1970) 105–112.

[7] P. V. Galkin, 'On the Solvability of the Problem of Periodic Spline Interpolation', *Mat. Zametki* **8** (1970).

[8] M. G. Krein, 'Integral Equations on the Half-Axis with a Kernel Depending on the Difference of the Arguments', *Usp. Mat. Nauk* **13** (1958) (83) 3–120.

[9] F. Krinzessa, *Zur periodischen Spline-Interpolation* (doctoral dissertation). Ruhr-Universität, Bochum 1969.

[10] V. S. Riaben'kii, *On the Stability of Finite-Difference Schemes and the Application of the Method of Finite Differences to the Solution of the Cauchy Problem for a System of Equations with Partial Derivatives*, Author's Thesis, Mos. Gos. Univ. 1952.

[11] V. S. Riaben'kii and A. F. Filippov, *On the Stability of Difference Equations*, Moskva 1956.

[12] V. S. Riaben'kii, 'Necessary and Sufficient Conditions for the Good Conditionary of Systems', *Ž. Vyčisl. Mat. i Matem. Fiz.* **4** (1964) 242–255.

[13] P. L. J. van Rooy and F. Schurer, *A Bibliography on Spline Functions*, 1971.

[14] I. J. Schoenberg, 'Contributions to Problem of Approximation of Equidistant Data by Analytic Functions', *Quart. Appl. Math.* **4** (1946) 112–141.

[15] I. J. Schoenberg, 'Cardinal Interpolation and Spline Functions', *J. Approx. Theory* **2** (1969) 167–206.

[16] S. L. Sobolev, *Lectures on the Theory of Cubature Formulas*, Novosibirsk 1965, 170.

[17] S. B. Stečkin and Yu. N. Subbotin, 'Contributions to the book: J. N. Ahlberg, E. N. Nilson and J. L. Walsh, *The Theory of Splines and their Applications*', Moskva 1972.

[18] Yu. N. Subbotin, 'On the Connection Between Finite Differences and the Corresponding Derivatives', *Trudy Mat. Inst. Akad. Nauk SSSR* **78** (1965) 24–42.

[19] Yu. N. Subbotin, 'On the Piecewise Polynomial Interpolation', *Mat. Zametki* **1** (1967) 63–70.

[20] Yu. N. Subbotin, 'The Diameter of the Class $w^r L$ in $L(0, 2\pi)$ and the Approximation by Spline Functions', *Mat. Zametki* **7** (1970) 43–52.

[21] Yu. N. Subbotin, 'Functional Interpolation in Mean with the Least $n$-th Derivative', *Trudy Mat. Inst. Akad. Nauk SSSR* **88** (1967) 30–60.

[22] Yu. N. Subbotin, 'On the Spline Approximation and Smooth Bases in C(0, 2π)', *Mat. Zametki* **12** (1972).

[23] Yu. N. Subbotin, 'On the Approximation to Functions of the Class $w^r H^p$ by Splines of Order $m$', *Dokl. Akad. Nauk SSSR* **195** (1970) 1039–1041.

[24] B. Swartz, '$O(h^{2r-1})$ Bounds on Some Spline Interpolation Errors', *Bull. Amer. Math. Soc.* **74** (1969) 1072–1078.

[25] R. S. Varga, '*Error Bounds for Spline Interpolation. Approximations with Special Emphasis on Spline Functions*, edited by I. J. Schoenberg, 367–388, Acad. Press, New York 1969.

# ON THE BEST APPROXIMATION OF FACTORIZED FOURIER SERIES

ISTVÁN SZALAY

*Szeged*

**1. Introduction.** Let $f(x)$ be a periodic function with period $2\pi$ and let $|f(x)|^p$ $(1 < p < \infty)$ be integrable in the sense of Lebesgue over $(0, 2\pi)$. Let the Fourier series of $f(x)$ be given by

$$(1) \qquad f(x) \sim \frac{a_0}{2} + \sum_{n=1}^{\infty} (a_n \cos nx + b_n \sin nx) \equiv \sum_{n=0}^{\infty} A_n(x).$$

By $s_n(x)$ we denote the $n$th partial sum of the series (1).

Denote $E_n^{(p)}(f)$ the best approximation of $f(x)$ in the metric of $L^p(0, 2\pi)$ by trigonometric polynomials of order $n$.

Finally $\omega_j^{(p)}(t, f)$ $(j \geqslant 1)$ means the modulus of continuity of order $j$ of $f(x)$, that is

$$\omega_j^{(p)}(t, f) = \sup_{|h| \leqslant t} \left\| \sum_{k=0}^{j} (-1)^{j-k} \binom{j}{k} f(x+kh) \right\|_p .$$

We prove the following theorems:

THEOREM 1. *Let $f(x) \in L^p(0, 2\pi)$ $(1 < p < \infty)$ and let $q$ be a real number such that $p \leqslant q < \infty$. Let $\{\lambda_n\}_0^\infty$ be a monotonic increasing, convex (or concave) sequence of the real numbers. If*

$$(2) \qquad \sum_{n=1}^{\infty} n^{1/p-1/q} \varrho_n E_n^{(p)}(f) < \infty,$$

*where $\varrho_n = \max(|\Delta \lambda_n|, n^{-1} \lambda_n)$, then the series*

$$(3) \qquad \sum_{n=0}^{\infty} \lambda_n A_n(x)$$

*is the Fourier series of a function* $f_\lambda(x)$ *belonging to the class* $L^q(0, 2\pi)$ *and*

$$(4) \quad E_n^{(q)}(f_\lambda) \leqq G(\lambda)\Big[(n+1)^{1/p-1/q}\lambda_{n+1}E_n^{(p)}(f) + \sum_{k=n+1}^{\infty} k^{1/p-1/q}\varrho_k E_k^{(p)}(f)\Big]$$

$$(n = 0, 1, \ldots),$$

*where the constant* $G(\lambda)$ *is independent of* $f(x)$, $n$, $p$ *and* $q$.

In the case $\lambda_n = (n+1)^\alpha$ $(\alpha > 0)$ and $p = q$ this theorem reduces to a theorem of Konjuškov ([1], Theorem 1).

The sequence of ideas of our proof is similar to that of Konjuškov.

THEOREM 2. *Let* $f(x) \in L^p(0, 2\pi)$ $(1 < p < \infty)$ *and let* $q$ *be a real number such that* $p \leqq q < \infty$. *Let* $\{\lambda_n\}_0^\infty$ *be a monotonic non-increasing sequence of the positive numbers. If*

$$\sum_{n=1}^{\infty} n^{1/p-1/q-1}\lambda_n E_n^{(p)}(f) < \infty,$$

*then the series*

$$\sum_{n=0}^{\infty} \lambda_n A_n(x)$$

*is the Fourier series of a function* $f_\lambda(x)$ *belonging to the class* $L^q(0, 2\pi)$ *and*

$$E_n^{(q)}(f_\lambda) \leqq G\left[(n+1)^{1/p-1/q}\lambda_n E_n^{(p)}(f) + \sum_{k=n+1}^{\infty} k^{1/p-1/q-1}\lambda_k E_k^{(p)}(f)\right]$$

$$(n = 0, 1, \ldots),$$

*where* $G$ *is an absolute constant.*

If $\lambda_n = (n+1)^\alpha$ $(\alpha < 0)$ and $p = q$, or if $\lambda_n = 1$ this theorem reduces to theorems of Konjuškov ([1], Theorems 1 and 2).

The proof of Theorem 2 is similar to the proof of Theorem 1, so we omit it.

It is of some interest to remark the following corollary to Theorem 2.

COROLLARY 1. *Let* $f(x) \in L^p(0, 2\pi)$ $(1 < p < \infty)$. *If*

$$\sum_{n=1}^{\infty} n^{-1}E_n^{(p)}(f) < \infty,$$

*then for any $q$ ($> p$) there exists a function $f_{p,q}(x)$ belonging to the class $L^q(0, 2\pi)$ such that*

$$f_{p,q}(x) \sim \frac{a_0}{2} + \sum_{n=1}^{\infty} n^{1/q-1/p} A_n(x),$$

$$E_n^{(q)}(f_{p,q}) \leqslant G\Big[E_n^{(p)}(f) + \sum_{k=n+1}^{\infty} k^{-1} E_k^{(p)}(f)\Big] \quad (n = 0, 1, \dots)$$

*where $G$ is an absolute constant.*

Using a result of A. Timan and M. Timan (see Lemma 3) we get the following corollaries.

COROLLARY 2. *Under the conditions of Theorem 1 the following estimation holds*

$$\omega_j^{(q)}\left(\frac{1}{n}, f_\lambda\right) \leqslant G(j, \lambda)\left[\frac{1}{n^j} \sum_{k=1}^{n} k^{j+1/p-1/q} \varrho_{k-1} E_{k-1}^{(p)}(f) + \right.$$
$$\left. + \sum_{k=n+1}^{\infty} k^{1/p-1/q} \varrho_k E_k^{(p)}(f)\right].$$

COROLLARY 3. *Under the conditions of Theorem 2 the following estimation holds*

$$\omega_j^{(q)}\left(\frac{1}{n}, f_\lambda\right) \leqslant G(j)\left[\frac{1}{n^j} \sum_{k=1}^{n} k^{j+1/p-1/q-1} \lambda_{k-1} E_{k-1}^{(p)}(f) + \right.$$
$$\left. + \sum_{k=n+1}^{\infty} k^{1/p-1/q-1} \lambda_k E_k^{(p)}(f)\right].$$

**2. Lemmas.** We require the following lemmas:

LEMMA 1. *If $f(x) \in L^p(0, 2\pi)$ ($1 < p < \infty$), then*

$$\|f(x) - s_n(x)\|_p = O\big(E_n^{(p)}(f)\big) \quad (n = 0, 1, \dots).$$

Lemma 1 follows from a theorem of M. Riesz (see e.g. [5], § 7.21).

LEMMA 2 (Nikol'skii [2], p. 256). *If $T_n(x)$ is a trigonometric polynomial of order $n$ and $1 \leqslant p < q < \infty$, then*

$$\|T_n(x)\|_q \leqslant 2n^{1/p-1/q} \|T_n(x)\|_p.$$

LEMMA 3 (A. Timan and M. Timan [4]). *If $f(x)\in L^p(0,2\pi)$ $(1<p<\infty)$, then*

$$\omega_j^{(p)}\left(\frac{1}{n},f\right)\leqslant\frac{G(j)}{n^j}\sum_{k=1}^{n}k^{j-1}E_{k-1}^{(p)}(f)\qquad(n=1,2,\ldots).$$

LEMMA 4. *Let $f(x)\in L^p(0,2\pi)$ $(1<p<\infty)$ and let $\{\lambda_n\}_0^\infty$ be a sequence of real numbers. If*

$$\sum_{n=1}^{\infty}|\Delta\lambda_n|\,E_n^{(p)}(f)<\infty,$$

*then $\sum_{n=0}^{\infty}\lambda_n A_n(x)$ is the Fourier series of a function of class $L^p(0,2\pi)$.*

The proof of Lemma 4 runs similarly to that of Lemma 6 of [3].

**3. Proof of Theorem 1.** By condition (2) and Lemma 4 we have that the series (3) is the Fourier series of a function $f_\lambda(x)$, and $f_\lambda(x)$ belongs to the class $L^p(0,2\pi)$.

Let $s_n(\lambda,x)$ denote the $n$th partial sum of the series (3). Withou loss of generality we may assume that $\lambda_0=1$. Applying the Abel's transformation we have

$$s_n(\lambda;x)-f(x)=\sum_{k=0}^{n-1}\big(s_k(x)-f(x)\big)\Delta\lambda_k+\big(s_n(x)-f(x)\big)\lambda_n$$

and so

$$s_{2l}(\lambda;x)-s_l(\lambda;x)=\sum_{k=l}^{2l-1}\big(s_k(x)-f(x)\big)\Delta\lambda_k+\big(s_{2l}(x)-f(x)\big)\lambda_{2l}-$$
$$-\big(s_l(x)-f(x)\big)\lambda_l.$$

By using of Lemma 1 we get

$$\|s_{2l}(\lambda;x)-s_l(\lambda;x)\|_p\leqslant G_1\Big(\lambda_l E_l^{(p)}(f)+\sum_{k=l}^{2l-1}|\Delta\lambda_k|\,E_k^{(p)}(f)\Big)$$

and by Lemma 2 we have

$$(5)\quad \|s_{2l}(\lambda;x)-s_l(\lambda;x)\|_q\leqslant G_2\, l^{1/p-1/q}\Big(\lambda_l E_l^{(p)}(f)+\sum_{k=l}^{2l-1}|\Delta\lambda_k|\,E_k^{(p)}(f)\Big),$$

where $G_1$ and $G_2$ are different absolute constants.

If the sequence $\{\lambda_n\}_0^\infty$ is monotonic increasing and convex, then the sequence $\{|\Delta\lambda_n|\}_0^\infty$ is monotonic non-decreasing, so for any $l = 1, 2, \ldots,$ we have

$$\lambda_{2l} = \lambda_0 + \sum_{k=0}^{2l-1} |\Delta\lambda_k| \leqslant G'(\lambda) \sum_{k=0}^{2l-1} |\Delta\lambda_k| \leqslant 2G'(\lambda) \sum_{k=l+1}^{2l} |\Delta\lambda_k| .$$

If the sequence $\{\lambda_n\}_0^\infty$ is monotonic increasing and concave, then the sequence $\{n^{-1}\lambda_n\}_1^\infty$ is monotonic non-increasing, so for any $l = 1, 2, \ldots$ we have

$$\lambda_{2l} \leqslant 2 \sum_{k=l+1}^{2l} k^{-1}\lambda_k .$$

Therefore for both cases we have

$$\lambda_{2l} E_{2l} \leqslant G_3(\lambda) \sum_{k=l+1}^{2l} \varrho_k E_k^{(p)}(f) \qquad (l = 1, 2, \ldots). \tag{6}$$

Applying (5) and (6) we have for any $n = 1, 2, \ldots$

$$\begin{aligned}
(7)\quad & \sum_{m=0}^{\infty} \|s_{2^{m+1}n}(\lambda, x) - s_{2^m n}(\lambda; x)\|_q \\
& \leqslant G_4(\lambda) \sum_{m=0}^{\infty} (2^m n)^{1/p-1/q} \Big( \lambda_{2^m n} E_{2^m}^{(p)}(f) + \sum_{k=2^m n}^{2^{m+1}n-1} |\Delta\lambda_k| E_k^{(p)}(f) \Big) \\
& \leqslant G_5(\lambda) \Big[ n^{1/p-1/q} \lambda_n E_n^{(p)}(f) + \sum_{m=0}^{\infty} (2^m n)^{1/p-1/q} \lambda_{2^{m+1}n} E_{2^{m+1}n}^{(p)}(f) + \\
& + \sum_{k=n}^{\infty} k^{1/p-1/q} |\Delta\lambda_k| E_k^{(p)}(f) \Big] \leqslant G_6(\lambda) \Big[ n^{1/p-1/q} \lambda_n E_n^{(p)}(f) + \\
& + \sum_{m=0}^{\infty} (2^m n)^{1/p-1/q} \sum_{k=2^m n+1}^{2^{m+1}n} \varrho_k E_k^{(p)}(f) + \sum_{k=n}^{\infty} k^{1/p-1/q} |\Delta\lambda_k| E_k^{(p)}(f) \Big] \\
& \leqslant G_6(\lambda) \Big[ n^{1/p-1/q} \lambda_{n+1} E_n^{(p)}(f) + \sum_{k=n+1}^{\infty} k^{1/p-1/q} \varrho_k E_k^{(p)}(f) \Big].
\end{aligned}$$

$G_3(\lambda)$, $G_4(\lambda)$, $G_5(\lambda)$ and $G_6(\lambda)$ are different constants depending on $\lambda$.

An easy computation shows that

$$n^{1/p-1/q}\lambda_{n+1}E_n^{(p)}(f) \leqslant G_7(\lambda)\sum_{k=1}^{n} k^{1/p-1/q}\varrho_k E_k^{(p)}(f)$$

thus the series (7) converges for any $n = 1, 2, \ldots$

Hence we get that the sequence $\{s_{2^m}(\lambda; x)\}$ converges in the sense of the metric of $L^q(0, 2\pi)$, that is there exists a function $\varphi(x)$ belonging to the class $L^q(0, 2\pi)$ such that

$$\lim_{m\to\infty} \|s_{2^m}(\lambda; x)-\varphi(x)\|_q = 0.$$

By Hölder's inequality we have

$$\lim_{m\to\infty} \|s_{2^m}(\lambda; x)-\varphi(x)\|_p = 0.$$

On the other hand using Lemma 1 we have

$$\lim_{m\to\infty} \|s_{2^m}(\lambda; x)-f_\lambda(x)\|_p = 0$$

therefore $f_\lambda(x) = \varphi(x)$ (in the sense of $L^p(0, 2\pi)$) and so $f_\lambda(x) \in L^q(0, 2\pi)$.

By Minkowski inequality we get

$$\begin{aligned} E_n^{(q)}(f_\lambda) &\leqslant \|f_\lambda(x)-s_n(\lambda; x)\|_q \\ &\leqslant \|f_\lambda(x)-s_{2^{m_n}}(\lambda; x)\|_q+\|s_{2^{m_n}}(\lambda; x)-s_n(\lambda; x)\|_q \\ &\leqslant \|f_\lambda(x)-s_{2^{m_n}}(\lambda; x)\|_q+\sum_{k=0}^{m-1}\|s_{2^{k+1}n}(\lambda; x)-s_{2^k n}(\lambda; x)\|_q \end{aligned}$$

if $m \to \infty$, then we have

$$(8)\quad E_n^{(q)}(f_\lambda) \leqslant \|f_\lambda(x)-s_n(\lambda; x)\|_q \leqslant \sum_{k=0}^{\infty}\|s_{2^{k+1}n}(\lambda; x)-s_{2^k n}(\lambda; x)\|_q.$$

Considering (7) we have the proof of (4).

Finally it is easy to verify that

$$\left\|s_1(\lambda; x)-\frac{a_0}{2}\right\|_q \leqslant 16\pi\, \lambda_1 E_0^{(p)}(f).$$

By (8) in the case $n = 1$, we have (4) for $n = 0$, so our proof is complete.

## REFERENCES

[1] A. A. Konjuškov, 'Best Approximation by Trigonometric Polynomials and Fourier Coefficients', *Mat. Sbornik* **44** (1958) 53–84.
[2] C. M. Nikolskii, 'Inequalities Concerning Entire Functions of a Finite Degree and their Application in the Theory of Differentiable Functions of Several Variables', *Trud. Mat. Inst. Stekloff* (1951) 244–278.
[3] I. Szalay, 'On the Absolute Summability of Fourier Series', *Tôhoku Math. J.* **21** (1959) 523–531.
[4] A. F. Timan and M. F. Timan, 'On the Generalized Modulus of Continuity and the Best Approximation in Mean', *Dokl. Akad. Nauk SSSR* **71** (1950) 17–20.
[5] A. Zygmund, *Trigonometric series*, Cambridge 1959.

# ON THE RIESZ MEANS OF FOURIER-BESSEL SERIES


ROMAN TABERSKI

*Poznań*


**1. Introduction.** Let $J_\nu(z)$ be a Bessel function of order $\nu > -1/2$, with the successive positive zeros $j_1, j_2, \ldots$, and let

$$\varphi_\nu(z) = (\tfrac{1}{2}\pi z)^{1/2} J_\nu(z) \quad \text{for } z > 0, \quad \varphi_\nu(0) = \lim_{z\to 0+} \varphi_\nu(z).$$

Denote by $L^*$ the class of all real-valued functions $f(t)$ Lebesgue-integrable over every finite subinterval of $\langle 0, \infty)$.

Given arbitrary $f \in L^*$ and $r \geqslant 0$, we set

$$a_k = \frac{2}{l\varphi_{\nu+1}^2(j_k)} \int_0^l f(t)\varphi_\nu\left(j_k \frac{t}{l}\right) dt, \qquad A_n = (n+\tfrac{1}{2}\nu+\tfrac{1}{4})\pi$$

and

$$\sigma_n^l(x; f, r) = \sum_{k=1}^n \left(1 - \frac{j_k^2}{A_n^2}\right)^r a_k \varphi_\nu\left(j_k \frac{x}{l}\right)$$

for $x \geqslant 0$, $l > 0$, $n = 1, 2, \ldots$ These Riesz means of Fourier–Bessel series of $f$ in $\langle 0, l \rangle$, with respect to the orthogonal system $\{\varphi_\nu(j_k x/l)\}$, will be examined here.

We shall present some theorems concerning the convergence of $\sigma_n^l(x; f, r)$, for positive integers $r$, assuming that $l \to \infty$, $n \to \infty$ and $n^{-1} l \to 0$ or $n^{-1} l^{\nu+3/2} \to 0$, etc. Our considerations will be based on the integral formula

$$\sigma_n^l(x; f, r) = \frac{1}{l} \int_0^l f(t) K_n^r\left(\frac{t}{l}, \frac{x}{l}\right) dt \qquad (x \geqslant 0,\ l > 0, n = 1, 2, \ldots),$$

where

$$K_n^r(u, v) = \sqrt{uv} \sum_{k=1}^n \left(1 - \frac{j_k^2}{A_n^2}\right)^r \frac{2J_\nu(j_k u) J_\nu(j_k v)}{J_{\nu+1}^2(j_k)}.$$

The suitable positive constants depending only on $\nu$ and $r$, which are not necessary the same in different estimates, will be denoted by $C$.

**2. Auxiliary results.** By 2.1, 2.2 and 2.3 of [2], 674–675, we obtain the following three lemmas.

LEMMA 1. *If $r$ is an arbitrary positive integer, then*

$$\text{(i)} \qquad \left|\frac{1}{l} K_n^r\left(\frac{t}{l}, \frac{x}{l}\right)\right| \leqslant C\frac{n}{l},$$

$$\text{(ii)} \qquad \left|\frac{1}{l} K_n^r\left(\frac{t}{l}, \frac{x}{l}\right)\right| \leqslant \frac{Cl^r}{n^r|t-x|^{r+1}} \qquad (t \neq x)$$

*for $t, x \in \langle 0, l\rangle$, $l > 0$, $n = 1, 2, \ldots$*

*In particular, these estimates imply*

$$\text{(1)} \qquad \frac{1}{l}\int_0^l \left|K_n^r\left(\frac{t}{l}, \frac{x}{l}\right)\right| dt \leqslant C$$

*for $x \in \langle 0, l\rangle$, $l > 0$, $n = 1, 2, \ldots$*

LEMMA 2. *Given any integer $r \geqslant \nu + 3/2$, the inequality*

$$\left|\frac{1}{l}\int_0^l K_n^r\left(\frac{t}{l}, \frac{x}{l}\right) dt - 1\right| \leqslant C\left\{\frac{l}{nx} + \frac{l^{r+1}}{n^{r+1}(l-x)^{r+1}}\right\}$$

*holds, whenever*

$$\frac{l}{n} \leqslant x \leqslant l - \frac{l}{n}, \qquad l > 0,\ n \geqslant 2.$$

LEMMA 3. *Considering positive integers $r$, we have*

$$\left|\frac{1}{l}\int_0^l t^{\nu+1/2} K_n^r\left(\frac{t}{l}, \frac{x}{l}\right) dt - x^{\nu+1/2}\right| \leqslant C\frac{l^{r+\nu+3/2}}{n^{r+1}(l-x)^{r+1}}$$

*for $x \in \langle 0, l - l/n\rangle$, $l > 0$, $n = 1, 2, \ldots$*

Now, some other estimates will be deduced.

LEMMA 4. *Let $r$ be a positive integer, and let $x$, $l$, $n$ be as in Lemma 2. Then,*

$$\left|\frac{1}{l}\int_0^x t^{\nu+1/2} K_n^r\left(\frac{t}{l}, \frac{x}{l}\right) dt - \frac{x^{\nu+1/2}}{2}\right| \leqslant C\left\{\frac{l^{\nu+3/2}}{nx} + \frac{l^{\nu+3/2}}{n(l-x)}\right\}.$$

PROOF. It is enough to show that

$$\left|\int_0^z u^{\nu+1/2}K_n^r(u,z)\,du-\frac{z^{\nu+1/2}}{2}\right|\leqslant C\left\{\frac{1}{nz}+\frac{1}{n(1-z)}\right\}$$

when $n^{-1}\leqslant z\leqslant 1-n^{-1}$.

As in [3], 586,

$$\int_0^z u^{\nu+1/2}K_n^r(u,z)\,du=\sqrt{z}\sum_{k=1}^{n}\left(1-\frac{j_k^2}{A_n^2}\right)^r\frac{2z^{\nu+1}J_\nu(j_kz)J_{\nu+1}(j_kz)}{j_kJ_{\nu+1}^2(j_k)}$$

for $z\in\langle 0,1\rangle$. Therefore

$$\int_0^z u^{\nu+1/2}K_n^r(u,z)\,du=\int_0^z u^{\nu+1/2}K_n^0(u,z)\,du+$$

$$+\sqrt{z}\sum_{p=1}^{r}(-1)^p\binom{r}{p}\frac{1}{A_n^{2p}}\sum_{k=1}^{n}j_k^{2p-1}\frac{2z^{\nu+1}J_\nu(j_kz)J_{\nu+1}(j_kz)}{J_{\nu+1}^2(j_k)}.$$

Since

$$j_k^{2p-1}\frac{J_\nu(j_kz)J_{\nu+1}(j_kz)}{J_{\nu+1}^2(j_k)}=-\operatorname*{res}_{w=j_k}\frac{w^{2p-1}J_\nu(wz)J_{\nu+1}(wz)}{J_\nu(w)J_{\nu+1}(w)},$$

we have

$$\int_0^z u^{\nu+1/2}K_n^r(u,z)\,du=\int_0^z u^{\nu+1/2}K_n^0(u,z)\,du-$$

$$-\frac{1}{\pi i}\sum_{p=1}^{r}(-1)^p\binom{r}{p}\frac{z^{\nu+3/2}}{A_n^{2p}}\lim_{B\to\infty}\int_{A_n-Bi}^{A_n+Bi}w^{2p-1}\frac{J_\nu(wz)J_{\nu+1}(wz)}{J_\nu(w)J_{\nu+1}(w)}\,dw$$

(see [3], § 18.51).

The inequality

$$\left|\frac{J_\nu(wz)}{J_\nu(w)}\right|\leqslant\frac{C}{\sqrt{z}}e^{-(1-z)|v|},$$

which is true when $0 < z < 1$, $w = A_n + iv$ and $n > N(\nu)$ ([3], § 18.51), implies

$$\left| \int_{A_n - Bi}^{A_n + Bi} w^{2p-1} \frac{J_\nu(wz) J_{\nu+1}(wz)}{J_\nu(w) J_{\nu+1}(w)} dw \right|$$

$$\leqslant \frac{(\sqrt{2})^{2p-1} C}{z} \left\{ A_n^{2p-1} \int_0^{A_n} e^{-2(1-z)v} dv + \right.$$

$$\left. + \int_{A_n}^{B} v^{2p-1} e^{-2(1-z)v} dv \right\} \qquad \text{if } B \geqslant A_n .$$

Hence

$$\left| \frac{z^{\nu+3/2}}{A_n^{2p}} \int_{A_n - Bi}^{A_n + Bi} w^{2p-1} \frac{J_\nu(wz) J_{\nu+1}(wz)}{J_\nu(w) J_{\nu+1}(w)} dw \right|$$

$$\leqslant \frac{(\sqrt{2})^{2p-1} C}{2} \left\{ \frac{1}{A_n(1-z)} + \frac{\Gamma(2p)}{A_n^{2p}(1-z)^{2p}} \right\} \qquad (p = 1, 2, \ldots, r) .$$

Consequently,

$$\left| \int_0^z u^{\nu+1/2} K_n^r(u, z) du - \frac{z^{\nu+1/2}}{2} \right| \leqslant \left| \int_0^z u^{\nu+1/2} K_n^0(u, z) du - \frac{z^{\nu+1/2}}{2} \right| +$$

$$+ \frac{C}{n(1-z)} \qquad \text{if } n(1-z) \geqslant 1 .$$

Now, by an argument similar to that of [3], 586–587, we easily get the desired assertion.

LEMMA 5. *Given a positive integer $r$, an arbitrary $\varepsilon > 0$ and any interval $\langle a, b \rangle \subset (0, l)$, there is a positive $\eta \leqslant a$ such that, for every positive $\delta \leqslant \eta$, a corresponding positive $\lambda$ exists for which*

(i) $$\left| \frac{1}{l} \int_{x-\delta}^{x} K_n^r\left(\frac{t}{l}, \frac{x}{t}\right) dt - \tfrac{1}{2} \right| < \varepsilon ,$$

(ii) $$\left| \frac{1}{l} \int_{x}^{x+\delta} K_n^r\left(\frac{t}{l}, \frac{x}{l}\right) dt - \tfrac{1}{2} \right| < \varepsilon ,$$

*whenever $n^{-1} l^{\nu+3/2} \leqslant \lambda$ ($l \geqslant b+1$) and $a \leqslant x \leqslant b$.*

PROOF OF (i). In the case $0 < \delta \leqslant a \leqslant x \leqslant b \leqslant l-\delta$ $(n = 1, 2, \ldots)$ estimate (1) ensures that

$$\left|\frac{1}{l}\int_{x-\delta}^{x} \{x^{\nu+1/2} - t^{\nu+1/2}\} K_n^r\left(\frac{t}{l}, \frac{x}{l}\right) dt\right|$$

$$\leqslant \{x^{\nu+1/2} - (x-\delta)^{\nu+1/2}\} \frac{1}{l}\int_{0}^{l} \left|K_n^r\left(\frac{t}{l}, \frac{x}{l}\right)\right| dt < \frac{\varepsilon}{2} a^{\nu+1/2}$$

for sufficiently small $\delta$, say $\delta \leqslant \eta_1 \leqslant \min(a, 1)$.

By Lemma 4 and Lemma 1 (ii),

$$\left|\frac{1}{l}\int_{x-\delta}^{x} t^{\nu+1/2} K_n^r\left(\frac{t}{l}, \frac{x}{l}\right) dt - \frac{x^{\nu+1/2}}{2}\right|$$

$$\leqslant \left|\frac{1}{l}\int_{0}^{x} t^{\nu+1/2} K_n^r\left(\frac{t}{l}, \frac{x}{l}\right) dt - \frac{x^{\nu+1/2}}{2}\right| +$$

$$+\left|\frac{1}{l}\int_{0}^{x-\delta} t^{\nu+1/2} K_n^r\left(\frac{t}{l}, \frac{x}{l}\right) dt\right| \leqslant C\frac{l^{\nu+3/2}}{n}\left\{\frac{1}{x} + \frac{1}{l-x}\right\} + C\frac{l^r}{n^r r \delta^r}.$$

Consequently,

$$\left|\frac{x^{\nu+1/2}}{l}\int_{x-\delta}^{x} K_n^r\left(\frac{t}{l}, \frac{x}{l}\right) dt - \frac{x^{\nu+1/2}}{2}\right| \leqslant \frac{\varepsilon}{2} a^{\nu+1/2} + C\left(\frac{l^{\nu+3/2}}{2n\delta} + \frac{l^r}{n^r \delta^r}\right),$$

and part (i) of our thesis follows.

PROOF OF (ii). Let $0 < \delta \leqslant a \leqslant x \leqslant b \leqslant l-\delta$ $(n = 1, 2, \ldots)$. Then, by (1),

$$\left|\frac{1}{l}\int_{x}^{x+\delta} \{t^{\nu+1/2} - x^{\nu+1/2}\} K_n^r\left(\frac{t}{l}, \frac{x}{l}\right) dt\right|$$

$$\leqslant \{(x+\delta)^{\nu+1/2} - x^{\nu+1/2}\} C < \tfrac{1}{2}\varepsilon a^{\nu+1/2}$$

whenever $\delta$ are small enough, say $\delta \leqslant \eta_2 \leqslant \min(a, 1)$.

In view of Lemmas 3, 4 and Lemma 1 (ii),

$$\left|\frac{1}{l}\int_x^{x+\delta} t^{\nu+1/2}K_n^r\left(\frac{t}{l},\frac{x}{l}\right)dt-\frac{x^{\nu+1/2}}{2}\right| \leqslant \left|\frac{1}{l}\int_0^l t^{\nu+1/2}K_n^r\left(\frac{t}{l},\frac{x}{l}\right)dt-\right.$$

$$\left.-x^{\nu+1/2}\right|+\left|\frac{x^{\nu+1/2}}{2}-\frac{1}{l}\int_0^x t^{\nu+1/2}K_n^r\left(\frac{t}{l},\frac{x}{l}\right)dt\right|+$$

$$+\left|\frac{1}{l}\int_{x+\delta}^{l} t^{\nu+1/2}K_n^r\left(\frac{t}{l},\frac{x}{l}\right)dt\right|$$

$$\leqslant C\frac{l^{r+\nu+3/2}}{n^{r+1}(l-x)^{r+1}}+C\frac{l^{\nu+3/2}}{n}\left\{\frac{1}{x}+\frac{1}{l-x}\right\}+C\frac{l^{r+\nu+1/2}}{n^r r\delta^r}$$

$$\leqslant C\frac{l^{\nu+3/2}}{n}\left\{\frac{1}{\delta^{r+1}}+\frac{1}{a}+\frac{1}{\delta}+\frac{1}{r\delta^r}\right\}\leqslant 4C\frac{l^{\nu+3/2}}{n\delta^{r+1}}$$

if $0<\delta\leqslant\eta_2$, $0<l/n\leqslant\eta_2$, $a\leqslant x\leqslant b\leqslant l-\eta_2$.

Thus, under the above restrictions,

$$\left|\frac{x^{\nu+1/2}}{l}\int_x^{x+\delta} K_n^r\left(\frac{t}{l},\frac{x}{l}\right)dt-\frac{x^{\nu+1/2}}{2}\right| \leqslant \left|\frac{1}{l}\int_x^{x+\delta}\{x^{\nu+1/2}-\right.$$

$$\left.-t^{\nu+1/2}\}K_n^r\left(\frac{t}{l},\frac{x}{l}\right)dt\right|+\left|\frac{1}{l}\int_x^{x+\delta} t^{\nu+1/2}K_n^r\left(\frac{t}{l},\frac{x}{l}\right)dt-\frac{x^{\nu+1/2}}{2}\right|$$

$$\leqslant\frac{\varepsilon}{2}a^{\nu+1/2}+4C\frac{l^{\nu+3/2}}{n\delta^{r+1}},$$

and the desired result is established.

**3. Pointwise convergence of $\sigma_n^l(x;f,r)$.** Considering the case $l\to\infty$ and $n\to\infty$, we shall now give two theorems analogous to that of (1.21) in [4], 246.

THEOREM 1. *Let $f(t)$ be a function of class $L^*$, such that*

$$\int_1^\infty\frac{|f(t)|}{t^{r+1}}dt<\infty, \tag{2}$$

*where the integer* $r \geqslant \nu + 3/2$. *Then, for any fixed* $x > 0$, *the condition*

$$(3) \qquad \int_0^h |f(x \pm z) - f(x)|\, dz = o(h) \quad \text{as } h \to 0+$$

*implies*

$$(4) \qquad \lim_{n^{-1}l \to 0} \sigma_n^l(x; f, r) = f(x).$$

PROOF. Given a positive $x$ for which relation (3) holds, we have

$$(5) \quad \sigma_n^l(x; f, r) - f(x) = \frac{1}{l} \int_0^l \{f(t) - f(x)\} K_n^r\left(\frac{t}{l}, \frac{x}{l}\right) dt +$$

$$+ f(x) \left\{ \frac{1}{l} \int_0^l K_n^r\left(\frac{t}{l}, \frac{x}{l}\right) dt - 1 \right\} = U_n^l(x) + V_n^l(x).$$

By Lemma 2,

$$|V_n^l(x)| \leqslant C |f(x)| \left\{ \frac{l}{nx} + \frac{l^{r+1}}{n^{r+1}(l-x)^{r+1}} \right\}$$

if $l/n \leqslant x \leqslant l - l/n$. Hence

$$\lim_{n^{-1}l \to 0} V_n^l(x) = 0.$$

Let us choose, for an arbitrary $\varepsilon > 0$, a positive $\eta < x$ such that

$$\frac{1}{h} \int_0^h |f(x \pm z) - f(x)|\, dz < \varepsilon \quad \text{when } 0 < h \leqslant \eta.$$

Suppose that

$$0 \leqslant x - \frac{l}{n} \leqslant x + \frac{l}{n} \leqslant x + \eta \leqslant l,$$

and put

$$g(t) = \int_x^t |f(s) - f(x)|\, ds \quad \text{for } t \in \langle x - \eta, x + \eta \rangle.$$

Write

$$U_n^l(x) = \frac{1}{l} \left( \int_0^{x-\eta} + \int_{x-\eta}^{x-l/n} + \int_{x-l/n}^{x} + \int_x^{x+l/n} + \int_{x+l/n}^{x+\eta} + \int_{x+\eta}^{l} \right) \{f(t) -$$

$$- f(x)\} K_n^r\left(\frac{t}{l}, \frac{x}{l}\right) dt = \sum_{i=1}^{6} W_i.$$

In view of Lemma 1 (ii),

$$|W_6| \leqslant \frac{Cl^r}{n^r} \int_{x+\eta}^{l} \frac{|f(t)-f(x)|}{(t-x)^{r+1}}\, dt \leqslant \frac{Cl^r}{n^r}\left(1+\frac{x}{\eta}\right)^{r+1} \int_{x+\eta}^{l} \frac{|f(t)|+|f(x)|}{t^{r+1}}\, dt;$$

whence, by (2), $|W_6| < \varepsilon$ for sufficiently small $l/n$.

Applying Lemma 1 (ii), we obtain

$$|W_5| \leqslant \frac{Cl^r}{n^r} \int_{x+l/n}^{x+\eta} \frac{|f(t)-f(x)|}{(t-x)^{r+1}}\, dt = \frac{Cl^r}{n^r} \int_{x+l/n}^{x+\eta} \frac{1}{(t-x)^{r+1}}\, dg(t).$$

Consequently,

$$|W_5| \leqslant \frac{Cl^r}{n^r}\left\{\frac{g(x+\eta)}{\eta^{r+1}} + \frac{g(x+l/n)}{(l/n)^{r+1}} + \int_{x+l/n}^{x+\eta} g(t)\, \frac{r+1}{(t-x)^{r+2}}\, dt\right\}$$

$$\leqslant \frac{Cl^r\varepsilon}{n^r}\left\{\frac{1}{\eta^r} + \frac{1}{(l/n)^r} + \int_{x+l/n}^{x+\eta} \frac{r+1}{(t-x)^{r+1}}\, dt\right\} \leqslant \left(2+\frac{1}{r}\right)C\varepsilon.$$

A similar calculation shows that $|W_2| \leqslant (2+1/r)C\varepsilon$.

By Lemma 1 (i),

$$|W_4| \leqslant \frac{1}{l} \int_{x}^{x+l/n} \left|K_n^r\left(\frac{t}{l}, \frac{x}{l}\right)\right| dg(t) \leqslant C\frac{n}{l}\, g(x+l/n) < C\varepsilon.$$

Analogously, $|W_3| < C\varepsilon$.

Arguing as in [1], 241–242, and using Lemma 1 (ii), we obtain

$$\lim_{n^{-1}l\to 0} \frac{1}{l} \int_0^{x-\eta} f(t)\, K_n^r\left(\frac{t}{l}, \frac{x}{l}\right) dt = 0$$

for any $f$ continuous in $\langle 0, x-\eta\rangle$. The last relation is also true for all $f$'s of class $L^*$; it follows at once from density of continuous functions in the Lebesgue space $L\langle 0, x-\eta\rangle$ and from Lemma 1 (ii). Hence $|W_1| < \varepsilon$ if $l/n$ are small enough.

Thus the proof is completed.

REMARK 1. If, instead of (3),

$$\int_0^h |f(z)|\, dz = o(h) \quad \text{as } h \to 0+$$

and if $f(0) = 0$, Theorem 1 remains valid for $x = 0$, too.

REMARK 2. In the case when $f$ is continuous at every $x$ on $\langle a, b\rangle$ $(0 < a < b < \infty)$, relation (4) holds uniformly in this interval.

THEOREM 2. *Suppose that $f \in L^*$ and that the integral* (2), *with a positive integer $r \geqslant 1$, is finite. Consider positive $x$ for which*

$$\int_0^h |f(x-t)-f_1(x)|\,dt = o(h) \quad \textit{and} \quad \int_0^h |f(x+t)-f_2(x)|\,dt = o(h)$$

*as $h \to 0+$, where $f_1(x), f_2(x)$ denote the suitable real numbers. Then,*

$$\lim_{n^{-1}l^{\nu+3/2}\to 0} \sigma_n^l(x; f, r) = \tfrac{1}{2}\{f_1(x)+f_2(x)\}.$$

PROOF. By the assumption and Lemma 5, for arbitrary $\varepsilon > 0$ and $x \in (0, l)$, there is a positive $\eta < x$ such that

$$\frac{1}{h}\int_0^h |f(x-t)-f_1(x)|\,dt < \varepsilon, \qquad \frac{1}{h}\int_0^h |f(x+t)-f_2(x)|\,dt < \varepsilon$$

and

$$\left|\frac{1}{l}\int_{x-\eta}^{x} K_n^r\left(\frac{t}{l}, \frac{x}{l}\right)dt-\tfrac{1}{2}\right| < \varepsilon, \qquad \left|\frac{1}{l}\int_{x}^{x+\eta} K_n^r\left(\frac{t}{l}, \frac{x}{l}\right)dt-\tfrac{1}{2}\right| < \varepsilon$$

when $0 < h \leqslant \eta$, $n^{-1}l^{\nu+3/2} < \lambda$ $(l \geqslant x+1)$. Condition (2) implies

$$\int_{x+\Delta}^{\infty} \frac{|f(t)|}{(t-x)^{r+1}}\,dt < \varepsilon$$

for sufficiently large $\Delta > \eta$.

Clearly,

$$\sigma_n^l(x; f, r) = \frac{1}{l}\left(\int_0^x + \int_x^l\right) f(t) K_n^r\left(\frac{t}{l}, \frac{x}{l}\right)dt = P+Q$$

and

$$P = \tfrac{1}{2}f_1(x)+f_1(x)\left\{\frac{1}{l}\int_{x-\eta}^{x} K_n^r\left(\frac{t}{l}, \frac{x}{l}\right)dt-\tfrac{1}{2}\right\}+$$

$$+\frac{1}{l}\int_{x-\eta}^{x} \{f(t)-f_1(x)\}K_n^r\left(\frac{t}{l}, \frac{x}{l}\right)dt+\frac{1}{l}\int_0^{x-\eta} f(t)K_n^r\left(\frac{t}{l}, \frac{x}{l}\right)dt,$$

$$Q = \tfrac{1}{2} f_2(x) + f_2(x) \left\{ \frac{1}{l} \int_x^{x+\eta} K_n^r\left(\frac{t}{l}, \frac{x}{l}\right) dt - \tfrac{1}{2} \right\} +$$

$$+ \frac{1}{l} \int_x^{x+\eta} \{f(t) - f_2(x)\} K_n^r\left(\frac{t}{l}, \frac{x}{l}\right) dt + \frac{1}{l} \int_{x+\eta}^{l} f(t) K_n^r\left(\frac{t}{l}, \frac{x}{l}\right) dt.$$

In view of Lemma 1 (ii),

$$\left| \frac{1}{l} \int_{x+\Delta}^{l} f(t) K_n^r\left(\frac{t}{l}, \frac{x}{l}\right) dt \right| \leqslant C \frac{l^r}{n^r} \int_{x+\Delta}^{l} \frac{f(t)}{(t-x)^{r+1}} dt < C\varepsilon$$

if $l/n \leqslant 1$; moreover,

$$\lim_{n^{-1}l \to 0} \frac{1}{l} \left( \int_0^{x-\eta} + \int_{x+\eta}^{x+\Delta} \right) f(t) K_n^r\left(\frac{t}{l}, \frac{x}{l}\right) dt = 0,$$

by a corresponding theorem of the Riemann–Lebesgue type. Hence

$$|\sigma_n^l(x; f, r) - \tfrac{1}{2}\{f_1(x) + f_2(x)\}| \leqslant \{|f_1(x)| + |f_2(x)| + C + 1\}\varepsilon +$$

$$+ \frac{1}{l} \left| \int_{x-\eta}^{x} \{f(t) - f_1(x)\} K_n^r\left(\frac{t}{l}, \frac{x}{l}\right) dt \right| + \frac{1}{l} \left| \int_x^{x+\eta} \{f(t) - f_2(x)\} K_n^r\left(\frac{t}{l}, \frac{x}{l}\right) dt \right|$$

for sufficiently small $n^{-1} l^{\nu+3/2}$ $(l \geqslant x+1)$.

Proceeding now as in the previous proof (estimates for $W_6 - W_2$), we easily get the desired assertion.

**4. Mean approximation of continuous functions.** Let $H^\alpha$ $(0 < \alpha \leqslant 1)$ be the class of all real-valued functions $f$, uniformly continuous in $\langle 0, \infty)$, satisfying the Hölder condition

$$|f(u) - f(v)| \leqslant L|u-v|^\alpha \quad \text{for } u, v \in \langle 0, \infty),$$

with a suitable positive constant $L$ (depending on $f$).

Assuming that $p > 1$ and $2 \leqslant l \leqslant n$ $(l \to \infty, n \to \infty)$, we shall give two estimates for the deviation

$$D_{l,n}^{p;r}(f) = \left\{ \frac{1}{l} \int_0^l |\sigma_n^l(x; f, r) - f(x)|^p dx \right\}^{1/p} \quad \text{when } f \in H^\alpha.$$

THEOREM 3. *Suppose that r is a positive integer and* $r \geqslant \nu+3/2$. *Then, for any* $f \in H^{\alpha}\,(0 < \alpha \leqslant 1)$,

$$D_{l,n}^{p;r}(f) = \begin{cases} O\left(\dfrac{l^{\alpha}}{n^{\alpha}}\right) & \text{if } \alpha p \leqslant 1, \\ O\left(\dfrac{l^{\alpha}}{n^{1/p}}\right) & \text{if } \alpha p > 1. \end{cases} \tag{6}$$

PROOF. By identity (5),

$$D_{l,n}^{p;r}(f) \leqslant \left\{\frac{1}{l}\int_0^l |U_n^l(x)|^p dx\right\}^{1/p} + \left\{\frac{1}{l}\int_0^l |V_n^l(x)|^p dx\right\}^{1/p}. \tag{7}$$

Considering $f \in H^{\alpha}$ $(0 < \alpha \leqslant 1)$, we can find two positive constants $L$, $M$ such that

$$|f(x)| \leqslant Lx^{\alpha}+M \qquad \text{for } x \geqslant 0.$$

In view of estimate (1) and Lemma 2,

$$\int_0^l |V_n^l(x)|^p dx \leqslant (C+1)^p\left(\int_0^{l/n} + \int_{l-l/n}^{l}\right)|f(x)|^p dx + \\ + C^p \int_{l/n}^{l-l/n} \left| f(x)\left\{\frac{l}{nx} + \frac{l^{r+1}}{n^{r+1}(l-x)^{r+1}}\right\}\right|^p dx.$$

Hence

$$\frac{1}{l}\int_0^l |V_n^l(x)|^p dx \leqslant \frac{2^p(C+1)^p}{l}\left(\int_0^{l/n} + \int_{l-l/n}^{l}\right)\{L^p x^{\alpha p}+M^p\}dx + \\ + \frac{(4CL)^p}{l}\left\{\int_{l/n}^{l-l/n} \frac{l^p}{n^p} x^{\alpha p-p} dx + \int_{l/n}^{l-l/n} \frac{l^{rp+p}}{n^{rp+p}} \cdot \frac{x^{\alpha p}}{(l-x)^{rp+p}} dx\right\} + \\ + \frac{(4CM)^p}{l}\left\{\int_{l/n}^{l-l/n} \frac{l^p}{n^p} x^{-p} dx + \int_{l/n}^{l-l/n} \frac{l^{rp+p}}{n^{rp+p}} \cdot \frac{1}{(l-x)^{rp+p}} dx\right\},$$

and, by simple calculation, we obtain

$$\frac{1}{l}\int_0^l |V_n^l(x)|^p dx = O\left(\frac{l^{\alpha p}}{n}\right) \qquad \text{as } l, n \to \infty. \tag{8}$$

Write

$$\Phi_n^l(x) = \frac{1}{l}\int_0^x \{f(t)-f(x)\}K_n^r\left(\frac{t}{l},\frac{x}{l}\right)dt\,,$$

$$\Psi_n^l(x) = \frac{1}{l}\int_x^l \{f(t)-f(x)\}K_n^r\left(\frac{t}{l},\frac{x}{l}\right)dt\,.$$

Then

$$\int_0^l |U_n^l(x)|^p dx \leqslant 2^p\left\{\int_0^l |\Phi_n^l(x)|^p dx+\int_0^l |\Psi_n^l(x)|^p dx\right\}.$$

In the case $0 \leqslant x \leqslant l/n$, Lemma 1 yields

$$|\Phi_n^l(x)| \leqslant L\int_0^x |t-x|^\alpha C\frac{n}{l}\,dt \leqslant \frac{CL}{\alpha+1}\left(\frac{l}{n}\right)^\alpha \leqslant CL\left(\frac{l}{n}\right)^\alpha,$$

$$|\Psi_n^l(x)| \leqslant L\left\{\int_x^{x+l/n} (t-x)^\alpha C\frac{n}{l}\,dt+\int_{x+l/n}^l (t-x)^\alpha\frac{Cl^r}{n^r(t-x)^{r+1}}\,dt\right\}$$

$$\leqslant L\left\{\frac{C}{\alpha+1}\left(\frac{l}{n}\right)^\alpha+\frac{C}{r-\alpha}\left(\frac{l}{n}\right)^\alpha\right\} \leqslant 2CL\left(\frac{l}{n}\right)^\alpha.$$

If $l/n \leqslant x \leqslant l$, Lemma 1 leads to

$$|\Phi_n^l(x)| \leqslant L\left\{\int_0^{x-l/n} (x-t)^\alpha\frac{Cl^r}{n^r(x-t)^{r+1}}\,dt+\int_{x-l/n}^x (x-t)^\alpha C\frac{n}{l}\,dt\right\}$$

$$\leqslant L\left\{\frac{C}{r-\alpha}\left(\frac{l}{n}\right)^\alpha+\frac{C}{\alpha+1}\left(\frac{l}{n}\right)^\alpha\right\} \leqslant 2CL\left(\frac{l}{n}\right)^\alpha.$$

Assuming that $l/n \leqslant x \leqslant l-l/n$, we have

$$|\Psi_n^l(x)| \leqslant L\left\{\int_x^{x+l/n} (t-x)^\alpha C\frac{n}{l}\,dt+\int_{x+l/n}^l (t-x)^\alpha\frac{Cl^r}{n^r(t-x)^{r+1}}\,dt\right\}$$

$$\leqslant L\left\{\frac{C}{\alpha+1}\left(\frac{l}{n}\right)^\alpha+\frac{C}{r-\alpha}\left(\frac{l}{n}\right)^\alpha\right\} \leqslant 2CL\left(\frac{l}{n}\right)^\alpha;$$

in the case $l-l/n \leqslant x \leqslant l$,

$$|\Psi_n^l(x)| \leqslant L\int_x^l (t-x)^\alpha C\frac{n}{l}\,dt = LC\frac{n}{l}\,\frac{(l-x)^{\alpha+1}}{\alpha+1} \leqslant CL\left(\frac{l}{n}\right)^\alpha .$$

Thus,

$$\frac{1}{l}\int_0^l |U_n^l(x)|^p dx \leqslant 2(4CL)^p\left(\frac{l}{n}\right)^{\alpha p} \tag{9}$$

and combining this with (8), we get (6).

REMARK. It is easy to see that for $p = 1$, the left-hand side of (6) is

$$O\left(\frac{l^\alpha}{n^\alpha}\right) \quad \text{if } \alpha < 1 \quad \text{and} \quad O\left(\frac{l+\log n}{n}\right) \quad \text{if } \alpha = 1.$$

THEOREM 4. *Let $r$ be a positive integer, and let $2 \leqslant r < \nu+3/2$. Then for any $f \in H^\alpha$ $(0 < \alpha \leqslant 1)$ such that $f(x) = O(x^{\nu-1/2})$ as $x \to 0+$,*

$$D_{l,n}^{p;r}(f) = O\left(\frac{l^{\nu+1/2}}{n^{1/p}} + \frac{l^\alpha}{n^\alpha}\right).$$

PROOF. As previously, we start with inequality (7) and we observe that estimate (9) remains valid.

If $0 \leqslant x \leqslant l/n$ or $l-l/n \leqslant x \leqslant l$, the Hölder condition

$$|f(x)-f(0)| \leqslant Lx^\alpha \qquad (f(0) = 0)$$

and estimate (1) imply

$$|V_n^l(x)| \leqslant \left| f(x)\left\{\frac{1}{l}\int_0^l K_n^r\left(\frac{t}{l},\frac{x}{l}\right)dt - 1\right\}\right| \leqslant (C+1)Lx^\alpha .$$

Consequently,

$$\frac{1}{l}\int_0^l |V_n^l(x)|^p dx \leqslant 2(C+1)^p L^p\frac{l^{\alpha p}}{n} + \frac{2^p}{l}\{I_1+I_2\},$$

where

$$I_1 = \int_{l/n}^{l-l/n}\left|\frac{f(x)}{x^{\nu+1/2}}\left\{\frac{1}{l}\int_0^l t^{\nu+1/2}K_n^r\left(\frac{t}{l},\frac{x}{l}\right)dt - x^{\nu+1/2}\right\}\right|^p dx,$$

$$I_2 = \int_{l/n}^{l-l/n}\left|\frac{f(x)}{x^{\nu+1/2}}\left\{\frac{1}{l}\int_0^l (x^{\nu+1/2}-t^{\nu+1/2})K_n^r\left(\frac{t}{n},\frac{x}{l}\right)dt\right\}\right|^p dx .$$

In view of Lemma 3,

$$I_1 \leqslant \left(\int_{l/n}^{1} + \int_{1}^{l-l/n}\right)\left|\frac{f(x)}{x^{\nu+1/2}}\, C\,\frac{l^{r+\nu+3/2}}{n^{r+1}(l-x)^{r+1}}\right|^p dx = Y_1+Y_2\,.$$

By the assumption, there is a constant $M$ such that

$$|f(x)| \leqslant Mx^{\nu-1/2} \qquad \text{for } x \subset \langle 0, 1\rangle.$$

Hence

$$Y_1 \leqslant (CM)^p \int_{l/n}^{1} \left|\frac{l^{r+\nu+3/2}}{xn^{r+1}(l-x)^{r+1}}\right|^p dx \leqslant (CM)^p \left(\frac{l^{r+\nu+3/2}}{n^{r+1}}\right)^p \int_{l/n}^{1} \frac{1}{x^p}\, dx$$

$$\leqslant \frac{(CM)^p}{p-1}\left(\frac{l^{r+\nu+1/2}}{n^r}\right)^p \cdot \frac{l}{n} \leqslant \frac{(CM)^p}{p-1}\,\frac{l^{(\nu+1/2)p+1}}{n}.$$

Analogously,

$$Y_2 \leqslant (CL)^p \int_{1}^{l-l/n} \left|\frac{x^{\alpha} l^{r+\nu+3/2}}{x^{\nu+1/2} n^{r+1}(l-x)^{r+1}}\right|^p dx$$

$$\leqslant (CL)^p \left(\frac{l^{r+\nu+3/2}}{n^{r+1}}\right)^p \int_{1}^{l-l/n} \frac{1}{(l-x)^{rp+p}}\, dx \leqslant (CL)^p\, \frac{l^{(\nu+1/2)p+1}}{n}.$$

Therefore

$$\frac{1}{l} I_1 = O\left(\frac{l^{(\nu+1/2)p}}{n}\right). \tag{10}$$

Further,

$$I_2 \leqslant \int_{l/n}^{1} \left|\frac{M}{x}\cdot\frac{1}{l}\int_{0}^{l} (x^{\nu+1/2} - t^{\nu+1/2}) K_n^r\left(\frac{t}{l}, \frac{x}{l}\right) dt\right|^p dx +$$

$$+ \int_{1}^{l-l/n} \left|\frac{Lx^{\alpha}}{x^{\nu+1/2}}\,\frac{1}{l}\int_{0}^{l} (x^{\nu+1/2} - t^{\nu+1/2}) K_n^r\left(\frac{t}{l}, \frac{x}{l}\right) dt\right|^p dx = Z_1+Z_2.$$

By the mean-value theorem and Lemma 1,

$$Z_1 \leqslant \int_{l/n}^{1} \left|\frac{(\nu+1/2)M}{xl}\left\{x^{\nu-1/2}\int_{0}^{x} (x-t)\left|K_n^r\left(\frac{t}{l}, \frac{x}{l}\right)\right| dt +\right.\right.$$

$$+l^{\nu-1/2}\int\limits_{x}^{l}(t-x)\left|K_n^r\left(\frac{t}{l},\frac{x}{l}\right)\right|dt\Bigg\}\Bigg|^p dx$$

$$\leqslant (2\nu+1)^p M^p\int\limits_{l/n}^{1}\Bigg|x^{\nu-3/2}\Bigg\{\int\limits_{0}^{x-l/n}\frac{Cl^r}{n^r(x-t)^r}\,dt+$$

$$+\int\limits_{x-l/n}^{x}(x-t)C\frac{n}{l}\,dt\Bigg\}\Bigg|^p dx+$$

$$+(2\nu+1)^p M^p\int\limits_{l/n}^{1}\Bigg|\frac{l^{\nu-1/2}}{x}\Bigg\{\int\limits_{x}^{x+l/n}(t-x)C\frac{n}{l}\,dt+$$

$$+\int\limits_{x+l/n}^{l}\frac{Cl^r}{n^r(t-x)^r}\,dt\Bigg\}\Bigg|^p dx;$$

whence $Z_1 = O(n^{-1}l^{(\nu-1/2)p+1})$.

Also, the mean-value theorem and Lemma 1 lead to

$$Z_2\leqslant\int\limits_{1}^{l-l/n}\Bigg|\frac{(\nu+1/2)Lx^\alpha}{x^{\nu+1/2}l}\Bigg\{x^{\nu-1/2}\int\limits_{0}^{x}(x-t)\left|K_n^r\left(\frac{t}{l},\frac{x}{l}\right)\right|dt+$$

$$+l^{\nu-1/2}\int\limits_{x}^{l}(t-x)\left|K_n^r\left(\frac{t}{l},\frac{x}{l}\right)\right|dt\Bigg\}\Bigg|^p dx$$

$$\leqslant (2\nu+1)^p L^p\int\limits_{1}^{l-l/n}\Bigg|x^{\alpha-1}\Bigg\{\int\limits_{0}^{x-l/n}\frac{Cl^r}{n^r(x-t)^r}\,dt+$$

$$+\int\limits_{x-l/n}^{x}(x-t)C\frac{n}{l}\,dt\Bigg\}\Bigg|^p dx+$$

$$+(2\nu+1)^p L^p\int\limits_{1}^{l-l/n}\Bigg|\frac{l^{\nu-1/2}}{x^{\nu-\alpha+1/2}}\Bigg\{\int\limits_{x}^{x+l/n}(t-x)\frac{Cn}{l}\,dt+$$

$$+\int\limits_{x+l/n}^{l}\frac{Cl^r}{n^r(t-x)^r}\,dt\Bigg\}\Bigg|^p dx.$$

Consequently, $Z_2 = O\left(\frac{l^{(\nu+1/2)p+1}}{n^p}\right)$. Thus

$$\frac{1}{l} I_2 = O\left(\frac{l^{(\nu+1/2)p}}{n}\right)$$

and, in view of (10), we have

$$\frac{1}{l}\int_0^l |V_n^l(x)|^p dx = O\left(\frac{l^{(\nu+1/2)p}}{n}\right).$$

Our thesis is now evident.

## REFERENCES

[1] I. P. Natanson, *Theory of Functions of a Real Variable*, Moskva–Leningrad 1950.
[2] R. Taberski, 'Some Properties of Fourier–Bessel Series, IV', *Bull. Acad. Polon. Sci., Sér. sci. math., astr. et phys.* **14** (1966) 673–680.
[3] G. N. Watson, *A Treatise on the Theory of Bessel Functions*, Cambridge 1922.
[4] A. Zygmund, *Trigonometric Series*, II, Cambridge 1959.

# POLYNOMIAL APPROXIMATION AND STRUCTURAL PROPERTIES OF FUNCTIONS DEFINED ON SUBSETS OF THE COMPLEX PLANE

P. M. TAMRAZOV

*Kiev*

In the first part of this paper we are concerned with the study of relations between the order of polynomial approximation of functions and their structural properties; in the second part we present results concerning the connection between properties of a holomorphic function with respect to its domain and analogous properties with respect to the boundary of this domain. Those results find an application in the approximation theory.

Let $\mathfrak{B}$ denote the class of all bounded closed subsets of the complex plane with connected complement.

Let $K \in \mathfrak{B}$ and let $f(\zeta)$ be a function continuous in $K$ and holomorphic in the interior of $K$.

For a point $z \in \partial K$ and a positive integer $n$ let $d_n(z)$ denote the distance from $z$ to the level line $\{\zeta : g(\infty, \zeta) = 1/n\}$ of the generalized Green function $g(\infty, \zeta)$ of the complement of $K$.

Moduli of continuity, local moduli of continuity and derivatives of the function $f(\zeta)$ taken with respect to $K$ will be denoted by $\omega_K(f, \delta)$, $\omega_K(f, \zeta, \delta)$, $f'_K(\zeta)$, and their analogues taken with respect to the boundary $\partial K$ will be denoted by $\omega_{\partial K}(f, \delta)$, $\omega_{\partial K}(f, \zeta, \delta)$, $f'_{\partial K}(\zeta)$.

We shall denote by $\mu(\delta)$ any function of the modulus of continuity type (i.e. any positive non-decreasing semi-continuous function defined on the half-axis $\delta > 0$ with $\mu(+0) = 0$).

The symbol $P_n(\zeta)$ will denote a polynomial of degree not exceeding $n$.

Suppose that for $r \geqslant 0$ the function $f(\zeta)$ admits a polynomial approximation on $K$ with the order

$$(1) \qquad |f(z) - P_n(z)| \leqslant [d_n(z)]^r \mu(d_n(z)), \qquad z \in \partial K, \qquad n = 1, 2, \ldots$$

Then there arises the problem of expressing the structural properties of $f(\zeta)$ on $K$ in terms of its moduli of continuity and its derivatives. This problem will be called the *inverse problem* of polynomial approximation. We shall use the expression $\alpha$-problem if the moduli and derivatives in question are taken with respect to the boundary $\partial K$ and $\eta$-problem if they are taken with respect to $K$.

In papers [3], [4] Dziadyk deals with the inverse $\alpha$-problem of polynomial approximation for a set of type (A), i.e. a continuum $K \in \mathfrak{B}$ whose boundary consists of a finite number of smooth arcs with continuous curvature (and with some additional restrictions).

In paper [9] Lebedev and Tamrazov solve the inverse $\alpha$-problem of polynomial approximation for an arbitrary continuum and even for an arbitrary compact set $K \in \mathfrak{B}$ regular with respect to the Dirichlet problem. Gorbaichuk and Tamrazov [7] solved the $\alpha$-problem for compact sets $K \in \mathfrak{B}$ whose every non-empty portion has positive logarithmic capacity (sets with this property form, in a sense, the largest class of compact sets for which the inverse $\alpha$-problem of polynomial approximation may be formulated in the above form).

What concerns the $\eta$-problem, it has been solved only in the case (cf. Dziadyk [3]) when $\mu(\delta) = c\delta^{\alpha}$, $0 < \alpha < 1$ and the set $K \in \mathfrak{B}$ is of type (A). In this case the solution of the inverse $\eta$-problem is a consequence of the solution of the $\alpha$-problem in virtue of certain results of Warschawski, Walsh and Sewell (see below, and also [11], 25–30). However, under more general assumptions, e.g. under the assumptions of papers [3], [4], [7], [9], the results of [11] cannot be applied. In particular, for an arbitrary function $\mu(\delta)$ of the modulus of continuity type the inverse $\eta$-problem has been open even for a set of type (A); it is precisely in this form that this problem was posed by Dziadyk a few years ago.

At the end of 1969 the inverse $\eta$-problem of polynomial approximation was solved for an arbitrary continuum $K \in \mathfrak{B}$ (a solution of Dziadyk's problem hence results as a particular case) and even for an arbitrary compact set $K \in \mathfrak{B}$ regular with respect to the Dirichlet problem and for any function $\mu(\delta)$ of the modulus of continuity type. This result has been announced in [12], [13] and fully exposed in [14].

We first formulate the following particular case.

THEOREM 1. *Let $f(\zeta)$ be a function defined on a continuum $K \in \mathfrak{B}$; suppose that $f(\zeta)$ admits a uniform (on $K$) approximation by a sequence of polynomials $P_1(\zeta), P_2(\zeta), \ldots$ and that the degree of approximation is given (on $\partial K$) by inequality* (1), *where r is a fixed non-negative number and $\mu(\delta)$ is any function of the modulus of continuity type. Then*

$$\omega_K(f, \delta) \leqslant c\delta \int_{\delta}^{d} t^{r-2}\mu(t)dt, \qquad 0 < \delta \leqslant \tfrac{1}{2}d,$$

*where d is the diameter of K and c is a constant independent of δ. Moreover, for all integers ν, with $1 \leqslant \nu \leqslant r-1$ ($1 \leqslant \nu \leqslant r$ in the case when $\int_0^1 t^{-1}\mu(t)dt < +\infty$) there exists the νth derivative $f_K^{(\nu)}(\zeta)$, which can be represented by the series*

$$f_K^{(\nu)}(\zeta) = P_1^{(\nu)}(\zeta) + \sum_{n=1}^{\infty} [P_{n+1}^{(\nu)}(\zeta) - P_n^{(\nu)}(\zeta)],$$

*uniformly convergent on K. This derivative satisfies also the inequality*

$$\omega_K(f_K^{(\nu)}, \delta) \leqslant c\left[\int_0^{\delta} t^{r-\nu-1}\mu(t)dt + \delta\int_{\delta}^{d} t^{r-\nu-2}\mu(t)\,dt\right],$$

$$0 < \delta < \tfrac{1}{2}d, \qquad c = \text{const},$$

*and the estimations*

$$|f_K^{(\nu)}(z) - P_n^{(\nu)}(z)| \leqslant c \int_0^{d_n(z)} t^{r-\nu-1}\mu(t)dt, \qquad z \in \partial K, \ n = 1, 2, \ldots$$

COROLLARY. *If the right-hand side of* (1) *is replaced by the quantity $n^{-2(k+\alpha)}$, where k is a non-negative integer and $0 < \alpha \leqslant 1$, then there exists the kth derivative $f_K^{(k)}$ satisfying the inequality*

$$\omega_K(f_K^{(k)}, \delta) \leqslant \begin{cases} c\delta & \text{for } 0 < \alpha < 1, \\ c(\delta + \delta \cdot |\ln \delta|) & \text{for } \alpha = 1. \end{cases} \tag{2}$$

The above corollary is a sharpening and an extension of two results due to Mergelyan [10] asserting the existence of the continuous $k$th derivative (without estimation (2)), obtained under the additional supposition that either $k = 1$ or $k > 1$ and $K$ is the closure of a domain.

Let $\mathfrak{B}_0$ denote the class of all bounded simply connected domains. Let $G \in \mathfrak{B}_0$ and let $\mu(\delta)$ be a function of the modulus of continuity type. Consider the following four hypotheses concerning a function $f(\zeta)$ continuous on $G$ and holomorphic in $\bar{G}$.

H1. If $\omega_{\partial G}(f, \delta) \leqslant \mu(\delta)$, then[1] $\omega_{\bar{G}}(f, \delta) \leqslant c\mu(\delta)$, $c = \text{const}$.

H2. If at some point $z_0 \in \partial G$ the inequality $\omega_{\partial G}(f, z_0, \delta) \leqslant \mu(\delta)$ holds, then we have at this point the estimation $\omega_{\bar{G}}(f, z_0, \delta) \leqslant c\mu(\delta)$, $c = \text{const}$.

H3. If at some point $z_0 \in \partial G$ the derivative $f'_{\partial G}(z_0)$ exists, then there exists also the derivative $f'_{\bar{G}}(z_0)$.

H4. If the derivative $f'_{\partial G}(z)$ is continuous on $\partial G$ and if $\omega_{\partial G}(f, \delta) \leqslant \delta \cdot \cdot \text{const}$, then the derivative $f'_{\bar{G}}(\zeta)$ is continuous on $\bar{G}$.

We shall be concerned with determining conditions on $G$ and $\mu(\delta)$ under which hypotheses H1–H4 hold true.

Hardy and Littlewood [8] proved that H1 is true in the case when $G$ is a disc and $\mu(\delta) = \delta^\alpha$, $\alpha = \text{const} > 0$.

In the case when $G$ is a Jordan domain and $\mu(\delta) = \delta^\alpha$, $\alpha > 0$, the validity of H2 has been proved by Warschawski [19] and that of H1 by Walsh and Sewell [17]; in both those results $c = 1$ (analogous results have been obtained also for the function $\mu(\delta) = \delta|\ln \delta|$, see [11], 23–24, and for $\mu(\delta) = \delta|\ln \delta|^2$, see [18]).

Walsh and Sewell [17] also proved that H3 and H4 are true if $G$ is a Jordan domain.

In connection with the above results Sewell (1942) posed in his monograph [11], 31–32, the following problems:

A. Extend the results of Warschawski, Walsh, Sewell to a more general class of domains than the Jordan domains.

B. Prove the analogues of the results of Warschawski, Walsh, Sewell with other majorants than $\delta^\alpha$ or $\delta|\ln \delta|$.

C. Find the most general form of the modulus of continuity $\mu(\delta)$ for which H1 holds true in the case when $G$ is a Jordan domain (or even a Jordan domain with analytic boundary).

In recent years several authors obtained partial results under restrictive additional assumptions on the domain $G$ and the function $\mu(\delta)$.

(1) Here and throughout the sequel the constants are supposed to be independent of $\delta$.

In the case when $G$ is a Jordan domain and $\mu(\delta)$ is a power function or a product of logarithms and iterated logarithms of $\delta$, the validity of H2 has been shown by Gagua; he also proved, that if, moreover, the conformal mapping of $G$ onto the disc and the inverse mapping satisfy the Hölder condition on the boundaries of the corresponding domains, then H1 holds true (see [5], [6]).

Brudnyi and Gopengauz [1] proved that if $G$ is the unit disc and if $\int_0^\delta t^{-1}\mu(t)\,dt = O(\mu(t))$ as $\delta \to 0$, then H1 is true.

The problem of relation between weak derivatives with respect to the domain and to its boundary has been, for Jordan domains with rectifiable boundary, investigated by Dolženko [2].

In 1971 the following results were obtained by the author [15], [16].

THEOREM 2. *Hypotheses* H3 *and* H1 *hold true for any domain* $G \in \mathfrak{B}_0$.

THEOREM 3. *Hypotheses* H1 *and* H2 *hold true for any domain* $G \in \mathfrak{B}_0$ *and for any function* $\mu(\delta)$ *of the modulus of continuity type.*

THEOREM 4. *The constant c occurring in* H1 *and* H2 *is absolute (independent of* $G, f$ *and* $\mu(\delta)$*); for instance,* $c = 108$ *is good.*

THEOREM 5. *The following conditions are equivalent:*

(i) *hypothesis* H2 *holds true with the constant* $c = 1$ *for any domain* $G \in \mathfrak{B}_0$ *and any function* $f(\zeta)$ *continuous on* $\bar{G}$ *and holomorphic in* $G$;

(ii) *the function* $\ln \mu(\exp \tau)$ *is concave.*

Let us mention here that also certain essentially more general results have been obtained by similar methods. Those results do not require the boundedness and simple connectedness of $G$, or the continuity of $f(\zeta)$ on $\partial G$; some of the remaining assumptions are also weakened. As an illustration we give here just one particular statement:

THEOREM 6. *Let* $F(\zeta)$ *be a function bounded and holomorphic in the open upper half-plane and continuous in the closed half-plane. Suppose that for* $x \to \pm\infty$, $x$ *real,* $F(x)$ *tends to zero with rapidity not smaller (greater) than* $\mu(|x|^{-1})$, *where* $\mu(t)$ *is a function of the modulus of continuity type. Then for* $\zeta \to \infty$ *through the upper half-plane the function* $F(\zeta)$ *decreases not more slowly (resp., more rapidly) than the function* $\mu(|\zeta|^{-1})$ const.

The above results can be applied in the study of direct and inverse problems of the approximation theory, in problems concerning Cauchy type integrals, in certain boundary problems, in singular integral equations, in elliptic differential equations and in other problems of analysis.

We shall give an example of a most direct application. Consider the following Cauchy type integral

$$\Phi(\zeta) = \frac{1}{2\pi i} \int_{\Gamma} \frac{\varphi(w)\, dw}{w-\zeta}, \tag{3}$$

where $\Gamma$ is a contour and $\varphi(w)$ is an integrable on $\Gamma$ density whose modulus of continuity is majorized by a function $\mu(\delta)$ of the modulus of continuity type. As is well known (Zygmund, Magnaradze), under these conditions the moduli of continuity of the singular (in the sense of the main value) Cauchy integral

$$\frac{1}{2\pi \mathrm{i}} \int_{\Gamma} \frac{\varphi(w)\, dw}{w-z}, \qquad z \in \Gamma,$$

and of the limit values of the integral (3)

$$\Phi_{+}(z) = \lim_{\zeta \to z} \Phi(\zeta), \qquad z \in \Gamma,$$

are majorized by the quantity

$$\left[ \int_{0}^{\delta} \frac{\mu(t)}{t} dt + \delta \int_{\delta}^{\pi} \frac{\mu(t)}{t^2} dt \right] \cdot \mathrm{const.} \tag{4}$$

It follows from the results of this paper that the modulus of continuity of the integral (3) itself can also be majorized by the expression (4) (formerly this fact has been known only for functions $\mu(\delta)$ of a very special form).

## REFERENCES

[1] Yu. A. Brudnyi and I. E. Gopengauz, *Mat. Sb.* **52** (94) (1960) 891–894.
[2] E. P. Dolženko, *Izv. Akad. Nauk SSSR, Ser. Mat.* **29** (1965) 1069–1084.
[3] V. K. Dziadyk, *ibidem* **23** (1959) 697–736.
[4] V. K. Dziadyk, *Ukr. Mat. Ž.* **15** (1963) 365–375.
[5] M. B. Gagua, *Soobšč. Akad. Nauk Gruz. SSSR* **10** (1949) 451–456.
[6] M. B. Gagua, *Usp. Mat. Nauk.* **8** (53) (1953) 121–125.

[7] V. I. Gorbaičuk and P. M. Tamrazov, *Ukr. Mat. Ž.* **24** (1972) 147–161.
[8] G. H. Hardy and J. E. Littlewood, *Math. Z.* **34** (1962) 403–439.
[9] N. A. Lebedev and P. M. Tamrazov, *Izv. Akad. Nauk SSSR, Ser. Mat.* **34** (1970) 1340–1390.
[10] S. N. Mergelyan, *Usp. Mat. Nauk* **7** (48) (1952) 31–132.
[11] W. E. Sewell, *Degree of Approximation by Polynomials in the Complex Domain*, Princeton 1942.
[12] P. M. Tamrazov, *Congres International des Mathematiciens*, Nice 1970, "265 communications individuelles", 176–177.
[13] P. M. Tamrazov, *Dokl. Akad. Nauk SSSR* **198** (1971) 540–542.
[14] P. M. Tamrazov, 'On the Inverse $\eta$-Problem of the Polynomial Approximation of Functions on a Regular Compact Set', IM-72-I (preprint), *Inst. Mat. Akad. Nauk Ukr. SSR*, Kiev 1972.
[15] P. M. Tamrazov, *Proceedings of the Conference on the Theory of Functions of a Complex Variable*, Kharkov 1971, 206–208.
[16] P. M. Tamrazov, *Dokl. Akad. Nauk SSSR* **204** (1972).
[17] J. L. Walsh and W. E. Sewell, *Duke Math. J.* **6** (1940) 658–705.
[18] J. L, Walsh, W. E. Sewell and H. M. Elliott, *Trans. Amer. Math. Soc.* **67** (1949) 341–420.
[19] S. Warschawski, *Math. Z.* **38** (1934) 669–683.

# TRIGONOMETRIC INTERPOLATION POLYNOMIALS WHICH GIVES BEST ORDER OF APPROXIMATION AMONG CONTINUOUSLY DIFFERENTIABLE FUNCTIONS OF CERTAIN ORDER

A. K. VARMA

*Gainesville, Fl.*

Let $f(x)$ be a $2\pi$ periodic continuous function on the real line. In earlier work [1] we have introduced the trigonometric polynomials

$$(1.1)\qquad R_n[f, x] \equiv R_n(x) = \sum_{k=0}^{n-1} f(x_{kn}) F_M(x - x_{kn}),$$

where

$$(1.2)\qquad x_{kn} = \frac{2k\pi}{n}, \quad K = 0, 1, \ldots, n-1,$$

$$(1.3)\qquad F_M(x) = \frac{1}{n}\left[1 + 2\sum_{j=1}^{n-1} \alpha_j \cos jx\right],$$

$$(1.4)\qquad \alpha_j = \begin{cases} \dfrac{(n-j)^M}{(n-j)^M + j^M} & \text{for } M \text{ — odd integer,} \\[2ex] \dfrac{(n-j)^M}{(n-j)^M - j^M} & \text{for } M \text{ — even integer.} \end{cases}$$

$R_n(x)$ is an interpolatory trigonometric polynomial of order $n-1$. It satisfies the interpolatory conditions

$$(1.5)\qquad R_n[f, x_i] = f(x_i), \quad R_n^{(M)}[f, x_i] = 0, \quad i = 0, 1, \ldots, n-1.$$

Concerning $R_n(x)$, the following theorems are known.

THEOREM 1 (A. Sharma and A. K. Varma [1]). *Let $M$ be odd fixed positive integer. If $f(x) \in C_{2\pi}$, then $R_n(x)$ as defined by* (1.1) *converges uniformly to $f(x)$ on the real line.*

THEOREM 2 (A. Sharma and A. K. Varma [1]). *Let $M$ be even fixed positive integer. If $f(x) \in C_{2\pi}$ and satisfies the Zygmund condition*

$$f(x+h)-2f(x)+f(x-h) = o(h),$$

*then $R_n(x)$ as defined by* (1.1) *converges uniformly to $f(x)$ on the real line.*

The object of this paper is to obtain the estimate $R_n(x)-f(x)$ in terms of generalized modulus of continuity of $f(x)$. Our result is somewhat analogous to Zygmund's well-known theorem concerning the approximation of functions by typical means of their Fourier series. We aim to prove the following:

THEOREM 3. *Let $M$ be an odd positive fixed integer $> 1$, and let $f(x) \in C_{2\pi}$ having $w_{M-1}(1/n, f)$ as its modulus of continuity of $(M-1)$th order; then we have*

$$|R_n(x)-f(x)| \leqslant C_1 w_{M-1}\left(\frac{1}{n}, f\right). \tag{1.6}$$

*In the case $f(x)$ is $M$ times continuously differentiable we have*

$$|R_n(x)-f(x)| \leqslant \frac{C_2}{n^M}(\log n) \max_x |f_x^{(M)}(x)|. \tag{1.7}$$

Professor A. Zygmund [8] has considered the problem of the degree of approximation of $f$ by the typical means

$$X_n^k(x) = \tfrac{1}{2}a_0 + \sum_{p=1}^{n-1}(a_p \cos px + b_p \sin px)\left(\frac{1-p^k}{n^k}\right) \tag{1.8}$$

of the Fourier series of $f$. He proved in 1945 the following:

THEOREM 4 (A. Zygmund). *Let $k$ be positive and odd and suppose that $f^{(h)}(x)$ exists for some $h \leqslant k$, and that $|f^{(h)}(x)| \leqslant L$. Then for $h < k$ we have*

$$|f(x)-X_n^k(x)| \leqslant LA_{h,k}n^{-h}, \tag{1.9}$$

*but for $h = k$*

$$|f(x)-X_n^k(x)| \leqslant LB_k n^{-k} \log(n+2), \tag{1.10}$$

*and the logarithm on the right cannot be removed. If $f^{(h)}(x)$, $h < k$ exists and satisfies a Lipschitz condition of order $\alpha$, $0 < \alpha < 1$, so that $|f^{(h)}(x+t)$*

$-f^{(h)}(x)| \leqslant L|t|^{\alpha}$, *then*

(1.11) $$|f(x)-X_n^k(x)| \leqslant C_{k,\alpha} L n^{-(\alpha+h)}.$$

*The restriction* $\alpha < 1$ *in* (1.9) *cannot be omitted, unless we assume that* $h < k-1$.

The analogy between Theorem 3 and Theorem 4 is apparent once we use the properties of generalized modulus of continuity (see [3], p. 103, equation (1)).

In later part of this paper we will strengthen the results of my earlier work [4] which we now describe. Linear process of approximation starting out with interpolation (trigonometric) was developed by S. N. Bernstein and S. M. Lozinski [3]. Let

(1.12) $$x_k = x_{kn} = \frac{2k\pi}{2n+1}, \quad k = 0, 1, \ldots, 2n,$$

and

(1.13) $$T_n(x) = a_0^{(n)} + \sum_{m=1}^{n} (a_m^{(n)} \cos mx + b_m^{(n)} \sin mx)$$

be a trigonometric polynomial which coincides at the nodes (1.12). It is known that

(1.14) $$a_0^{(n)} = \frac{1}{2n+1} \sum_{k=0}^{2n} f(x_k),$$
$$a_m^{(n)} = \frac{2}{2n+1} \sum_{k=0}^{2n} f(x_k) \cos mx,$$
$$b_m^{(n)} = \frac{2}{2n+1} \sum_{k=0}^{2n} f(x_k) \sin mx_k.$$

Consider

(1.15) $$L_n(f, x) = \lambda_0^{(n)} a_0^{(n)} + \sum_{m=1}^{n} \lambda_m^{(n)} (a_m^{(n)} \cos mx + b_m^{(n)} \sin mx)$$

which is a linear process of approximation starting with (1.13). In my earlier work [4] we proved the following:

THEOREM 5 (A. K. Varma [4]). *Let the triangular matrix $\lambda_j^{(n)}$ of bounded numbers satisfy the following conditions*

$$\lambda_0^{(n)} = 1, \quad \lambda_j^{(n)} = 0, \quad j \geqslant n+1, \tag{1.16}$$

$$\lambda_n^{(n)} = \left(\frac{1}{n}\right), \quad 1-\lambda_1^{(n)} = O\left(\frac{1}{n}\right),$$

$$|\lambda_{j+1}^{(n)} - 2\lambda_j^{(n)} + \lambda_{j-1}^{(n)}| = O\left(\frac{1}{n^2}\right), \quad j = 1, 2, \ldots, n, \tag{1.17}$$

*then we have*

$$|L_n(f, x) - f(x)| \leqslant c w_f\left(\frac{1}{\sqrt{n}}\right), \tag{1.18}$$

*and*

$$|L_n(f, x) - f(x)| \leqslant c_1(\log n) w\left(\frac{1}{n}\right), \tag{1.19}$$

*where $c$ and $c_1$ are positive constants independent of $n$ and $x$, $w(\delta)$ being the modulus of continuity of $f(x)$.*

Now we aim to impose some more simple restrictions on the triangular matrix so that $L_n(f, x)$ gives best order of approximation among continuously differentiable functions of certain order. Let us denote

$$\mu_i = \begin{cases} \dfrac{1-\lambda_i}{i^M}, & i = 1, 2, \ldots, n, \quad 1-\lambda_1 = O\left(\dfrac{1}{n^M}\right), \\ 0, & i = 0. \end{cases} \tag{1.20}$$

Let $\mu_i$'s be such that

$$|\mu_{i+1} - 2\mu_i + \mu_{i-1}| = O\left(\frac{1}{n^{M+2}}\right), \quad i = 1, 2, \ldots, n-1, \tag{1.21}$$

$$|\mu_{i+1} - \mu_i| = O\left(\frac{1}{n^{M+1}}\right), \quad i = 1, 2, \ldots, n-1. \tag{1.21a}$$

These conditions are satisfied by some examples of $\lambda_j$ given in [4]. We now aim to prove the following

THEOREM 6. *Let* $\lambda_j^{(n)}$ *satisfy* (1.16), (1.17), (1.20), (1.21) *and* (1.21a); *then we have*

$$|L_n[f, x]-f(x)| \leqslant c_2 w_M\left(\frac{1}{n}\right) \quad \textit{for } M \textit{ even}, \tag{1.22}$$

$$|L_n[f, x]-f(x)| \leqslant c_3 w_{M-1}\left(\frac{1}{n}\right) \quad \textit{for } M \textit{ odd}. \tag{1.23}$$

**2. Preliminaries.** Following theorem is due to S. B. Stečkin [2].

THEOREM 7 (S. B. Stečkin). *Let* $p$ *be a natural number and* $u_n$ ($n = 1, 2, \ldots$) *be a linear method of approximation of functions, having the following properties:*

(i) *for any function* $f(x) \in C_{2\pi}$

$$\|u_n(f)\| \leqslant M_0\|f\|. \tag{2.1}$$

(ii) *For any function* $f(x) \in C_{2\pi}$ *for which* $f^{(p)}(x) \in C_{2\pi}$

$$\|f-u_n(f)\| \leqslant \frac{M_p\|f^{(p)}\|}{n^p}, \quad n = 1, 2, \ldots \tag{2.2}$$

*Then for any function* $f(x) \in C_{2\pi}$ *we have*

$$\|f-u_n(f)\| \leqslant B_p(M_0+M_p)w_p\left(\frac{1}{n}, f\right). \tag{2.3}$$

In the case $u_n(f) = X_n^M(x)$ we see that in view of A. Zygmund's Theorem 4, conditions (2.1) and (2.2) are satisfied. On using Theorem 7 we have for $f(x) \in C_{2\pi}$

$$|f(x)-X_n^M(x)| \leqslant c_4 w_{M-1}\left(\frac{1}{n}, f\right) \quad (M \text{ odd}). \tag{2.4}$$

Following known results and using (1.2) we have

$$\sum_{k=0}^{n-1} \sin ix_{kn} \cos jx_{kn} = 0 \quad \text{for } i, j = 0, 1, \ldots, n-1. \tag{2.5}$$

Also we have

$$\left.\begin{aligned} &\sum_{k=0}^{n-1} \cos ix_{kn} \cos jx_{kn} = 0 \\ &\sum_{k=0}^{n-1} \sin ix_{kn} \sin jx_{kn} = 0 \end{aligned}\right\} \quad \begin{aligned} &\text{for } i \neq j, \text{ or } i \neq n-j, \\ &i, j = 0, 1, \ldots, n-1, \end{aligned} \tag{2.6}$$

and for $i = 1, 2, \ldots, n-1$

$$(2.7)\quad \begin{aligned} &\sum_{k=0}^{n-1} \cos^2 ix_{kn} = \sum_{k=0}^{n-1} \sin^2 ix_{kn} = \sum_{k=0}^{n-1} \cos ix_{kn} \cos(n-i)\, x_{kn} = \frac{n}{2}, \\ &\sum_{k=0}^{n-1} \sin ix_{kn} \sin(n-i)\, x_{kn} = \frac{-n}{2}. \end{aligned}$$

On using these properties along with the definition of $R_n[f, x]$ as given by (1.1) we obtain

$$R_n(\cos it, x) = \alpha_i \cos ix + \alpha_{n-i} \cos(n-i)x, \quad i = 1, 2, \ldots, n-1.$$

But from (1.4) it follows that

$$\alpha_i + \alpha_{n-i} \equiv 1.$$

Thus we obtain

$$(2.8)\quad R_n[\cos it, x] - \cos ix = \alpha_{n-i}\big(\cos(n-i)x - \cos ix\big).$$

Similarly we have

$$(2.9)\quad R_n[\sin it, x] - \sin ix = -\alpha_{n-i}\big(\sin ix + \sin(n-i)x\big),$$

and

$$(2.10)\quad R_n[1, x] \equiv 1.$$

We proved in earlier work ([1], p. 347, equation (21)) the following inequality concerning the fundamental function $F_m(x)$ as defined in (1.3)

$$(2.11)\quad \sum_{k=0}^{n-1} |F_M(x - x_{kn})| \leqslant 1 + M(M^2-1)2^{3M-1} = c_7 \quad (M \text{ odd}).$$

On using (2.11) and (1.1) we obtain for $f(x) \in C_{2\pi}$ and for $M$ being odd fixed integer

$$(2.12)\quad |R_n[f, x]| \leqslant c_7 \|f\|.$$

We denote the Fejér kernel by

$$(2.13)\quad t_i(u) = \begin{cases} 1 + \dfrac{2}{i} \displaystyle\sum_{j=1}^{i-1} (i-j) \cos ju \geqslant 0 & \text{for } i > 1, \\ 1 & \text{for } i = 1. \end{cases}$$

It follows from (2.13) that

$$\int_0^{2\pi} |t_i(u)|\,du = 2\pi, \qquad i = 1, 2, \ldots \tag{2.14}$$

It is well known that

$$2\cos iu = (i+1)\,t_{i+1}(u) - 2i t_i(u) + (i-1)\,t_{i-1}(u). \tag{2.15}$$

We also need the following theorem concerning trigonometric polynomials of order $n$.

THEOREM 8. *If $T(x)$ is a trigonometric polynomial of order $\leqslant n-1$, then we have*

$$|T'(x)| \leqslant n \max_x |T(x)|, \qquad |\tilde{T}'(x)| \leqslant n \max_x |T(x)|, \tag{2.16}$$

$$|\tilde{T}(x)| \leqslant c_8 \log n \max_x |T(x)|. \tag{2.17}$$

**3. Proof of Theorem 3.** Let $\varphi_{n-1}(x)$ be any arbitrary trigonometric polynomial of order $\leqslant n-1$ given by

$$\varphi_{n-1}(x) = e_0 + \sum_{i=1}^{n-1} (e_i \cos ix + f_i \sin ix).$$

On using (2.8)–(2.10) we have

$$\varphi_{n-1}(x) - R_n[\varphi_{n-1}(t); x] = (1-\cos nx) \sum_{i=1}^{n-1} \alpha_{n-i}(e_i \cos ix + f_i \sin ix) +$$

$$+ \sin nx \sum_{i=1}^{n-1} \alpha_{n-i}(e_i \sin ix - f_i \cos ix).$$

Let $\varphi_{n-1}(x) = X_n^M(x)$ (as defined by 1.8). Then

$$e_0 = \tfrac{1}{2} a_0, \quad e_i = \left(1 - \frac{i^M}{n^M}\right) a_i, \quad f_i = \left(1 - \frac{i^M}{n^M}\right) b_i,$$

$$i = 1, 2, \ldots, n-1.$$

Here $a_i$ and $b_i$ are the Fourier coefficients of $f(x)$. Let us denote

$$\delta_i = \frac{1 - i^M/n^M}{(n-i)^M + i^M}, \qquad i = 1, 2, \ldots, n. \tag{3.1}$$

Therefore we may write

$$X_n^M(x)-R_n[X_n^M(t),x] = \frac{(1-\cos nx)}{\pi}\sum_{i=1}^{n-1}\delta_i i^M\int_0^{2\pi} f(t)\cos i(t-x)\,dt+$$

$$+\frac{\sin nx}{\pi}\sum_{i=1}^{n-1} i^M\delta_i\int_0^{2\pi} f(t)\sin i(x-t)\,dt$$

$$= (1-\cos nx)I_1(x)+\sin nx\, I_2(x) \quad \text{(say)}.$$

We set

$$F(x) = \frac{1}{\pi}\int_0^{2\pi} f(t+x)\sum_{i=1}^{n-1} i^{M-1}\delta_i\cos it\,dt, \tag{3.3}$$

and observe that it is a trigonometric polynomial of order $\leqslant n-1$. From (3.2) and (3.3) it follows that

$$X_n^M(x)-R_n^M[X_n^M(t),x] = (1-\cos nx)\tilde{F}'(x)+\sin nx F'(x). \tag{3.4}$$

Here $\tilde{F}'(x)$ denotes the conjugate polynomial corresponding to $F'(x)$. Suppose $f(x)$ be $M-1$ times continuously differentiable function of $x$. Integrating by parts $M-1$ times we obtain ($M$ being odd positive integer)

$$F(x) = \frac{(-1)^{(M-1)/2}}{\pi}\int_0^{2\pi} f^{(M-1)}(t+x)\left(\sum_{i=1}^{n-1}\delta_i\cos it\right)dt. \tag{3.5}$$

On using (2.13) and (3.1) we obtain

$$F(x) = \frac{(-1)^{(M-1)/2}}{2\pi}\int_0^{2\pi} f^{(M-1)}(t+x)\sum_{i=1}^{n-1}(\delta_{i-1}-2\delta_i+\delta_{i+1})it_i(t)\,dt. \tag{3.6}$$

From (3.1) it follows that

$$|\delta_{i-1}-2\delta_i+\delta_{i+1}| = O\left(\frac{1}{n^{M+2}}\right), \quad i = 1, 2, \ldots, n-1. \tag{3.7}$$

On using (3.6), (3.7) and (2.14) we obtain

$$|F(x)| = O\left(\frac{1}{n^M}\right)\max_x|f^{(M-1)}(x)|. \tag{3.8}$$

Since $F(x)$ is a trigonometric polynomial of order $\leqslant n-1$, on using Theorem 8 we have

$$|F'(x)| = O\left(\frac{1}{n^{M-1}}\right) \max_x |f^{(M-1)}(x)|, \tag{3.9}$$

and

$$|\tilde{F}'(x)| = O\left(\frac{1}{n^{M-1}}\right) \max_x |f^{(M-1)}(x)|. \tag{3.10}$$

Thus, under the assumption that $f(x) \in C_{2\pi}^{M-1}$ we obtain from (3.4), (3.9) and (3.10)

$$|X_n^M(x) - R_n[X_n^M(t), x]| = O\left(\frac{1}{n^{M-1}}\right) \max_x |f^{(M-1)}(x)|. \tag{3.11}$$

From (2.12) and (3.11) it follows that

$$|R_n[X_n^M(t), x] - R_n[f, x]| = O\left(\frac{1}{n^{M-1}}\right) \max_x |f^{(M-1)}(x)|. \tag{3.12}$$

Since

$$f(x) - R_n[f, x] = f(x) - X_n^M(x) + X_n^M(x) - R_n[X_n^M(t), x] + \\ + R_n[X_n^M(t), x] - R_n[f, x]$$

it follows that for $f \in C_{2\pi}^{M-1}$ we have

$$|f(x) - R_n[f, x]| = O\left(\frac{1}{n^{M-1}}\right) \max_x |f^{(M-1)}(x)|. \tag{3.13}$$

On using (2.12) and (3.13) we observe that the conditions of Stečkin's theorem are satisfied for $p = M-1$. Therefore we may conclude that

$$|f(x) - R_n[f, x]| \leqslant c_1 w_{M-1}\left(\frac{1}{n}, f\right), \qquad M > 1.$$

It may be remarked that the case $M = 1$ and $M$ even integer is well settled in the paper of P. Vertesi [7].

Now we turn to prove (1.7). Let us assume that $f(x) \in C_{2\pi}^M$. From (3.2) we have

$$I_2(x) \equiv \frac{1}{\pi} \sum_{i=1}^{n-1} i^M \delta_i \int_0^{2\pi} f(t) \sin i(t-x)\, dt.$$

Integrating by parts $M$ times we have ($M$ being odd positive integer)

$$I_2(x) = \frac{(-1)^{(M-1)/2}}{\pi} \sum_{i=1}^{n-1} \delta_i \int_0^{2\pi} f^{(M)}(t) \cos i(t-x)\, dt$$

$$= \frac{(-1)^{(M-1)/2}}{\pi} \sum_{i=1}^{n-1} \delta_i \int_0^{2\pi} f^{(M)}(t+x) \cos it\, dt.$$

On using (2.15) we have

$$I_2(x) = \frac{(-1)^{(M-1)/2}}{2\pi} \sum_{i=1}^{n-1} \int_0^{2\pi} f^{(M)}(t+x)\, (\delta_{i-1} - 2\delta_i + \delta_{i+1})\, it_i(t)\, dt.$$

On using (2.14) and (3.7) we obtain

$$|I_2(x)| = \frac{1}{2\pi} \max_x |f^{(M)}(x)|\, O\left(\frac{1}{n^M}\right). \tag{3.14}$$

From Theorem 8, (3.14) and observing that $I_1(x)$ and $I_2(x)$ are trigonometric polynomials of order $\leqslant n-1$ such that

$$I_1(x) = \tilde{I}_2(x)$$

we obtain

$$|I_1(x)| = O\left(\frac{1}{n^M}\right) \log n \max_x |f^{(M)}(x)|. \tag{3.15}$$

Therefore, from (3.2), (3.14) and (3.15) we have

$$X_n^M(x) - R_n[X_n^M(t), x] = O\left(\frac{1}{n^M}\right) \log n \max_x |f^M(x)|.$$

On using (1.10) (for $h = k = M$) and (3.16) we have for $f(x) \in C_{2\pi}^M$

$$\begin{aligned} f(x) - R_n[f, x] &= f(x) - X_n^M(x) + X_n^M(x) - R_n[X_n^M(t), x] \\ &\quad + R_n[X_n^M(t), x] - R_n[f, x] \\ &= \Big[ B_M n^{-M} \log(n+2) + O\left(\frac{1}{n^M}\right) \log n + \\ &\quad + c_1 B_M n^{-M} \log(n+2) \Big] \max_x |f^{(M)}(x)|, \end{aligned} \tag{3.17}$$

which in turn proves (1.7).

**4. Proof of Theorem 6.** On using (1.14) and $\lambda_0^n = 1$ we get

$$L_n[f, x] = \sum_{k=0}^{2n} f(x_{kn}) A_n(x_k - x), \tag{4.1}$$

where

$$A_n(t) = \frac{1}{(2n+1)} \left[1 + 2 \sum_{m=1}^{n} \lambda_m^n \cos mt\right].$$

We proved in earlier work (Lemma 2, p. 186, [4]) that

$$\sum_{k=0}^{2n} |A_n(x - x_{kn})| \leqslant c_1. \tag{4.2}$$

From (4.1) and (4.2) it follows that if $f(x) \in C_{2\pi}$ we have

$$|L_n[f, x]| \leqslant c_1 \|f\|. \tag{4.3}$$

Let $\varphi_{n-1}(x)$ be any arbitrary trigonometric polynomial of order $\leqslant n-1$ given by

$$\varphi_{n-1}(x) = e_0 + \sum_{i=1}^{n-1} (e_i \cos ix + f_i \sin ix).$$

From the definition of $L_n[f, x]$ and using similar results as given in (2.5)–(2.7) we obtain

$$\varphi_{n-1}(x) - L_n[\varphi_{n-1}(t), x] = \sum_{i=1}^{n-1} (1 - \lambda_i)(e_i \cos ix + f_i \sin ix).$$

Let in particular $\varphi_{n-1}(x) = X_n^M(x)$; then

$$e_0 = \tfrac{1}{2} a_0, \quad e_i = \left(1 - \frac{i^M}{n^M}\right) a^i, \quad f_i = \left(1 - \frac{i^M}{n^M}\right) b_i, \quad i = 1, 2, \ldots, n-1.$$

We set as in (1.20)

$$\mu_i = \frac{1 - \lambda_i}{i^M} = 0, \quad i = 0,$$

then we obtain

$$X_n^M(x) - L_n[X_n^M(t), x] = \frac{1}{\pi} \sum_{i=1}^{n-1} i^M \mu_i \left(1 - \frac{i^M}{n^M}\right) \int_0^{2\pi} f(t) \cos i(t-x)\, dt. \tag{4.4}$$

Let $M$ be even positive integer; then assuming $f \in C_{2\pi}^M$ and integrating by parts $M$ times we obtain

$$(4.5)\quad X_n^M(x) - L_n[X_n^M(t), x] = \sum_{i=1}^{n-1} \mu_i i^M \left(1 - \frac{i^M}{n^M}\right) \int_0^{2\pi} f(t+x) \cos it\, dt$$

$$= \frac{(-1)^{M/2}}{\pi} \sum_{i=1}^{n-1} \mu_i \left(1 - \frac{i^M}{n^M}\right) \int_0^{2\pi} f^{(M)}(t+x) \cos it\, dt.$$

We set

$$(4.6)\quad \delta_i = \mu_i \left(1 - \frac{i^M}{n^M}\right), \quad i = 1, 2, \ldots, n-1,$$

$$(4.7)\quad |\delta_{i+1} - 2\delta_i + \delta_{i-1}| = O\left(\frac{1}{n^{M+2}}\right), \quad i = 2, 3, \ldots, n-1,$$

$$|\delta_2 - 2\delta_1| = O\left(\frac{1}{n^M}\right).$$

Now on using (2.15) and (4.5) we obtain

$$X_n^M(x) - L_n[X_n^M(t), x] = \frac{(-1)^{M/2}}{2\pi} \int_0^{2\pi} f^{(M)}(x+t) \left[\sum_{i=2}^{n-1} (\delta_{i-1} - \right.$$

$$\left. - 2\delta_i + \delta_{i+1}) i t_i(t) + (\delta_2 - 2\delta_1) t_1(t)\right] dt.$$

Therefore on using (4.7) and (2.14) we have

$$(4.8)\quad |X_n^M(x) - L_n[X_n^M(t), x]| = O\left(\frac{1}{n^M}\right) \max_x |f^{(M)}(x)|.$$

Now we use a known result of A. Zygmund [8] which asserts that if $f(x) \in C_{2\pi}^M$ ($M$ even), then

$$(4.9)\quad |f(x) - X_n^M(x)| = O\left(\frac{1}{n^M}\right) \max_x |f^{(M)}(x)|.$$

From (4.3) it follows that

$$(4.10)\quad L_n[f(t) - X_n^{(M)}(t), x] = O\left(\frac{1}{n^M}\right) \max_x |f^{(M)}(x)|.$$

Following as in (3.17) we obtain from (4.8)–(4.10) that if $f(x) \in C_{2\pi}^M$ ($M$ even) we obtain

$$|f(x)-L_n[f, x]| = O\left(\frac{1}{n^M}\right)\max_x |f^{(M)}(x)|.$$

Now we use Stečkin's Theorem 7 and obtain for $f(x) \in C_{2\pi}$

$$|f(x)-L_n[f, x]| \leqslant cw_M\left(\frac{1}{n}f\right).$$

Proof of (1.23) is similar as in Theorem 3 so we omit the details. This completes the proof of Theorem 6.

I would like to take this opportunity to express my thanks to Prof. A. Zygmund for some valuable conversation.

## REFERENCES

[1] A. Sharma and A. K. Varma, 'Trigonometric Interpolation', *Duke Math. J.* **32** (1965) 341–358.

[2] S. B. Stečkin, 'The Approximation of Periodic Functions by Fejér Sums', *Amer. Math. Soc. Trans.* **2** (28) (1963) 269–282.

[3] A. F. Timan, *Theory of Approximations of Functions of a Real Variable*, Pergamon Press (Chapter 8).

[4] A. K. Varma, 'Some Remarks on a Theorem of S. M. Lozinski Concerning Linear Process of Approximation of Periodic Functions', *Studia Math.* **41** (1972) 183–190.

[5] A. K. Varma, 'The Approximation of Functions by Certain Trigonometric Interpolation Polynomials', communicated in *J. Approx. Theory.*

[6] A. K. Varma, 'Simultaneous Approximation of Periodic Continuous Functions and their Derivatives', *Israel J. Math.* **6** (1968) 66–73.

[7] P. Vertesi, 'On the Convergence of the Trigonometric (0, $M$) Interpolation', *Acta Math. Acad. Sci. Hung.* **22** 11–126.

[8] A. Zygmund, 'The Approximation of Functions by Typical Means of Their Fourier Series', *Duke Math. J.* (1945) 695–704.

[9] A. Zygmund, *Trigonometric Series*, Vol. 2.

# HERMITE-FEJÉR INTERPOLATION ON THE JACOBI ABSCISSAS

P. O. H. VÉRTESI

*Budapest*

**1. Notations and preliminary results.** We define the uniquely determined Hermite–Fejér interpolating polynomial of degree $\leqslant 2n-1$ for a continuous function on the interval $[-1, 1]$ by

$$H_n^{(\alpha,\beta)}(f; x) = \sum_{k=1}^{n} f(x_{kn})\, v_k(x)\, l_k^2(x), \tag{1.1}$$

where (with $x_k = x_{kn}^{(\alpha,\beta)}$, $k = 1, 2, \ldots, n$)

$$-1 < x_{nn}^{(\alpha,\beta)} < x_{n-1,n}^{(\alpha,\beta)} < \ldots < x_{2n}^{(\alpha,\beta)} < x_{1n}^{(\alpha,\beta)} < 1 \tag{1.2}$$

are the roots of the Jacobi polynomial $P_n^{(\alpha,\beta)}(x)$ of degree $n$ defined by

$$(1-x)^{\alpha}(1+x)^{\beta}P_n^{(\alpha,\beta)}(x) = \frac{(-1)^n}{2^n n!}\frac{d^n}{dx^n}[(1-x)^{\alpha+n}(1+x)^{\beta+n}] \text{ (}^1\text{)}; \tag{1.3}$$

$$l_k(x) = l_{k,n}^{(\alpha,\beta)}(x) = \frac{P_n^{(\alpha,\beta)}(x)}{P_n'^{(\alpha,\beta)}(x_{kn})\,(x-x_{kn})}, \tag{1.4}$$

$$v_n(x) = v_{kn}^{(\alpha,\beta)}(x) \tag{1.5}$$

$$= \frac{1-x[\alpha-\beta+(\alpha+\beta+2)x_k]+(\alpha-\beta)x_k+(\alpha+\beta+1)x_k^2}{1-x_k^2},$$

It is well known that

$$H_n^{(\alpha,\beta)}(f; x_k) = f(x_k), \qquad H_n'^{(\alpha,\beta)}(f; x_k) = 0, \tag{1.6}$$

$$\sum_{k=1}^{n} v_k(x)\, l_k^2(x) = 1. \tag{1.7}$$

---

(1) Throughout this paper let $\alpha, \beta > -1$.

Further, supposing that $\alpha < 0$, $\beta < 0$, we have for every continuous function $f(x)$

$$(1.8) \qquad |H_n^{(\alpha,\beta)}(f; x) - f(x)| \to 0 \qquad (x \in [-1, 1])$$

uniformly in $x$ ([3], 14.1).

As for the rate of uniform convergence, we have for arbitrary $[a, b]$ contained in $(-1, 1)$ that

$$(1.9) \qquad |H_n^{(\alpha,\beta)}(f; x) - f(x)| = O(1) \sum_{i=1}^{n} \omega\left(f; \frac{i}{n}\right)\frac{1}{i^2},$$

$$x \in [a, b] \subset (-1, 1)$$

for arbitrary $\alpha, \beta > -1$, where $\omega(f; t) = \omega(t)$ is the modulus of continuity of $f(x)$, the $O$ sign depends on $\alpha, \beta$, $a$ and $b$ (see [4]).

For the whole interval $[-1, 1]$ we know only for the Čebyšev nodes the exact estimation

$$(1.10) \quad |H_n^{(-1/2,-1/2)}(f; x) - f(x)| = O(1) \sum_{i=1}^{n} \left[\omega\left(f; \frac{i\sqrt{1-x^2}}{n}\right) + \right.$$

$$\left. + \omega\left(f; \frac{i^2|x|}{n^2}\right)\right]\frac{1}{i^2} \qquad (x \in [-1, 1])$$

(see [1], [5]).

## 2. Uniform estimations on the whole interval $[-1, 1]$.

**2.1.** Our new results are the following:

THEOREM 2.1. *For every continuous function $f(x)$ we have*

$$(2.1) \qquad |H_n^{(\alpha,\beta)}(f; x) - f(x)| = O(1) \sum_{i=1}^{n} \left[\omega\left(f; \frac{i\sqrt{1-x^2}}{n}\right) + \right.$$

$$\left. + \omega\left(f; \frac{i^2|x|}{n^2}\right)\right] i^{2\gamma-1}$$

*for every $x \in [-1, 1]$. Here the $O$ sign depends on $\alpha$ and $\beta$, $\gamma = \max(\alpha, \beta, -\frac{1}{2})$.*

**2.2.** Let us consider some special cases.

**2.2.1.** If $f \in \operatorname{Lip} \varrho$ and $x \in (-1, 1)$ we have for the difference (1.8)

$$\sum_{i=1}^{n} \omega\left(\frac{i}{n}\right) i^{2\gamma-1} \sim \begin{cases} n^{\gamma 2} & \text{if } \varrho > -2\gamma, \\ n^{2\gamma} \log n & \text{if } \varrho = -2\gamma, \\ n^{-\varrho} & \text{if } \varrho < -2\gamma \end{cases}$$

(see (2.1)), i.e., in many cases we obtain the Jackson-order for the difference (1.8). ($f_n \sim g_n$ means that $f_n = O(g_n)$ and $g_n = O(f_n)$.)

THEOREM 2.2.

$$(2.2)\quad |H_n^{(\alpha,\beta)}(f; 1) - f(1)| = O(1) \sum_{i=1}^{n} \omega\left(f; \frac{i^2}{n^2}\right) i^{2\alpha-1} \qquad (\alpha > -1),$$

*from where for* Lip $\varrho$

$$(2.3)\qquad |H_n^{(\alpha,\beta)}(f; 1) - f(1)| = \begin{cases} O(1) n^{2\alpha} & \textit{if } \varrho > -\alpha, \\ O(1) n^{2\alpha} \log n & \textit{if } \varrho = -\alpha, \\ O(1) n^{-2\varrho} & \textit{if } \varrho < -\alpha, \end{cases}$$

*i.e., in the last case we obtain the Timan-order for the difference.*

## REFERENCES


[1] R. Bojanic, 'A Note on the Precision of Interpolation by Hermite–Fejér Polynomials', *Proceedings of the Conference on Constructive Theory of Functions* (Budapest 1971) 69–76.

[2] G. J. Natanson, 'A Two-Sided Estimate for the Lebesgue Function of the Lagrange Interpolation Process with Jacobi Knots', *Izv. Vysš. Učebn. Zaved.* **11** (66) (1967) 67–74.

[3] G. Szegö, 'Orthogonal Polynomials', *Amer. Math. Soc. Colloq. Publ.* **28** (New York 1959).

[4] P. O. H. Vértesi, 'Hermite–Fejér Interpolation Based on the Roots of Jacobi Polynomials', *Studia Sci. Math. Hung.* **5** (1971) 183–187.

[5] P. O. H. Vértesi, 'On the Convergence of Hermite–Fejér Interpolation', *Acta Math. Acad. Sci. Hung.* **22** (1971) 161–158.

[6] P. O. H. Vértesi, 'Notes on the Hermite–Fejér Interpolation Based on the Jacobi Abscissas', *Acta Math. Acad. Sci. Hung.* **24** (1973) 233–239

# PROBLEMS

## PROBLEM 1

Let $\alpha, 0 < \alpha < 1$, and the integers $d \geqslant 1$, $m \geqslant 0$ be given. Let $I = \langle 0, 1\rangle$ and let $C^{m+\alpha}(I^d)$ be the Banach space of $m$ times continuously differentiable functions in $I^d$ with the derivatives of order $m$ satisfying Hölder condition with the exponent $\alpha$ (Lipschitz $\alpha$ condition). Denote by $\|\cdot\|_{m+\alpha}$ the Hölder norm in $C^{m+\alpha}(I^d)$. Construct a basis $\{f_n\}$ in the Banach space $[C^{m+\alpha}(I^d), \|\ \|_{m+\alpha}]$ normalized, i.e. $\|f_n\|_{m+\alpha} = 1$ for all $n$, such that

$$f = \sum_n a_n f_n \in C^{m+\alpha}(I^d) \quad \text{iff } a_n = O(1).$$

REMARK. The answer is positive for $d = 1$, $m \geqslant 0$ and for $m = 0$ $d \geqslant 1$.

*Z. Ciesielski*

## PROBLEM 2

Let $X$ be a closed subspace of $C_{2\pi}$ such that for every $f \in X$ the Fourier series of $f$ converges. Prove—possibly under additional restrictions on the spectrum—that $X$ is finite-dimensional.

*V. Popov, S. Troianskii*

## PROBLEM 3

Find an estimate for $E_n((x-a)^s)$ on $[-1, 1]$, $s > 0$, uniform with respect to $a \geqslant 1$,

*B. Sendov*

## PROBLEM 4

Let $0 < p < 1$. If $\sum_{n=1}^{\infty} |s_n(x) - f(x)|^p \leqslant K$ for all $x$, where $s_n(x)$ the $n$th partial sum of Fourier expansion of $f(x)$, does $f \in \text{Lip}\,1$ hold?

*L. Leindler*

## PROBLEM 5

If $1 \leqslant r, s, \gamma \leqslant \infty$ and $\frac{1}{r}+\frac{1}{s}=1+\frac{1}{\gamma}$, then for every $a_n, b_n \geqslant 0$ $(n=0, \pm 1, \ldots)$ we have—with $C(r, s, \gamma) \equiv 1$

$$\max_{i,j}(a_i b_j)+\sum_{n=-\infty}^{\infty}\left(\sum_{k=-\infty}^{\infty} a_k^\gamma b_{n-k}^\gamma\right)^{1/\gamma} \geqslant C(r, s, \gamma)\left(\sum_{a=-\infty}^{\infty} a_a^r\right)^{1/r}\left(\sum_{a=-\infty}^{\infty} b_a^s\right)^{1/s}.$$

What is the best possible $C(r, s, \gamma)$?

P.S. If $\gamma=\infty$, then $C(r, s, \gamma)=r^{1/r}s^{1/s}$ is the best possible.

*L. Leindler*

## PROBLEM 6

Is it possible to represent every bounded $2\pi$-periodic function in the form of a sum of two functions in such a way that each of these two has a uniformly bounded subsequence of Fourier sums?

*K. I. Oskolkow*

## PROBLEM 7

Let $R_{n,m}(f)$ denote the distance of $f \in C[-1,+1]$ from the set of rational functions which are quotients of polynomials of order $n$ and order $m$ respectively. Is it possible to estimate $R_{n,m}(f)$ by properties of $f$ as the modulus of continuity?

*D. Braess*

## PROBLEM 8

Characterize the convex sets $V$ in $C(I)$, $I$ being an interval, in which the best approximation is always unique in the sense of Chebyshev. (Generalization of Haar's Theorem on linear subsets.)

*D. Braess*

## PROBLEM 9

Let $w^2$ denote the class of absolutely continuously differentiable $2\pi$-periodic functions whose second-derivative satisfies the inequality

$$\|f^{(2)}\|_{L_p(0,2\pi)} \leqslant 1.$$

Let $d_n(w^2, L_p)$ denote the $n$th Kolmogorov diameter, i.e.

$$d_n(w^2, L_p) = \inf_{H_n \in L_p} \sup_{f \in w^2} \inf_{s \in H_n} \|f-s\|_{L_p(0,2\pi)},$$

the infimum being taken with respect to $n$-dimensional subspaces $H_n$ of $L_p$.

Compute the quantity

$$\inf_{H_n \in L_p} \sup_{0 \leqslant i \leqslant k_0 < 2} \sup_{f \in w^2} \inf_{s \in H_n} \frac{\|f^{(i)}-s^{(i)}\|_{L_p(0,2\pi)}}{d_n(w^{2-i}, L_p)}.$$

*Yu. Subbotin*

## PROBLEM 10

Let $\{x_{kn}\}$ $(1 \leqslant k \leqslant n, n = 1, 2, \ldots)$ be a triangular matrix of elements $1 \geqslant x_{1n} > x_{2n} > \ldots > x_{nn} \geqslant -1$.

Find necessary and sufficient criteria concerning $\{x_{kn}\}$ so that for every $\varepsilon > 0$ there exists a sequence $\{A_n\}$ of linear operators mapping $C[-1, 1]$ in the set of polynomials $(1+\varepsilon)n$ interpolating at the points $x_{kn}$, i.e.

$$A_n(f; x_{kn}) = f(x_{kn}) \quad (k = 1, 2, \ldots, n)$$

and we have for every $f \in C[-1, 1]$

$$|f(x) - A_n(f; x)| \leqslant C(\varepsilon)\, \omega\left(f; \frac{\sqrt{1-x^2}}{n} + \frac{1}{n^2}\right),$$

$C(\varepsilon)$ independent of $f$ and $n$.

We know that zeros of Jacobi polynomials $P_n^{(\alpha,\beta)}$ $(\alpha, \beta > -1)$ have this property (joint result of G. Freud and A. Sharma).

*G. Freud*

## PROBLEM 11

Let $\Delta_r(p)$ be the set of positive decreasing sequences $\{\delta_k\}$ having the following property:

for arbitrary $0 < \beta < r$ and $2\pi$-periodic $f \in L_p$ the assumption

$$\omega_r(L_p; f; \delta_k) = O(\delta_k^\beta)$$

implies

$$\omega_r(L_p; f; h) = O(h^\beta).$$

Conjecture: $\{\delta_k\} \in \Delta_r(p)$ iff

$$(*) \qquad \overline{\lim_{k\to\infty}}\, \delta_k/\delta_{k+1} < \infty .$$

For $r = 2$ this conjecture was formulated by R. de Vore and proved for $p = \infty$. We proved the generalized conjecture for $p = 2$. Moreover, we proved that $\{\delta_k\} \in \Delta_r(p)$ implies $(*)$ for every $1 \leqslant p \leqslant \infty$.

*G. Freud*

## PROBLEM 12 (¹)

Let $H_n$ denote the Hermite function normalized in $L_2(R)$. Show that

$$\sup_{x\in R} |H_n(x)| \geqslant \sup_{x\in R} |H_{n+1}(x)| \quad \text{for all } n.$$

Similar question about the monotonicity of the $L_p(R)$ norm of $H_n$ as a function of $n$.

*Z. Ciesielski*

## PROBLEM 13

Let $f$ be a bounded function defined on an $n$-dimensional parallelepiped $Q$: $a_i \leqslant x_i \leqslant b_i$, $i \in \overline{1, n}$. Consider the class of functions defined on $Q$ which can be written as the sum of finitely many functions of a smaller number of variables: $\sum \varphi_\nu = \sum_{\nu=1}^{m} \varphi_\nu(x_{\overline{s\nu}})$, where $x_{\overline{s\nu}}$ are arbitrary proper subsets of $\{x_1, \ldots, x_n\}$. Let $E_f$ denote, as usual, the distance from $f$ to that class: $E_f = \inf_{\Sigma\varphi_\nu} \|f - \sum \varphi_\nu\|$, where $\|\cdot\|$ is the sup norm or the $L_p$ norm. A possibility of constructive expressing the effectiveness of an approximation apparatus and finding best approximations has not only theoretical, but also practical value: We thus regard as interesting answers to the following problems:

(*) For $p = \infty$ this problem was solved by O. Szas in the paper: 'On Relative Extrema of the Hermite Orthogonal Functions', *J. Indian Math. Soc.* **15** (1951) 129–154. In the case of $p = 4$ it has been solved recently by G. Freud and G. Németh.

I. Find a simple method of computing or precisely estimating $E_f$;

II. Find a method of construction or the best approximating functions;

III. Solve problems I and II in domains different from a parallelepiped.

Up to what extent can the domain of approximation be deformed without actually affecting the simplicity of the solution.

IV. Solve problems I–III in the complex domain.

P.S. Certain particular cases of problems I and II have been solved in the space $C$ (M.-B. A. Babaev, *Mat. Zametki* **12** (1972), I. I. Ibragimov and M.-B. A. Babaev, *Dokl. Akad. Nauk SSSR* **197** (1971)) and $L_2$ (V. M. Mordashev, *Dokl. Akad. Nauk SSSR* **183** (1968)).

*M.-B. A. Babaev*

## PROBLEM 14

Show that very few sets allow the possibility of approximation. For example, show that if $C$ is the class of compact sets $K$ in the plane having the property that every function continuous on $K$ and holomorphic in the interior is the uniform limit on $K$ of rational functions, then $C$ is of Baire category 1 in the space of all compact sets.

*P. M. Gauthier*